VOLUME ONE HUNDRED AND TWENTY FOUR

CURRENT TOPICS IN DEVELOPMENTAL BIOLOGY

Zebrafish at the Interface of Development and Disease Research

CURRENT TOPICS IN DEVELOPMENTAL BIOLOGY

"A meeting-ground for critical review and discussion of developmental processes"
A.A. Moscona and Alberto Monroy (Volume 1, 1966)

VOLUME ONE HUNDRED AND TWENTY FOUR

CURRENT TOPICS IN DEVELOPMENTAL BIOLOGY

Zebrafish at the Interface of Development and Disease Research

Edited by

KIRSTEN C. SADLER
New York University Abu Dhabi, United Arab Emirates

ELSEVIER

ACADEMIC PRESS
An imprint of Elsevier

Academic Press is an imprint of Elsevier
50 Hampshire Street, 5th Floor, Cambridge, MA 02139, United States
525 B Street, Suite 1800, San Diego, CA 92101-4495, United States
The Boulevard, Langford Lane, Kidlington, Oxford OX5 1GB, United Kingdom
125 London Wall, London, EC2Y 5AS, United Kingdom

First edition 2017

Notices
Knowledge and best practice in this field are constantly changing. As new research and experience broaden our understanding, changes in research methods, professional practices, or medical treatment may become necessary.

Practitioners and researchers must always rely on their own experience and knowledge in evaluating and using any information, methods, compounds, or experiments described herein. In using such information or methods they should be mindful of their own safety and the safety of others, including parties for whom they have a professional responsibility.

To the fullest extent of the law, neither the Publisher nor the authors, contributors, or editors, assume any liability for any injury and/or damage to persons or property as a matter of products liability, negligence or otherwise, or from any use or operation of any methods, products, instructions, or ideas contained in the material herein.

ISBN: 978-0-12-803308-1
ISSN: 0070-2153

Publisher: Zoe Kruze
Acquisition Editor: Zoe Kruze
Editorial Project Manager: Shellie Bryant
Production Project Manager: Vignesh Tamil
Cover Designer: Greg Harris

Typeset by SPi Global, India

CONTENTS

CONTRIBUTORS

Ryan M. Anderson
Indiana University School of Medicine, Indianapolis, IN, United States

Kathryn Bambino
Icahn School of Medicine at Mount Sinai, New York, United States

Rebecca D. Burdine
Princeton University, Princeton, NJ, United States

Erin V. Carter
Graduate School of Biomedical Sciences and Engineering, University of Maine, Orono, ME, United States

Jaime Chu
Icahn School of Medicine at Mount Sinai, New York, United States

Gokhan Dalgin
The University of Chicago, Chicago, IL, United States

Michelle F. Goody
School of Biology and Ecology, University of Maine, Orono, ME, United States

Meagan G. Grant
Princeton University, Princeton, NJ, United States

Daniel T. Grimes
Princeton University, Princeton, NJ, United States

Clarissa A. Henry
School of Biology and Ecology; Graduate School of Biomedical Sciences and Engineering, University of Maine, Orono, ME, United States

Stephanie Jerman
Yale University School of Medicine, New Haven, CT, United States

Elisabeth A. Kilroy
Graduate School of Biomedical Sciences and Engineering, University of Maine, Orono, ME, United States

Ela W. Knapik
Vanderbilt University Medical Center; Vanderbilt Brain Institute; Vanderbilt Genetics Institute, Vanderbilt University, Nashville, TN, United States

Wanda Kwan
Beth Israel Deaconess Medical Center, Harvard Medical School, Boston, MA, United States

Lauryn N. Luderman
Vanderbilt University Medical Center; Vanderbilt Brain Institute; Vanderbilt Genetics Institute, Vanderbilt University, Nashville, TN, United States

Samrah Masud
Institute of Biology, Leiden University, Leiden, The Netherlands

Lisa Maves
Center for Developmental Biology and Regenerative Medicine, Seattle Children's Research Institute; Division of Cardiology, University of Washington, Seattle, WA, United States

Annemarie H. Meijer
Institute of Biology, Leiden University, Leiden, The Netherlands

Sophie R. Miller
Danish Stem Cell Center (DanStem), University of Copenhagen, Copenhagen N, Denmark

Trista E. North
Beth Israel Deaconess Medical Center, Harvard Medical School, Boston; Harvard Stem Cell Institute, Cambridge, MA, United States

Elke A. Ober
Danish Stem Cell Center (DanStem), University of Copenhagen, Copenhagen N, Denmark

Victoria L. Patterson
Princeton University, Princeton, NJ, United States

Victoria E. Prince
The University of Chicago, Chicago, IL, United States

Kirsten C. Sadler
Icahn School of Medicine at Mount Sinai, New York, NY, United States; New York University Abu Dhabi, Abu Dhabi, United Arab Emirates

Zhaoxia Sun
Yale University School of Medicine, New Haven, CT, United States

Vincenzo Torraca
Institute of Biology, Leiden University, Leiden, The Netherlands

Gokhan Unlu
Vanderbilt University Medical Center; Vanderbilt Genetics Institute, Vanderbilt University, Nashville, TN, United States

Shuang Wang
Icahn School of Medicine at Mount Sinai, New York, NY, United States

PREFACE

The complexities of the developing embryo have captured the fascination of scientists for centuries. Cells must follow cues to achieve the proper position and function in the embryo and coordinate their activities with neighboring cells to form organs that function in concert in the newly formed organism. These processes are hidden from view in the developing mammalian embryo, but are on display in nonmammalian organisms, where embryos develop outside of the mother. In particular, fish embryos develop quickly and thus have been favorites of embryologists for centuries. The authors who contributed to this volume witness the marvel of vertebrate development as routine practice in their laboratories: all are developmental biologists who study how the small, transparent zebrafish embryo develops in a Petri dish from fertilized egg to fully formed larvae within a matter of days.

Why zebrafish? The combined attributes of this system make it an extremely powerful system for developmental genetics. Zebrafish are vertebrates and share nearly all organ systems with mammals, with the obvious exception of structures like gills and scales that are relevant to aquatic life. Most notably, zebrafish, like all animals, are susceptible to disease, and in the past decade, work of many zebrafish investigators has uncovered the genes and pathways that give rise to pathology, and found that nearly all of them are shared with mammals. Many disease pathways are reiterations of processes that are required for development; this volume describes how lessons learned from studying development in zebrafish have informed our understanding of a range of diseases.

During zebrafish development, formation of nearly all organs can be observed in real time as development is extremely fast: gastrulation is complete within 6 h postfertilization (hpf), nearly all organs are patterned within the first day and half, and by the end of the third day postfertilization (dpf), the embryo has transformed to a free swimming larvae. By 5 dpf, the larvae have nearly completed development. During this rapid development, organs such as the heart, craniofacial skeleton, muscle, brain, and eyes are large enough to be easily observed using a low power stereomicroscope, while others require specific markers for visualization. Illustrations of several of these structures are provided in figures that accompany most chapters in this volume. Since the zebrafish embryo is transparent, fluorescent markers—either as vital dyes or fluorescent proteins expressed under the control of cell-type-specific promoters—provide a straightforward means to visualize specific cell types in developing embryos and larvae. Indeed, using imaging to study how cells move and organs take shape and to track the dynamic behavior of intracellular structures within the developing embryo is a major strength of zebrafish. Studies described here include work to image how the heart loops (Chapter 1), how innate immune cells interact with pathogens (Chapter 8) and use of fluorescent reporters to signal the presence of toxins (Chapter 9) are some examples of how imaging has advanced the field. Of practical utility is that each chapter includes a table listing of commonly used markers for identifying specific cell types in each organ or system covered here. While the rapid advancement in the field will render these lists outdated by the time of publication, there is a great value for compiling these resources in single place.

The wealth of genetic tools that have been applied in forward and reverse genetic studies and in the generation of tissue-specific transgenics is the foundation of zebrafish research. Most of the genes governing development of all shared organ systems are highly conserved across vertebrates. In fact, several chapters highlight how work in zebrafish has both contributed to the understanding of human congenital diseases and provided information that has helped to diagnose some of them. In several chapters, a table summarizing the human diseases in the cardiovascular (Chapter 1), skeletal, (Chapter 3) immune (Chapter 4) and muscular (Chapter 6) systems which have been studied using zebrafish is provided as a resource. The early era of zebrafish developmental genetics was rooted in forward genetic screens in which the genome was randomly mutagenized either using chemicals (Chen et al., 1996; Geisler et al., 2007; Malicki et al., 1996; Pack et al., 1996; Ransom et al., 1996) or by viral insertion. Forward genetics is based on selecting a specific phenotype for which the investigator wants to identify the genes that govern it. This approach has been extremely powerful, as mutants that emerged from the several large-scale screens carried out decades ago are still being studied intensely. Screens identifying mutants with phenotypes that resemble human disease of the heart (Chapter 1), kidney (Chapter 2), skeleton (Chapter 3), blood (Chapter 4), liver (Chapter 5), muscle (Chapter 6), and pancreas (Chapter 7) are discussed in detail in this volume.

The revolution in gene editing technology has transformed the zebrafish field. With the advent of zinc finger nucleases, TALENs, and now the CRISPR/Cas9 system, nearly any gene of interest can be mutated. The wide application of these approaches in zebrafish, most notably, CRISPR/Cas9, has led to innovative methods to target specific genes in specific tissues or at discrete developmental times. Use of mutants, whereby the genetic change is heritable and stable, is complemented by the use of morpholinos, whereby a specific RNA can be targeted acutely and transiently in the embryo. The controversy surrounding the use of morpholinos in zebrafish has been sensationalized by a few outspoken advocates for exclusive use of mutants; however, the power of using a knock-down approach is demonstrated throughout the work described in nearly all of the chapters in this volume. In fact, the combined use of both mutants and morphants has led to unexpected discoveries on how zebrafish adapt to heritable mutations through compensatory changes in other genes (Rossi et al., 2015).

The overarching theme of this volume is the interplay between developmental processes and disease. Developmental biologists as well as our colleagues in obstetrical and pediatric fields can attest that development is not always perfect. Congenital defects—i.e., the consequences of failed developmental processes—afflict millions of children each year (http://www.marchofdimes.org/mission/march-of-dimes-global-report-on-birth-defects.aspx) and affect every organ system. The genes that govern developmental processes are highly conserved and zebrafish with mutations in homologous genes that cause congenital defects in humans display phenotypes analogous to the human diseases; cardiomyopathies (Chapter 1), cystic kidneys (Chapter 2), anemia, and bone marrow deficiencies (Chapter 4) are discussed in detail here. Many of these disorders have a genetic basis, while a host of others are caused by environmental factors and both of these are covered in this volume.

The impact of studying development processes reaches beyond congenital disorders. Our collective work in zebrafish and other model systems has also provided instructions for manipulating development to benefit human disease. In no case is this more evident than in stem cell biology, where the factors that govern potency and differentiation are manipulated to replace or repair mature cells and tissues which have been rendered insufficient by pathology. In fact, the lifelong impact of defects in development is exemplified by blood disorders (Chapter 4), degenerative disease of the muscle (Chapter 6), skeletal malformation (Chapter 3), the failure of pancreatic beta cells in diabetes (Chapter 7), and the regeneration of the liver (Chapter 5) or kidney (Chapter 4) following injury. Many aspects of cancer are attributed to the cooption of developmental processes, and the relationship between development and cancer is illustrated in hematopoietic malignancies described in Chapter 4. Screening for chemical modifiers of disease-relevant phenotypes has been extremely powerful, and illustrations of how drug screens have been applied to identify novel compounds to treat defects in hematopoiesis, kidney cysts, beta cell division, and myopathies are described in the relevant chapters.

Several decades before zebrafish was first used to study development, fish embryos were the model of choice in the field of environmental toxicology. The impact of toxicants on fish and wildlife is a broad and important topic, and these have been reviewed elsewhere. Chapter 9 describes the attributes of zebrafish that make it a powerful tool to broaden our understanding of environmental toxicology and the acute, long-term, and transgenerational effects of early toxicant exposures.

This volume provides both a unique perspective and a much needed resource. Work in zebrafish disease models has expanded beyond genetic disorders and developmental defects, and there is now an active and growing community of researchers studying a range of diseases in zebrafish. The recent establishment of the Zebrafish Disease Models Society (zdmsociety.org) and the well-attended ZDMS annual meeting attest to the growing impact of zebrafish in disease modeling.

The authors who contributed to each chapter are scientists working at the cutting edge of both disease and developmental fields. Combined, these researchers who have participated in pioneering screens for zebrafish developmental mutants, those whose research has uncovered the molecular basis for the disease-relevant developmental defects caused by these mutations, and carried out drug screens to identify compounds to modify developmental defects, and clinicians apply their first hand understanding of the importance of disease research to making discoveries using zebrafish. I am indebted to these dedicated and talented researchers for bringing their expertise to these pages and for our collegial work during the editorial process. The contributions of their trainees and other colleagues who share authorship are invaluable. The editorial input of members of my research team, Drs. Bhavani Madakashira and Anjana Ramdas Nair, anonymous reviewers, and the expert support of the publishing team at Elsevier have promoted cohesiveness among chapters and has made this an exciting process. This is not a comprehensive review of development and disease research in zebrafish, which a decade ago could have been compiled, but now the field is too vast to cover within a single volume. Indeed, major areas of active research in zebrafish, including neuroscience, sensory perception, metabolism, and cancer, were omitted largely due to space constraints. We collectively apologize to those researchers whose work was not covered and are grateful for the strong community of zebrafish researchers who enable collaborative and exciting research in the field of zebrafish development and disease.

REFERENCES

Chen, J. N., Haffter, P., Odenthal, J., Vogelsang, E., Brand, M., van Eeden, F. J., et al. (1996). Mutations affecting the cardiovascular system and other internal organs in zebrafish. *Development, 123*, 293–302.

Geisler, R., Rauch, G. J., Geiger-Rudolph, S., Albrecht, A., van Bebber, F., Berger, A., et al. (2007). Large-scale mapping of mutations affecting zebrafish development. *BMC Genomics, 8*, 11.

Malicki, J., Neuhauss, S. C., Schier, A. F., Solnica-Krezel, L., Stemple, D. L., Stainier, D. Y., et al. (1996). Mutations affecting development of the zebrafish retina. *Development, 123*, 263–273.

Pack, M., Solnica-Krezel, L., Malicki, J., Neuhauss, S. C., Schier, A. F., Stemple, D. L., et al. (1996). Mutations affecting development of zebrafish digestive organs. *Development, 123*, 321–328.

Ransom, D. G., Haffter, P., Odenthal, J., Brownlie, A., Vogelsang, E., Kelsh, R. N., et al. (1996). Characterization of zebrafish mutants with defects in embryonic hematopoiesis. *Development, 123*, 311–319.

Rossi, A., Kontarakis, Z., Gerri, C., Nolte, H., Holper, S., Kruger, M., et al. (2015). Genetic compensation induced by deleterious mutations but not gene knockdowns. *Nature, 524*(7564), 230–233.

Kirsten C. Sadler
New York University Abu Dhabi,
United Arab Emirates

CHAPTER ONE

Modeling Syndromic Congenital Heart Defects in Zebrafish

Meagan G. Grant[1], Victoria L. Patterson[1], Daniel T. Grimes, Rebecca D. Burdine[2]
Princeton University, Princeton, NJ, United States
[2]Corresponding author: e-mail address: rburdine@Princeton.EDU

Contents

Abstract

Cardiac development is a dynamic process regulated by spatial and temporal cues that are integrated to effect molecular, cellular, and tissue-level events that form the adult heart. Disruption of these highly orchestrated events can be devastating for cardiac form and function. Aberrations in heart development result in congenital heart defects (CHDs), which affect 1 in 100 infants in the United States each year. Zebrafish have

[1] These authors contributed equally.

Current Topics in Developmental Biology, Volume 124
ISSN 0070-2153
http://dx.doi.org/10.1016/bs.ctdb.2016.11.010

proven informative as a model organism to understand both heart development and the mechanisms associated with CHDs due to the similarities in heart morphogenesis among vertebrates, as well as their genetic tractability and amenability to live imaging. In this review, we discuss the mechanisms of zebrafish heart development and the utility of zebrafish for understanding syndromic CHDs, those cardiac abnormalities that occur in the context of multisystem disorders. We conclude with avenues of zebrafish research that will potentially inform future therapeutic approaches for the treatment of CHDs.

1. INTRODUCTION

1.1 Overview of Congenital Heart Defects

During human embryogenesis, the fetal heart undergoes a series of dynamic morphogenetic events. Cardiac progenitor cells (CPCs) are specified and differentiated based on location in order to migrate and form a primitive heart tube. Multiple signaling and patterning events occurring throughout this process ensure the heart is positioned appropriately and that the embryonic heart tube transforms into a contractile, multichambered organ capable of pumping blood throughout the body. The processes driving human heart development are myriad and complex. Disruption of these processes results in congenital heart defects (CHDs), the leading cause of infant mortality arising from structural birth defects in the United States. CHDs differ in prevalence, severity, and the heart tissues affected, existing in isolation or in the context of syndromic disorders affecting multiple organ systems, and can arise as a result of multigenic causes or mutations in single genes (Mozaffarian et al., 2016). Animal models are critical to studying these pathologies, with zebrafish emerging as a premier tool for understanding CHDs.

1.2 Zebrafish as a Model of Vertebrate Cardiogenesis and CHDs

The zebrafish is an excellent system for studying heart development and the aberrations in morphogenesis that result in cardiac defects. The human heart is comprised of two upper chambers (the left and right atria) and two lower chambers (the left and right ventricles). A septum separates the left atrium from the right atrium and the left ventricle from the right ventricle. Despite the fact that the zebrafish heart consists of a single atrium and ventricle that are not septated, the morphogenetic behaviors required for heart development in humans and zebrafish are strikingly similar. Moreover, many of the genes involved in heart development (Table 1) are conserved (for example, see Bruneau, 2008; Richards & Garg, 2010; Szeto et al., 2002), and are useful in visualizing specific cell types in the heart (Table 2). Other facets making zebrafish ideal for cardiovascular studies include their fecundity,

Table 1 Human and Zebrafish Cardiac Phenotypes Caused by Mutations in Single Genes Associated With Nonsyndromic Heart Defects

Gene	Patient CHD	Zebrafish Heart Defects	References
GATA4	ASD, VSD	Cardiomyopathy, defective cardiac looping, chamber expansion, blood circulation, and heart tube displacement	Bruneau (2008), Holtzinger and Evans (2005), and Richards and Garg (2010)
GATA6	AVSD, PTA, TOF	Cardia bifida, impaired cardiac looping, and heart tube fusion defects	Holtzman, Schoenebeck, Tsai, and Yelon (2007) and Kodo and Yamagishi (2010)
MYH6	ASD, HCM	Dilated atrium, weakened atrial contractility, defective myofibrillar organization, thickening of ventricular wall, and narrowing of ventricular lumen	Auman et al. (2007), Berdougo, Coleman, Lee, Stainier and Yelon (2003), Bruneau (2008), and Richards and Garg (2010)
NKX2.5	ASD, VSD, conduction defects, TOF, TGA, valve defects	Defective proliferation of SHF progenitors, misshapen atria, smaller ventricle, loss of SHF-derived ventricular myocardium and OFT smooth muscle	Bruneau (2008), Guner-Ataman et al. (2013), Richards and Garg (2010), and Targoff, Schell, and Yelon (2008)
TBX20	ASD, VSD	Abnormal contractility, no blood circulation, edema, abnormal cardiac looping and chamber morphology	Bruneau (2008) and Szeto, Griffin, and Kimelman (2002)
HAND2	TOF, PS, AVSD, VSD-DORV	Reduced myocardium, endocardium fails to form cone	Palencia-Desai et al. (2015), Shen et al. (2010), and Yelon et al. (2000)

ASD, atrial septal defect; *AVSD*, atrioventricular septal defect; *DCM*, dilated cardiomyopathy; *DORV*, double outlet right ventricle; *HCM*, hypertrophic cardiomyopathy; *PS*, pulmonary stenosis; *PTA*, persistent truncus arteriosus; *TGA*, transposition of the great arteries; *TOF*, tetralogy of fallot; *VSD*, ventricular septal defect.

Table 2 Markers for Cell Types in the Zebrafish Heart

Cell Type	Transgenic Marker	Gene Marker	Antibody or Fluorescent Indicator	Functional Assay	References
Myocardial	*Tg(myl7:GFP)*, *Tg(cmlc2:dsred2-nuc)*	*myl7*[a], *vmhc*, *amhc*	MF20, S46	Cardiac conduction: *Tg(cmlc2:gCaMP)*^{s878}	Miura and Yelon (2011)
				Timing of differentiation: *Tg(cmlc2:Kaede)*	
Endocardial and endothelial	*Tg(flk1:EGFP)*, *Tg(fli1:EGFP)*	*flk1* *(kdrl/vegfr2)*, *fli1*	Kdrl	Angioblast migration and vascular tube formation: *Tg(flk1:EGFP)*	Miura and Yelon (2011) and Poon, Liebling, Kondrychyn, Garcia-Lecea, and Korzh (2010)
Epicardial	*Tg(tcf21:DsRed2)*, *Tg(wt1b:eGFP)*	*tcf21*	Raldh2 (also in endocardium)	Cellular contributions of the epicardium: *tcf21:CreER*; *gata5:RnG*	Kikuchi et al. (2011) and Perner, Englert, and Bollig (2007)
Second heart field	*Tg(ltbp3:TagRFP2Acre)*; *Tg(cmlc2:CSY)*, *Tg(nkx2.5::ZsYellow)*; *Tg(cmlc2::CSY)*	*ltbp3*	Eln2 and DAF-2DA (BA), Isl1	Contributions to SHF-derived structures: photoconversion of *Tg(nkx2.5:Kaede)* transgenics	de Pater et al. (2009), Grimes, Stadt, Shepherd, and Kirby (2006), Guner-Ataman et al. (2013), and Zhou et al. (2011)
	BA smooth muscle cells: *Tg(eln2:CSY)*			Cellular contributions to OFT SMCs: *Tg(eln2:CSY)* reporter strain crossed with a driver strain [*Tg(ltbp3::TagRFP2Acre)*, *Tg(gata4:ERCreER)*, *or* *Tg(nkx2.5:ERCreER)*, for example]	

Erythrocyte	*Tg(Gata1:DsRed)*, *Tg(Gata1:EGFP)*	*gata1*	–	Blood flow through vessels: *Tg(Gata1:Dsred); Tg(Fli1: EGFP)*	De Domenico et al. (2007), Miura and Yelon (2011), and Poon et al. (2010)
				Hemoglobin production: o-dianisidine staining	
				Tissue iron delivery: Prussian blue staining	

[a] *myl7* was formerly referred to as *cmlc2*.

external fertilization, rapid development, and transparency. These qualities allow in vivo time-lapse imaging, and examination of multiple stages and aspects of heart development. Additionally, the small size of zebrafish embryos allows for passive diffusion of oxygen, permitting survival of the organism for several days despite cardiovascular defects (Stainier, 2001). Thus, mutants with abrogated cardiovascular function can be studied for longer than is possible in mammalian organisms where cardiac defects cause early lethality. Indeed, many zebrafish cardiovascular mutants, some discovered decades ago (Haffter et al., 1996; Stainier et al., 1996; Stainier, Weinstein, Detrich, Zon, & Fishman, 1995), have provided important insights into the molecular and cellular processes underlying vertebrate heart formation. Finally, methodologies for manipulating the genome in order to generate tools and disease models are rapidly advancing in zebrafish. These aspects make zebrafish a versatile model for furthering our understanding of human CHDs. Here we discuss how this system has been successfully used to understand the genetic basis of heart development and to identify the mechanisms of syndromic heart defects in humans (Table 3).

Table 3 Syndromic Heart Defects: Causative Genes, Human Heart Defects, and Zebrafish Phenotypes

Syndrome	Example Genes Involved	Patient CHDs Include	Zebrafish Phenotypes
Ciliopathies and Heterotaxy	*PKD2*, *PKD1L1*	Dextrocardia	Aberrant heart looping laterality
RASopathies	*PTPN11*, *RIT1*, *NF1*, *RAF1*, *KRAS*, *HRAS*	HCM, CoA, ASD, VSD, PVS, PS	Impaired cardiac jogging and looping, reduced cardiac function, edema, defects in heart size, hypoplastic chambers, valve defects, thickening of heart walls, delays in heart morphogenesis
Cohesinopathies	*NIPBL*, *RAD21*	VSD, ASD, PS, TOF, HLHS	Reduced heart size, impaired cardiac looping, valve defects
CHARGE	*CHD7*, *SEMA3A*, *SEMA3E*	TOF,PDA, DORV, VSD, ASD, AVSD, LVOTO, RVOTO	Dysmorphic chambers, edema, reduced blood flow, weak heartbeat, abnormal narrowing of dorsal aorta

Table 3 Syndromic Heart Defects: Causative Genes, Human Heart Defects, and Zebrafish Phenotypes—cont'd

Syndrome	Example Genes Involved	Patient CHDs Include	Zebrafish Phenotypes
DiGeorge	*TBX1*	TOF, PA-VSD, truncus arteriosus, IAA, aortic arch anomalies, VSD	Impaired cardiac jogging and looping, aortic arch defects, reduced proliferation in FHF, reduced incorporation of SHF cells at arterial pole, impaired OFT development
Williams-Beuren	*ELN*	SVAS, pulmonary arterial stenosis, STA, PPS, OFT obstruction	Hypoplasia and reduced contraction of the BA, aberrant cardiac/smooth muscle differentiation
Holt–Oram	*TBX5*	ASD, VSD, TOF, arrhythmias, HLHS, PDA	Impaired cardiac looping, arrested differentiation, reduced contractility, stretching (and ripping) of atrium, smaller ventricle

ASD: atrial septal defect; *AVSD*: atrioventricular septal defect; *BA*: bulbus arteriosus; *CoA*: coarctation of the aorta; *DORV*: double outlet right ventricle; *HCM*: hypertrophic cardiomyopathy; *HLHS*: hypoplastic left heart syndrome; *IAA*: interrupted aortic arch; *LVOTO*: left ventricular outflow tract obstruction; *OFT*: outflow tract; *PA-VSD*: pulmonary atresia with VSD; *PDA*: persistent ductus arteriosus; *PPS*: peripheral pulmonary stenosis; *PS*: pulmonary stenosis; *PVS*: pulmonary valve stenosis; *RVOTO*: right ventricular outflow tract obstruction; *STA*: stenosis of the thoracic aorta; *SVAS*: supravalvular aortic stenosis; *TOF*: tetralogy of fallot; *VSD*: ventricular septal defect.

2. ZEBRAFISH HEART DEVELOPMENT

Although occurring on a much faster scale, the processes driving zebrafish heart formation are very similar to those described for other vertebrates. In brief, in vertebrates the heart is formed from mesodermal cells that are specified during gastrulation on both the left and right sides of the embryo. These cells undergo similar processes including differentiation, medial migration to the midline, and fusion of both left and right populations at the midline to form a contractile heart tube. Additional cells are subsequently recruited from a secondary heart field to the poles of the heart tube, where they contribute to structures including the myocardium and the outflow tract (OFT). The heart tube bends rightward in a process known as cardiac looping, which, in humans, is necessary for alignment and septation of the cardiac chambers (Harvey & Rosenthal, 1999). The bending of the linear tube during cardiac looping in zebrafish is preceded by an asymmetric process termed "jogging," whereby the atrial cells are placed to the anterior and left of ventricular cells during generation of the linear tube. Despite differences, the genes and mechanisms involved in cardiac development are conserved between zebrafish and humans, and are presented in more detail below.

2.1 Specification and Differentiation of Cardiac Progenitor Cells

The first step in cardiac development is the specification of cells as CPCs. Labeling of individual blastomeres using laser-mediated activation of caged fluorescein revealed that by 5 h postfertilization (hpf), myocardial populations are specified as either ventricular or atrial precursors (Fig. 1A; Keegan, Meyer, & Yelon, 2004). Both populations reside in the lateral marginal zone (LMZ), an area of developing mesendodermal cells at the border between the cells and the yolk, with ventricular precursors located more dorsally and marginally. Endocardial precursors reside in the LMZ alongside myocardial cells (Fig. 2A), but are not spatially segregated into ventricular and atrial populations at this stage (Keegan et al., 2004). After their involution during gastrulation, precursor cells migrate to the anterior lateral plate mesoderm (ALPM), where they form distinct bilateral sheets on either side of the midline by 15 hpf, with ventricular precursors located more medially than atrial progenitors (Fig. 1B–D; Yelon, Horne, & Stainier, 1999).

Multiple signaling pathways converge to regulate CPC specification. Hedgehog (Hh) signaling acts cell autonomously to promote both ventricular and atrial CPC specification, while retinoic acid (RA) acts to restrict the number of cardiac progenitors. Wnt signaling acts biphasically, with induction or repression of specification dependent on whether signaling is active before, or during, gastrulation, respectively. Subpopulation-specific effects are also evident; ventricular CPCs are more sensitive to regulation by Nodal and fibroblast growth factor (FGF) signaling than atrial CPCs, while atrial precursors are more affected by changes in bone morphogenetic protein (BMP) signaling than ventricular precursors (for a more in depth review of signaling in CPC development, see Staudt & Stainier, 2012).

Cardiomyocyte differentiation is driven by a variety of transcription factors. Both pools of myocardial CPCs express *nkx2.5* (Yelon et al., 1999), and

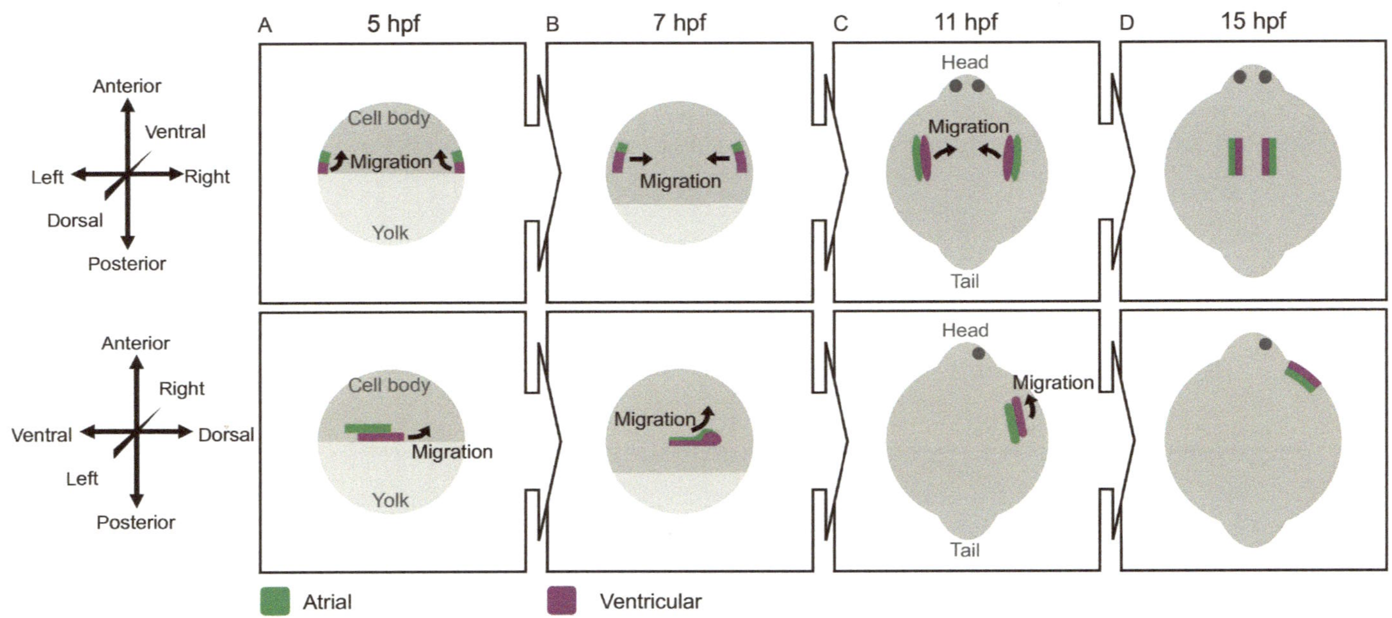

Fig. 1 Specification and migration of cardiac precursor cells. (A) Atrial and ventricular cardiac progenitor cells (CPCs) are specified on both the left and right side of the embryo. Ventricular precursors lie dorsal and marginal to atrial precursors. (B) CPCs maintain their relative spatial organization while migrating dorsally as cohesive sheets. (C) CPCs migrate toward the embryonic midline, such that atrial progenitors are lateral and ventral to ventricular precursors. (D) This organization is maintained during migration to the anterior lateral plate mesoderm.

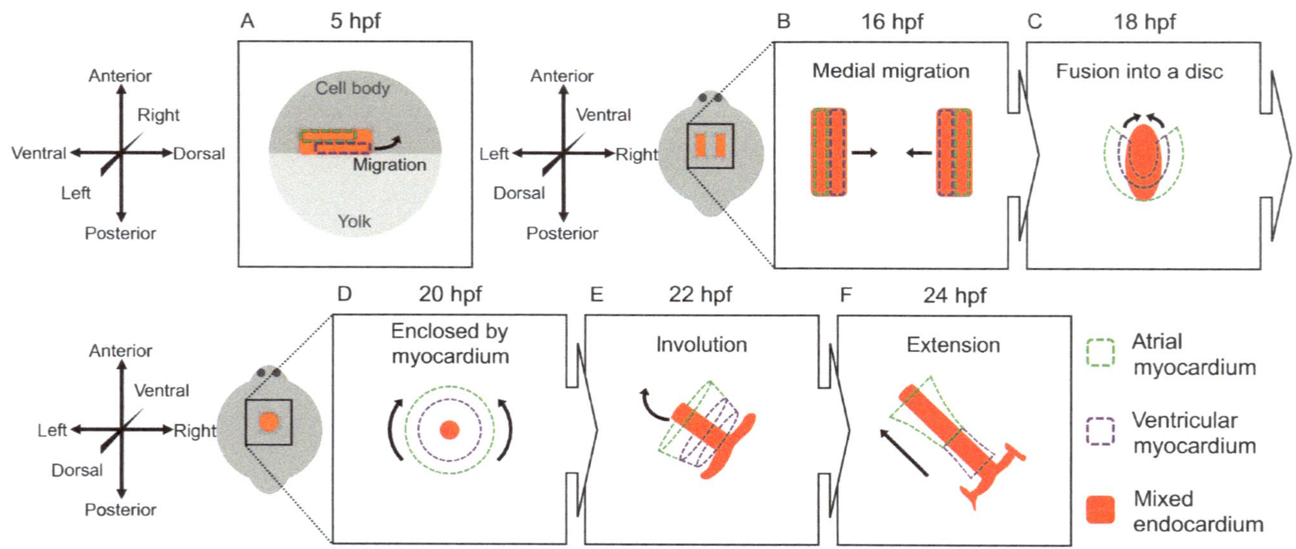

Fig. 2 Endocardial development and migration. (A) Endocardial precursors originate in the lateral marginal zone but show no separation into atrial and ventricular subpopulations. (B) Bilateral populations of endocardial cells initiate medial migration at 16 hpf. (C) Endocardium coalesces into a disk in the embryonic midline. (D) Myocardial fusion initiates at the posterior of the endocardial disk and proceeds to enclose the endocardium. (E) During jogging, the endocardium involutes with the myocardium. (F) The endocardium lines the lumen of the myocardial tube.

loss of *nkx2.5* and *nkx2.7* disrupts heart tube assembly, resulting in excess atrial cells at 26 hpf and reduced ventricular cardiomyocytes by 52 hpf (Targoff et al., 2008). Loss of both Gata5 and Gata6 results in reduced CPCs (Holtzinger & Evans, 2007) and minimal differentiated myocardial tissue develops in mutants for *hand2*, the basic helix-loop-helix transcription factor (Yelon et al., 2000). All CPCs express *myosin light chain 7* (*myl7*), but subpopulations can be distinguished since ventricular cardiomyocytes express *ventricular myosin heavy chain* (*vmhc*) while atrial cardiomyoctyes express *atrial myosin heavy chain* (*amhc*) (Yelon et al., 1999).

2.2 Heart Tube Formation and Extension

Heart tube development requires the medial migration of bilateral populations of endocardial and myocardial CPCs. The endocardial cells initiate migration at 16 hpf (14 somites, Fig. 2B) and move posteriorly (Bussmann, Bakkers, & Schulte-Merker, 2007). Fusion of the populations is initiated at 16.5 hpf (15 somites) and completed by 18 hpf (18 somites), when the endocardium forms a disk at the embryonic midline (Fig. 2C). Endocardium development is incompletely understood, but the transcription factor Tal1 is involved. Tal1 gain-of-function leads to expansion of the endothelial population, while loss-of-function causes aggregation of endocardial cells at the arterial pole of the heart, inducing ventricular stenosis (Bussmann et al., 2007; Gering, Yamada, Rabbitts, & Patient, 2003). *cloche*, a bHLH-PAS transcription factor, acts upstream of Tal1 as a master regulator of endothelial specification and its mutation leads to loss of endocardium (Reischauer et al., 2016). Medial migration is also influenced by vascular endothelial growth factor. Increasing the response to this important factor by modulating Slit/Robo signaling leads to multiple heart lumens as a result of faster movement of individual cells at the expense of collective migration. Decreasing the response to this factor causes unfused heart fields as a consequence of inhibited migration (Fish et al., 2011).

Endocardial CPC migration is essential in forming the endocardium, but also facilitates morphogenesis of the myocardium. Endocardial loss causes myocardial dysmorphia, as in the *cloche* mutant where angular cardiomyocyte migration is incorrectly executed (Holtzman et al., 2007; Stainier et al., 1995). This crosstalk is reciprocal, as when myocardial cells are reduced in number by deletion of *hand2*, endocardial cells migrate properly but fail to form the cardiac cone, seemingly as a consequence of disrupting BMP signaling (Garavito-Aguilar, Riley, & Yelon, 2010;

Palencia-Desai et al., 2015). Similarly, myocardial Tmem2 promotes migration of both myocardial and endocardial CPCs, with loss delaying medial migration of both populations and endocardial populations ultimately failing to fuse (Totong et al., 2011).

Myocardial CPCs begin medial migration to the midline later than endocardial cells, at 17.5 hpf (17 somites), as revealed through a combination of in situ hybridization, fate mapping, and live cell imaging experiments using fluorescently tagged cell type markers (Holtzman et al., 2007; Stainier et al., 1996; Yelon et al., 1999). Fusion initiates at the posterior end of the myocardium at 18 hpf (18 somites), with ventricular CPCs making contact prior to atrial CPCs (Fig. 3B). Myocardial fusion occurs immediately anterior to the most posterior endocardial cells (Fig. 2C). A second phase of angular migration allows the cardiomyocytes to surround the central endocardial cells (Fig. 2D) (Holtzman et al., 2007). Anterior myocardial CPCs then fuse to form a shallow cone (Fig. 3C) where ventricular precursors form the apex and atrial precursors localize beneath them at the wider base by 20 hpf (22 somites). The central lumen is lined with those endocardial cells that will connect to the aortic arches, although most of the endocardium remains ventral to the myocardial cardiac cone (Holtzman et al., 2007; Stainier, Lee, & Fishman, 1993).

The myocardial cells are organized into a single-layered epithelium, where ventricular CPCs adopt a cuboidal morphology and atrial CPCs are more squamous (Rohr, Bit-Avragim, & Abdelilah-Seyfried, 2006; Rohr, Otten, & Abdelilah-Seyfried, 2008). Proper cone formation requires cells to maintain their epithelial integrity. Mutations disrupting the adherens or tight junctions, such as *prkci* in *heart-and-soul* (*has*) or *mpp5a* in *nagie oko* (*nok*), impact cone formation (Horne-Badovinac et al., 2001; Peterson, Mably, Chen, & Fishman, 2001; Rohr et al., 2006, 2008). In *has* mutants, myocardial CPCs exhibit delayed posterior fusion until after the anterior regions have coalesced, and this delay causes a failure of tube morphogenesis with ventricular tissue developing within the atrium. CPCs do fuse at the midline in *nok* mutants, but the myocardial layer loses epithelial coherence, forming partial multilayers, and failing to undergo cone rotation. Epithelial integrity is also important during subsequent stages of cardiac morphogenesis. Disruption of cell junctions results in nondirectional migration of single cells that have lost contact with the surrounding cells (Rohr et al., 2008). Thus, epithelial coherence ensures cells migrate collectively, move uniformly in the correct direction, and maintain contact with neighboring cells to prevent population mixing.

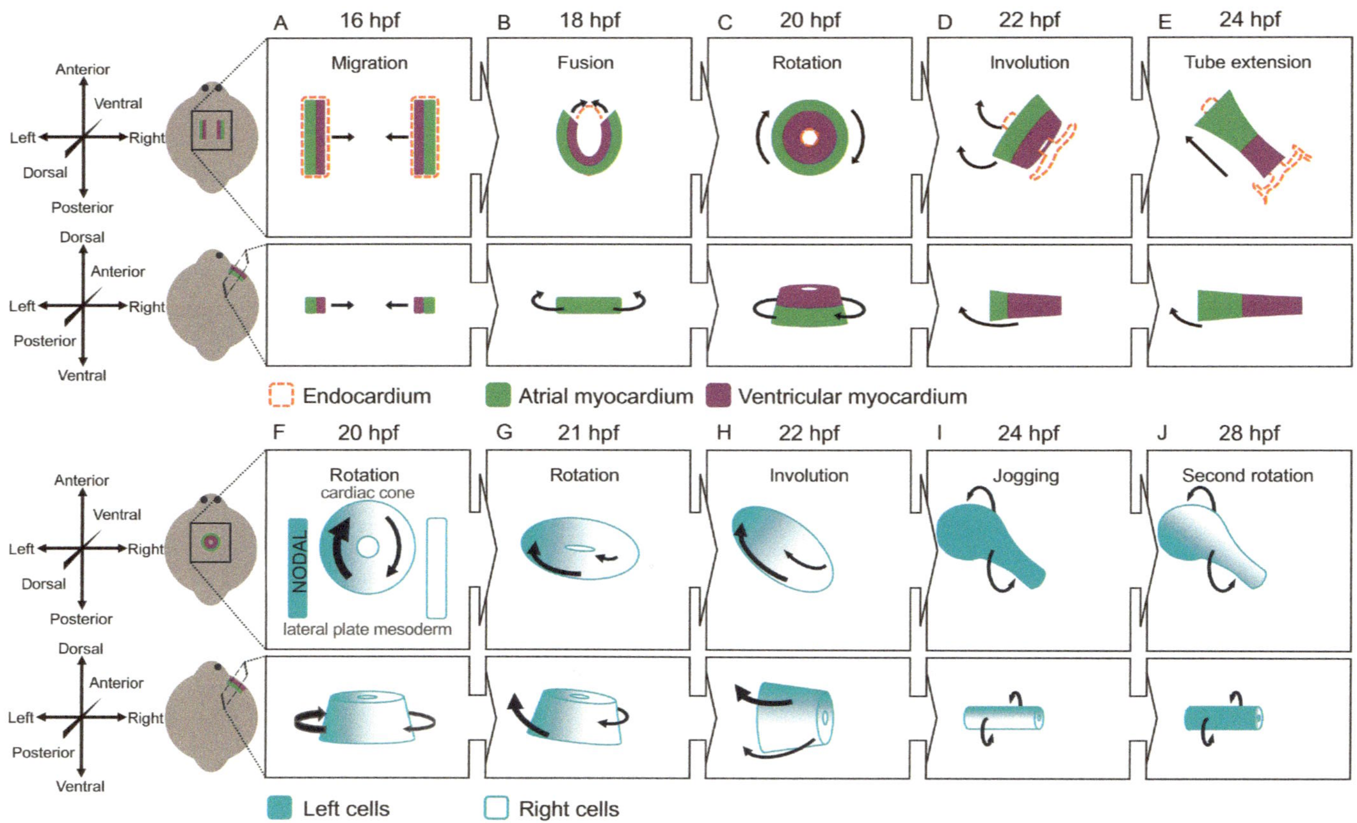

Fig. 3 See legend on next page.

Successful migration of CPCs additionally depends on input from the endoderm and extracellular matrix (ECM). The endoderm contributes a surface along which CPCs can migrate and actively signals to the myocardium, for example through sphingolipids (Fukushima, Ishii, Contos, Weiner, & Chun, 2001; Maceyka, Payne, Milstien, & Spiegel, 2002). Mutations in the transcription factors *gata5* and *sox32*, or the Nodal cofactor *oep*, result in loss of endoderm tissue and a failure of CPCs to migrate to the midline. Mutation of the *sphingosine-1 phosphate receptor* (*s1pr2*) or the sphingosine-1 phosphate transporter *spinster 2* (*spns2*) disrupts sphingolipid signaling and causes similar CPC migration phenotypes. The result is cardia bifida, in which two separate hearts develop on either side of the midline (Hisano, Ota, Takada, & Kawahara, 2013; Kawahara et al., 2009; Kupperman, An, Osborne, Waldron, & Stainier, 2000; Osborne et al., 2008; Stainier et al., 1996). The finding that cardia bifida also occurs in mutants where fibronectin deposition is altered, blocking migration of the CPCs toward the midline, demonstrates the importance of the ECM in directing the location and formation of the heart (Arrington & Yost, 2009; Sakaguchi, Kikuchi, Kuroiwa, Takeda, & Stainier, 2006; Trinh & Stainier, 2004; Yelon et al., 2000).

Once the cardiac cone forms it continues to migrate as a collective population culminating in the conversion of the cone into a linear heart tube. This process, called "jogging" in zebrafish, is influenced by left–right patterning cues generating the first morphological visceral asymmetry in the embryo. Jogging occurs over the span of 4 h and culminates in asymmetric positioning of the atrium to the left and anterior of the ventricle (Figs. 2E and F and 3D and E). The entire process can be visualized by live imaging of

Fig. 3 Formation and positioning of the linear heart tube. (A) Cardiac precursors are localized to the anterior lateral plate mesoderm (ALPM), with ventricular precursors situated more medially than atrial precursors. (B) Populations migrate toward the midline, initiating contact at the posterior. (C) Anterior fusion forms a shallow cone, with ventricular precursors forming the tip and atrial precursors forming the base. (D) The cone rotates, tilting such that atrial cells are to the left of ventricular cells. (E) Cone extension positions atrial cells to the left and anterior of ventricular populations. (F) Nodal signaling from the left LPM increases the migratory velocity of cells on the left of the cardiac cone, resulting in clockwise rotation. (G) Left cells migrate left anteriorly along the lateral edges of the cone while slower cells on the right migrate anteriorly around the lumen of the cone. (H) During later stages of rotation the cone tilts in addition to rotating. (I) Cells originating on the left are displaced to form the dorsal region of the extending heart tube. (J) An additional leftward rotation occurs to reposition cells originating from the left and right cells back to their respective sides of the extending tube.

fluorescently tagged cardiomyocytes and in situ hybridization (for example, see Baker, Holtzman, & Burdine, 2008; de Campos-Baptista, Holtzman, Yelon, & Schier, 2008; Rohr et al., 2008; Smith et al., 2008) providing detailed characterization of cell behavior. The jogging direction of the heart is regulated by asymmetric Nodal signaling (Fig. 3F). The zebrafish Nodal gene *southpaw* (*spaw*) is expressed in the left lateral plate mesoderm (LPM) but is absent from the right LPM (Baker et al., 2008; Long, Ahmad, & Rebagliati, 2003; Schier & Shen, 2000). Live imaging revealed that exposure to Nodal signaling increases the velocity of CPC migration in the left side of the cardiac cone, compared to the right (Baker et al., 2008; de Campos-Baptista et al., 2008; Lenhart, Holtzman, Williams, & Burdine, 2013; Smith et al., 2008). This left–right (L–R) asymmetry in CPC migration velocity causes the cardiac cone to rotate clockwise as the entire cone migrates anteriorly (Fig. 3G). This rotation is accompanied by involution of cells on the posterior right of the midline (Fig. 3H), leading to tilting of the cone along the anterior posterior axis (Rohr et al., 2008). Together, asymmetric cell migrations and involution cause the tube to extend toward the left into the typical leftward jog (Fig. 3I). In the absence of *spaw*, CPC migration velocities are significantly slowed and become more L–R symmetrical. In addition, the point of involution within the cone becomes randomized. Together this leads to loss of left jogging, and heart tube positioning along the L–R axis becomes random in direction (Lenhart et al., 2013; Rohr et al., 2008; Smith et al., 2008).

While it is clear Nodal signaling is the dominant laterality cue in the heart, BMP signaling also plays a role in jogging laterality though the exact role remains to be determined (Lenhart et al., 2013; Smith et al., 2008; Veerkamp et al., 2013). Migration of CPCs toward exogenous BMP protein suggests this molecule can function as a promigratory cue (Smith et al., 2008). However, Bmp activity is reported to be higher on the right in slower migrating cells (Veerkamp et al., 2013). Additionally, mutations in *bmp4* suggest a role for limiting cell velocity of CPCs (Lenhart et al., 2013). Given that heterozygous mutants for *bmp4* could modify the phenotypes observed in *spaw* morphants, the levels of BMP signaling may be critical for proper CPC migration (Lenhart et al., 2013). Interestingly, the response to BMP signaling, as visualized by phospho-SMAD 1/5/8 immunoreactivity, appears to be within the endocardium. The BMP response in this tissue is abolished with the Nodal transcription factor FoxH1 is mutated, further suggesting Nodal signaling lies upstream of the BMP response (Lenhart et al., 2013).

As a result of rotation and involution, cells from the left half of the cone that received Nodal signals from the LPM localize to the dorsal side of the heart tube (Baker et al., 2008; Rohr et al., 2008). A second, leftward rotation occurs to reposition the dorsally displaced cells back to the left of the heart tube by 48 hpf, prior to cardiac looping (Fig. 3J; Baker et al., 2008), but the mechanisms involved remain to be clarified.

Overall, the role of Nodal in directing asymmetric cardiac morphogenesis is conserved among vertebrates, although the exact event influenced by this pathway is organism specific. In mouse for example, *Nodal* is asymmetrically expressed just prior to cardiac looping and influences the directionality of this process. In zebrafish, Nodal is expressed lateral to the cardiac cone and influences cone rotation and jogging as described earlier. Additionally, cardiac cone rotation in zebrafish seems analogous to the slight rotation observed in the linear heart tube during looping in mouse and chick (Baker et al., 2008; Rohr et al., 2008; Smith et al., 2008).

Following its formation, the heart tube undergoes extension. Elongation is not a result of increased proliferation, as the cardiomyocyte proliferation rate is very low (de Pater et al., 2009; Rohr et al., 2006). At 20 hpf (22 somites) the cardiac cone comprises approximately 85 cells (R. Burdine, unpublished). By 24 hpf, the heart tube is approximately 150 cells, growing to 270 cells by 36 hpf and 310 cells by 48 hpf (de Pater et al., 2009; Rohr et al., 2006). Such low proliferation rates are insufficient to account for the increases in tube length. Instead, the cone is lengthened by remodeling myocardial cell morphology, such that cells—especially atrial precursors—adopt an extended squamous morphology, and by addition of cells from secondary sources (de Pater et al., 2009; Hami, Grimes, Tsai, & Kirby, 2011; Rohr et al., 2006).

2.3 The Second Heart Field

Two fields of CPCs form the vertebrate heart. The first heart field (FHF) gives rise to the linear heart tube (Fig. 3). The evolutionarily conserved second heart field (SHF) contributes CPCs that are progressively added to the poles of the heart tube. SHF progenitors remain undifferentiated in pharyngeal mesoderm until incorporated into heart tube myocardium between 24 and 48 hpf, where they are added to either the arterial (outflow) pole at the ventricular end of the heart tube (Fig. 4A and B) or to the venous pole (inflow tract) at the atrial end (de Pater et al., 2009; Hami et al., 2011). SHF cells contribute to smooth muscle in the OFT, also referred to as the bulbus

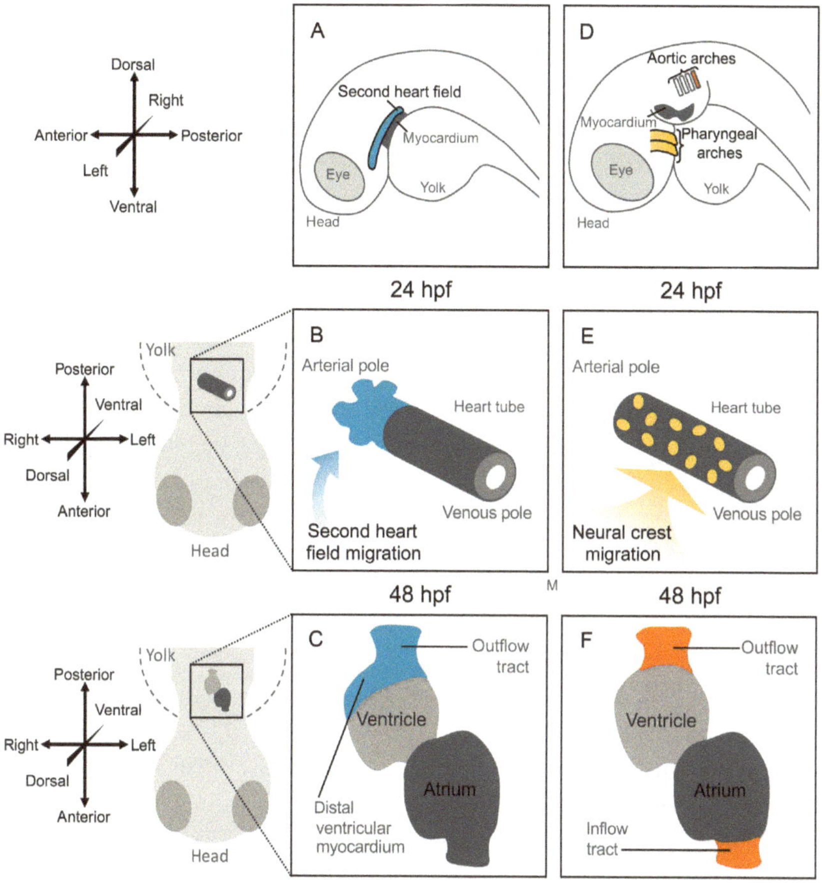

Fig. 4 Secondary cardiac precursor populations contribute cells to different regions of the developing heart. (A) The second heart field (SHF) overlies the proliferating myocardium, localizing to the pharyngeal arches. (B) SHF cells contribute to the arterial pole. (C) Later in development, SHF cells localize to distal ventricular myocardium and the outflow tract. (D) Cardiac neural crest cell (CNC) arise from between rhombomeres 1 and 6 (not shown). Early streaming CNC migrate through pharyngeal arches 1 and 2. Later streaming CNC migrates through aortic arch 6. (E) Early migrating CNC contribute cardiomyocytes throughout the linear heart tube. (F) Later migrating CNC populate the inflow and outflow tracts.

arteriosus (BA) in zebrafish, as well as the distal ventricular myocardium (Fig. 4C) (Grimes et al., 2006; Hami et al., 2011; Zhou et al., 2011).

Multiple transcription factors have been identified as SHF regulators in zebrafish. For example, knockdown of *tbx1*, *mef2c*, or *nkx2.5* impairs SHF cell proliferation and differentiation, and underdevelopment of

SHF-derived structures, including the OFT and distal ventricular myocardium (Guner-Ataman et al., 2013; Hinits et al., 2012; Lazic & Scott, 2011). SHF accretion to the arterial pole requires FGF and TGFβ signaling; pharmacological inhibition of FGF signaling or knockdown of latent TGFβ-binding protein 3 (Ltbp3) results in decreased accretion at the arterial pole and truncation of the OFT (Marques, Lee, Poss, & Yelon, 2008; Zhou et al., 2011). These defects are consistent with human conotruncal heart defects, which can arise from defective SHF development and comprise nearly 30% of all CHDs (Rochais, Mesbah, & Kelly, 2009). There are some interesting differences, however. For example, in mouse, the transcription factor Isl1 is required for the recruitment of cells to both the venous and arterial pole of the heart tube (Cai et al., 2003). In zebrafish, it is required for cardiomyocyte differentiation at the venous pole but not at the arterial pole (de Pater et al., 2009). Thus, since the SHF in zebrafish is akin in origin and function to that of mammals, studying SHF formation in zebrafish embryos can inform how SHF-related CHDs arise in humans.

2.4 Cardiac Neural Crest

Neural crest cells (NCCs) give rise to a large number of differentiated cell types and can be divided into five subtypes—cranial, vagal, sacral, trunk, and cardiac. In zebrafish, cardiac neural crest cells (CNCCs) invade the myocardium of many cardiac structures, including the OFT, atrium, ventricle, and atrioventricular canal (AVC) (Li et al., 2003; Sato & Yost, 2003). CNCCs originate in a broad region between rhombomeres 1 and 6, but their addition to the developing heart occurs in two waves as visualized by lineage tracing and fate mapping (Cavanaugh, Huang, & Chen, 2015; Sato & Yost, 2003). In the first stream, CNCCs migrate via pharyngeal arches 1 and 2 and are added throughout the heart tube between 24 and 30 hpf (Fig. 4D and E), where they differentiate into cardiomyocytes. In the second stream, CNCCs migrate via aortic arch 6 and are added to the ventral aorta and OFT around 80 hpf (Fig. 4D and F) (Cavanaugh et al., 2015). A variety of factors are involved in regulating CNCC contributions to the heart, including Wnt, FGF, and Semaphorin pathways (Cavanaugh et al., 2015; Sato, Tsai, & Yost, 2006; Sun, Zhang, Lin, & Xu, 2008).

Given the cellular contributions of CNCCs to different cardiac structures, it is unsurprising that defective CNCC development has been implicated in CHDs, including OFT malformations, ventricular septal defects

(VSDs), aortic arch anomalies, pulmonary stenosis (PS), and coarctation of the aorta (CoA) (Keyte & Hutson, 2012). Disrupting CNCCs in zebrafish also causes cardiac defects including compromised cardiac looping, depressed heart rate, smaller ventricles, and loss of SHF cell recruitment (Cavanaugh et al., 2015; Li et al., 2003), making fish a useful model for pursuing the role of CNCCs in cardiac disease.

2.5 Cardiac Looping, Ballooning, and Chamber Formation

The heart tube bends rightward in a process known as cardiac looping, which, in humans, is necessary for alignment and septation of the cardiac chambers (Harvey & Rosenthal, 1999). Though the zebrafish heart chambers are not septated, cardiac looping is still occurs. Looping begins at 30 hpf, forming a slight kink in the middle of the linear tube (Fig. 5A). Bending becomes more pronounced, with the tube becoming increasingly "S" shaped as development proceeds (Fig. 5B). Looping occurs in a defined L–R asymmetric fashion, resulting in a dextral loop that positions the ventricle to the right and anterior of the atrium. The events underlying looping morphogenesis remain poorly understood, although cellular migration and tissue-level forces exerted through the cytoskeleton are likely to be involved. Treatment of explanted linear heart tubes with cytochalasin B or blebbistatin, inhibitors of actin polymerization and myosin II, respectively, impairs cardiac looping and constriction at the AVC (Noel et al., 2013).

Following looping, heart chambers expand via cardiac ballooning (Fig. 5C). During this process, the chamber curvatures can be distinguished by the expression of *naturietic peptide precursor a* (*nppa*), which is regionally restricted to the myocardium of the convex outer curvature (OC) of both chambers, but absent from the concave inner curvature (IC) and AVC (Auman et al., 2007). OC cells appear elongated and flattened, while cells of the IC remain cuboidal, suggesting that regionalized differences in cell morphology bring about curvature formation and chamber expansion (Fig. 5C and D). The enlargement and elongation of OC cells are stimulated by blood flow but restricted by contractility (Auman et al., 2007). For example, *weak atrium* (*wea*) mutants exhibit decreased blood flow through the ventricle, resulting in OC cells that are smaller and less elongated than normal (Auman et al., 2007). Pharmacological reduction of blood flow has a similar effect on cardiomyocyte morphology. Conversely, ventricular cardiomyocytes in *half-hearted* (*haf*) mutants, in which ventricular contractility is defective, are overly enlarged and elongated.

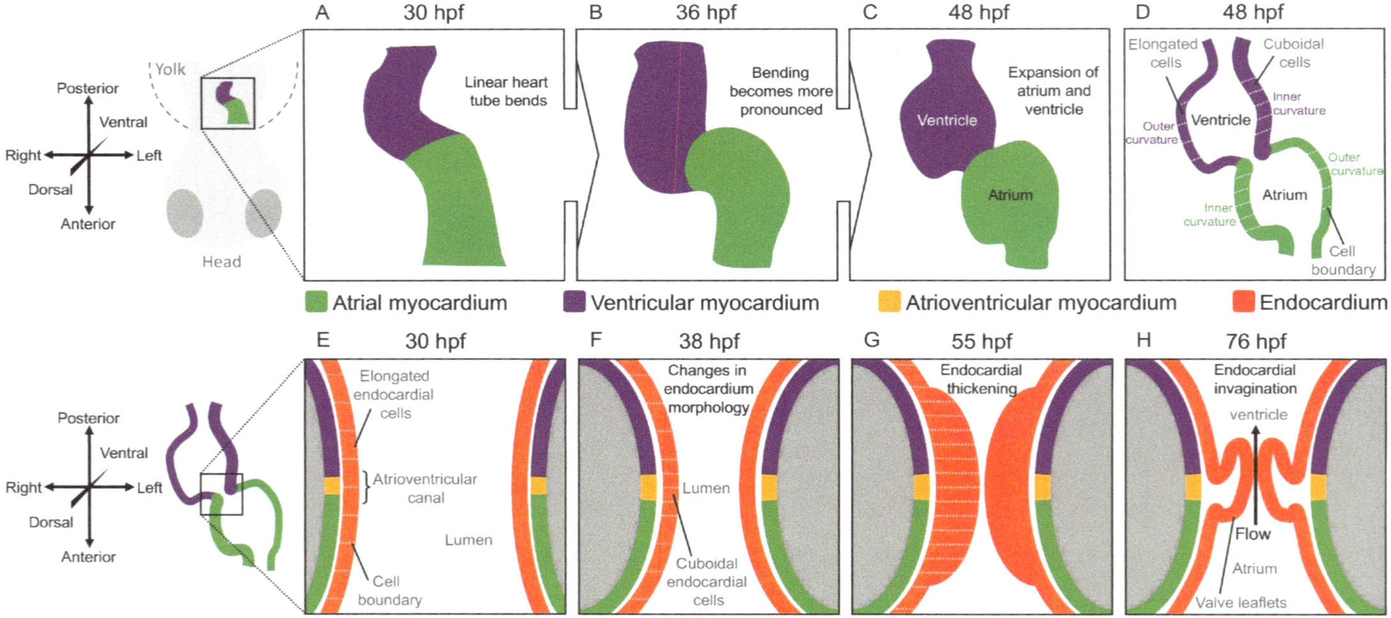

Fig. 5 Transforming the linear tube into a functional heart. (A) Looping begins with bending of the linear tube to form a slight "S" shaped tube. (B) Looping becomes more pronounced and the "S" shape more prominent. (C) The atrium and ventricle expand to become discernible chambers during ballooning. (D) Expansion requires changes in cell morphology, where cells in the inner curvature remain cuboidal while cells in the outer curvature are elongated and flattened. (E) During valve formation, the atrioventricular myocardium (AVC) forms the boundary between the ventricular and atrial regions. (F) Endocardial cells adjacent to the AVC change morphology from elongated to cuboidal. (G) The increasing cell morphology changes result in thickening of the endocardium in the AVC region. (H) Invagination of the endocardium forms unidirectional valves.

2.6 Valvulogenesis

The linear heart tube drives circulation by acting as a suction pump. Contractions of the linear heart tube result in systemic distribution of blood with little regurgitation, even though there are no valves. However, the looped heart is an ineffective suction pump with substantial backflow (Forouhar et al., 2006; Liebling et al., 2006). Thus, dextral looping and growth of the heart necessitate the development of valves in the AVC that reestablish unidirectional flow (Liebling et al., 2006; Scherz, Huisken, Sahai-Hernandez, & Stainier, 2008). The AVC arises as a constriction at the boundary between the chambers during cardiac looping (Fig. 5E). Cells within the AVC begin differentiating at 36 hpf, a time when most endocardial cells are squamous in shape, but a few at the boundary between the chambers appear cuboidal (Fig. 5F) and express different markers, such as DM-GRASP (Beis et al., 2005). By 55 hpf, endocardial cells lining the AVC form a pronounced layer of cuboidal cells (Fig. 5G) and then leaflets form through invagination of AVC endocardium (Fig. 5H). These leaflets ultimately function to prevent backflow (Scherz et al., 2008). At the same time, because of apical membrane constriction, AVC myocardial cells become trapezoidal in morphology (Chi et al., 2008).

Cardiac function is critical for proper valvulogenesis. In the *silent heart* (*sih*) mutant, where mutation of *cardiac troponin T* (*tnnt2*) causes lack of heartbeat and circulation, endocardial cells lack expression of DM-GRASP, remain squamous in morphology, and do not constrict at the AVC (Bartman et al., 2004; Beis et al., 2005). Analysis of *sih* and pharmacological reduction of myocardial force suggests that AVC development is more sensitive to defects in contraction and blood flow directionality, than to shear stress (Bartman et al., 2004; Vermot et al., 2009).

In humans, defective valve formation manifests as aortic valve stenosis, mitral valve prolapse (MVP), calcific aortic valve disease, pulmonary valve stenosis (PVS), Ebstein's anomaly, or bicuspid aortic valve (BAV). Studies in zebrafish have already proven useful in dissecting the genetics underlying these defects. Deregulated Notch or BMP signaling, or activation of RAS/mitogen-activated protein kinase (MAPK) signaling, contribute to valve defects in both human patients and zebrafish, demonstrating the promise of zebrafish models in understanding human valve disease (LaHaye, Lincoln, & Garg, 2014; Padmanabhan et al., 2009; Peal, Lynch, & Milan, 2011; Rose, Force, & Wang, 2010).

3. ZEBRAFISH MODELS OF SYNDROMIC CHD

Considering the complexity of heart development, it is unsurprising that CHDs are so common in humans; failure in any of the precise progressive steps can derail development, resulting in malformation and/or dysfunction. Mutations can impinge on heart development alone or generate a spectrum of defects that comprise developmental syndromes, arising from a monogenic change (Table 1 and discussed throughout the text) or a multigenic mechanism. We discuss a subset of developmental syndromes that include CHDs as a major or defining feature, and arise from a range of primary causes. Modeling these syndromes often recapitulates the pleiotropic symptoms observed in patients, but here we focus on the cardiac aspects of these syndromes.

3.1 Left–Right Asymmetry and CHDs

The vertebrate heart exhibits L–R asymmetry in both the placement of the heart and the pattern of chambers and vessels. L–R asymmetry originates at midline structures called L–R coordinators (LRCs); Kupffer's vesicle (KV) in zebrafish (Amack, 2014; Blum, Weber, Beyer, & Vick, 2009). Cells within LRCs bear motile cilia, small microtubule-based organelles protruding from the apical surface, which beat or rotate to drive asymmetric fluid flow and elicit asymmetries in gene expression (Yoshiba & Hamada, 2014).

In mouse, flow is detected by sensory cilia around the LRC in a mechanism that involves Polycystin proteins (Norris & Grimes, 2012). The polycystins Pkd1l1 and Pkd2 also play a role in L–R patterning in fish, although it is yet unknown whether cilia are required for flow sensation in KV (Bisgrove, Snarr, Emrazian, & Yost, 2005; Kamura et al., 2011; Schottenfeld, Sullivan-Brown, & Burdine, 2007). Flow sensation culminates in *nodal* (*spaw*) expression in the left LPM only, and Nodal signaling activates the homeodomain transcription factor Pitx2 in the left LPM (Hamada & Tam, 2014). While Pitx2 is a primary target of Nodal signaling on the left, the role of Pitx2 in early asymmetric cardiac morphogenesis appears minimal, although it may play later roles in asymmetries in the OFT in mice (Ai et al., 2006; Yashiro, Shiratori, & Hamada, 2007). Although in *Xenopus* and chick misexpression of *pitx2* in the right LPM reverses heart looping, in zebrafish *pitx2* mutants heart jogging and looping occurs normally (Ji, Buel, & Amack, 2016; Levin et al., 1997; Ryan et al., 1998; Sampath, Cheng, Frisch, & Wright, 1997). Looping also occurs normally in mouse mutants, suggesting *Pitx2* is dispensable for this event (Gage, Suh, & Camper, 1999; Lin et al., 1999; Lu, Pressman, Dyer, Johnson, & Martin, 1999).

Disruption of L–R asymmetry causes CHDs, including transposition of the great arteries (TGA), double outlet right ventricle (DORV) and persistent truncus arteriosus (Ramsdell, 2005). Given their importance in generating heart asymmetry, it is unsurprising that mutations disrupting ciliary structure or function, resulting in ciliopathies, commonly involve CHDs (Ramsdell, 2005). CHDs are particularly prevalent in primary ciliary dyskinesia (PCD), a subset of ciliopathies caused by cilia motility defects (Li et al., 2015; Zariwala, Omran, & Ferkol, 2011), and in heterotaxy, a rare condition characterized by discordant placement of organs resulting from L–R defects (Icardo & Sanchez de Vega, 1991; Nakhleh et al., 2012; Sutherland & Ware, 2009). Many potentially causative heterotaxy genes have been identified by sequencing patient cohorts and model organism genetic screens (Fakhro et al., 2011; Guimier et al., 2015; Li et al., 2015). Investigation of the specific role played by these genes in model organisms like zebrafish, where the stages of L–R patterning are particularly well understood, will be critical for understanding how human sequence variants cause disease (Fakhro et al., 2011). Interestingly, mutations in *PKD1L1* and *PKD2*, required for L–R patterning in zebrafish, cause heterotaxy with CHD in humans (Bataille et al., 2011; Bisgrove et al., 2005; Schottenfeld et al., 2007; Vetrini et al., 2016). Conversely, it is important to assess whether L–R patterning genes identified in zebrafish contribute to heterotaxy and/or CHD in humans.

3.2 CHARGE Syndrome

CHARGE syndrome (*C*oloboma, *H*eart defects, *A*tresia choanae, *R*etarded growth/development, *G*enital abnormalities, and *E*ar anomalies/deafness) affects 1 in 10,000 individuals (Blake & Prasad, 2006). CHDs are found in 75–85% of CHARGE patients and are a major cause of mortality (Zentner, Layman, Martin, & Scacheri, 2010).

Several genes associated with CHARGE syndrome have been discovered to be important for zebrafish heart development. For instance, *Chromodomain helicase DNA-binding protein-7* (*CHD7*) haploinsufficiency causes CHARGE syndrome (Vissers et al., 2004; Zentner et al., 2010), and knockdown of *chd7* in zebrafish results in defects reminiscent of those of CHARGE patients. *chd7* morphants display heart defects including dysmorphic cardiac chambers and pericardial edema (Balow et al., 2013; Patten et al., 2012). Blood flow and heartbeat are also reduced, resembling the cardiac rhythm abnormalities of CHARGE patients (Blake & Prasad, 2006; Roger et al., 1999). CHD7 localizes and remodels chromatin in human cell lines, associating with activating methylation marks, implicating epigenetic misregulation in the development of CHDs (Schnetz et al., 2009).

Roughly two-thirds of CHARGE syndrome patients have a mutation in CHD7. Although the genetic cause remains to be discovered for some patients, others have mutations in genes known to be important for heart development. For example, mutations in *SEMA3E* and *SEMA3A* were found in a subset of CHARGE patients, but the role of these Semaphorins in disease development is unclear (Lalani et al., 2004; Schulz et al., 2014). However, zebrafish studies suggest that Semaphorins govern vascular angioblast migration and dorsal aorta formation (Shoji, Isogai, Sato-Maeda, Obinata, & Kuwada, 2003). Overexpression or knockdown of *sema3a* causes heart swelling, loss of circulation, and narrowing of the dorsal aorta in zebrafish, reminiscent of CoA in CHARGE patients (Shoji et al., 2003; Wyse, al-Mahdawi, Burn, & Blake, 1993). Thus, studies in zebrafish substantiate the argument that mutations in Semaphorins may cause CHARGE syndrome. Studies such as these demonstrate how zebrafish can be useful for assessing variants of unknown function associated with CHD.

3.3 Holt–Oram Syndrome

Patients with Holt–Oram syndrome (HOS) exhibit a combination of CHDs and upper limb defects (Holt & Oram, 1960). Affecting 1 in 100,000 individuals, 85% of HOS patients present with CHDs, most commonly atrial septal defects (ASDs), VSD, and conduction defects, and prognosis depends on the severity of the associated CHDs (Basson et al., 1994; Chryssostomidis et al., 2014).

HOS is caused by haploinsufficiency of *TBX5*, a T-box family transcription factor expressed throughout cardiac development (Hatcher, Goldstein, Mah, Delia, & Basson, 2000; McDermott et al., 2005). In the zebrafish *tbx5* mutant *heartstrings* (*hst*), differentiation arrests after 33 hpf and cardiac looping is impaired (Garrity, Childs, & Fishman, 2002). Contractility progressively decreases with a smaller ventricle and the atrium stretching and tearing (Garrity et al., 2002). *hst* mutants also exhibit decreased *camk2b2* expression in the heart and lower activity of CaMK-II (a target of calcium signaling), phenocopying patients with dilated cardiomyopathy and conduction defects (Rothschild et al., 2009; Swaminathan, Purohit, Hund, & Anderson, 2012).

Although *TBX5* is the most frequently mutated gene in HOS, 25% of patients lack mutations in this gene. Mutations in genes regulated by TBX5 may cause HOS symptoms. For example, in some HOS patients, TBX5 mutations reduce expression of *MYH6*, a transcriptional target of TBX5 (Ching et al., 2005; Granados-Riveron et al., 2010). Moreover, mutations in *MYH6* itself cause several human heart defects, including ASD, although a direct link to HOS has not been established. Furthermore, zebrafish studies have demonstrated that Tbx5 synergizes with Mef2C, another transcription factor, to activate *myh6* transcription (Ghosh et al., 2009). Together with the fact that *mef2c* is essential for zebrafish heart development, these data suggest MEF2C function may also be critical in HOS pathology. In this way, probing molecular mechanisms in zebrafish may help inform the etiology of HOS by identifying new candidate genes for variant analyses in patient cohorts.

3.4 Cohesinopathies

Cohesin is a large multicomponent ring-shaped complex required for sister chromatid cohesion, whose function is governed by its ability to generate topological links between distant chromatin segments (Losada, 2014; Michaelis, Ciosk, & Nasmyth, 1997). Mutations in either core cohesin subunits or its regulators cause "cohesinopathies" (Watrin, Kaiser, & Wendt, 2016). Cohesinopathies encompass a broad spectrum of developmental abnormalities including CHDs such as VSD, ASD, PS, and tetralogy of fallot (TOF). Cornelia de Lange syndrome (CdLS), the best studied cohesinopathy, is frequently caused by heterozygous mutations in NIPBL, a protein critical for the loading of cohesin onto DNA, which cause misregulation of gene expression (Horsfield, Print, & Monnich, 2012; Krantz et al., 2004; Tonkin et al., 2004). *nipbl* depletion in zebrafish induces heart and gut defects, reminiscent of those observed in patients (Muto, Calof, Lander, & Schilling, 2011). The expression of genes controlling endodermal differentiation and L–R patterning is altered upon *nipbl* knockdown, and embryonic defects are caused by additive, synergistic interactions between misregulated genes (Muto et al., 2011).

The neural crest has also been implicated in heart defects in cohesinopathies. Depletion of the disease-associated cohesin subunit Rad21 led to smaller hearts, impaired looping, and valve defects in zebrafish (Deardorff et al., 2012; Schuster et al., 2015). Rather than contributing to the heart, NCCs exhibited a "wandering" behavior linked to dysregulation of Wnt, chemokine, and cadherin genes (Schuster et al., 2015). These studies in zebrafish support the hypothesis that cohesinopathies result from the collective effect of multiple quantitative changes in the expression of developmental genes, rather than defects in chromosome segregation, and further suggest that mild mutations in cohesin subunits or regulators might underlie a higher-than-previously appreciated fraction of human CHDs.

3.5 RASopathies

Components of the RAS signaling pathway, a kinase cascade that activates MAPK, are mutated in the developmental syndromes termed RASopathies (Jindal, Goyal, Burdine, Rauen, & Shvartsman, 2015; Tidyman & Rauen, 2009). Mutations are generally thought to activate the pathway, and indeed, some mutations found in RASopathy patients can also be found in cancer lesions. Collectively occurring in 1 in 1000 births, individual RASopathy incidences range from 1 in 1500 for Noonan syndrome (NS), to 1 in 810,000 and 1 in 1,290,000 for the rare cardio-facio-cutaneous (CFC) syndrome and Costello syndrome (CS), respectively (Abe et al., 2012; Rauen, 2013).

CHDs are prevalent among RASopathies, with NS-associated mutations considered the most frequent cause of CHDs arising from a monogenic mutation (Roberts et al., 2007). NS-associated cardiac defects include hypertrophic cardiomyopathy (HCM), PS, and ASD, although CoA, AVC, and mitral valve abnormalities, and VSD can occur. Noonan syndrome with multiple lentigines (NSML) patients present with HCM or PVS, ventricular OFT obstructions, valve abnormalities, ASD, and VSD. CFC patients commonly display PVS, HCM, and ASD, with some instances of CoA, subaortic stenosis and, rarely, arrhythmias. Arrythmias are more common in CS, alongside PS, HCM and, more infrequently, ASD, VSD, and aortic dilation. HCM is also prevalent in patients with neurofibromatosis type 1 syndrome, alongside CoA or PVS.

Zebrafish have been predominantly used to study NS and NSML, although more RASopathy models are being developed (Jindal et al., 2015). NS and NSML models, generated by variant RNA overexpression, recapitulate many features of the human syndrome, but heart defects manifest differently according to the specific mutation. Looping defects are observed upon mutation of *kras* and *ras-like without CAAX 1* (*rit1*), members of the RAS-family of GTPases. The heart tube is enlarged and fails to loop when *kras* function is lost, while expression of disease-associated mRNA causes a smaller heart with reduced ventricle thickness (Razzaque et al., 2012). *rit1* variants also compromise looping in zebrafish, resulting in hypoplastic chambers, and impaired cardiac function (Aoki, Niihori, Narumi, Kure, & Matsubara, 2008; Koenighofer et al., 2016).

Modeling RASopathy-associated mutations in zebrafish provides insight into mutation-phenotype correlations, especially for different mutations within a single gene. Studies with *protein tyrosine phosphatase, nonreceptor type 11* (*PTPN11*), a gene identified as mutated in patients with either NS or NSML, illustrate this elegantly. Injecting zebrafish embryos with *ptpn11* that contains certain patient-derived mutations induce heart defects, including randomized jogging laterality caused by defects in motile cilia in KV, looping failure, and reduced cardiac function (Bonetti et al., 2014; Jopling, van Geemen, & den Hertog, 2007). However, alternative mutations result in edematous embryos and reduced heart size in adult fish (Miura et al., 2013). This use of zebrafish to discover the phenotypic effects caused by disease-associated alleles clearly indicates that distinct mutations in *ptpn11* have different effects and can cause different diseases.

Studies in zebrafish are useful in informing therapeutic strategies; drugs that have been developed to inhibit Ras signaling in cancer have been effectively used in zebrafish models of CFC and NS (Anastasaki, Estep, Marais, Rauen, & Patton, 2009; Anastasaki, Rauen, & Patton, 2012; Chen et al., 2010; Lee et al., 2014; Wang et al., 2012). Indeed, assaying the strength of the causative mutation can be used to predict the required treatment dose (Jindal et al., 2017). Thus, using zebrafish has advanced both the functional relevance of disease causing mutations and provides a platform for testing therapeutic strategies for treating RASopathy-associated heart disease.

3.6 Williams–Beuren Syndrome

Williams–Beuren syndrome (WBS), also called Williams' syndrome, is a chromosomal microdeletion disorder occurring in 1 in 10,000 individuals (Pober, 2010). WBS is caused by deletion of a critical region (WBSCR), comprising 1.5–1.8 Mb of DNA on chromosome 7 (7q11.23) and containing 26–28 genes (Pober, 2010). Nonallelic homologous recombination between highly homologous blocks of low-copy repeat regions flanking the WBSCR mediates the deletion underlying WBS, resulting in hemizygosity for multiple genes. CHDs are the primary cause of death in WBS patients and 80% of patients have a cardiac abnormality, including VSD, MVP, and OFT obstruction, but arterial stenosis is most common, including supravalvar aortic stenosis (SVAS), pulmonary arterial stenosis, stenosis of the thoracic aorta, and peripheral pulmonary stenosis (PPS) (Collins, 2013).

Although many genes localize to the WBSCR, *Elastin* (*ELN*) hemizygosity has been demonstrated to contribute to the arteriopathy associated with WBS (Ewart et al., 1993). Elastin is an ECM component that allows the characteristic stretching and recoiling of arteries, and arterial stiffness may be increased in WBS patients as a result of reduced elastin in the ECM (Collins, 2013; Kozel et al., 2014). Indeed, cultured cells isolated from WBS patients show reduced deposition of elastin in the ECM (Urbán et al., 2002). Poor elastin deposition also contributes to increased proliferation of fetal smooth muscle cells, likely causing arterial stenosis in WBS, and may result from altered ECM signaling to cells (Kim, Turnbull, & Guimond, 2011; Moriyama et al., 2016). Collectively, these data suggest that cell fate may be altered by hemizygosity of *ELN* in WBS.

Elastin is critical for the form and function of the BA, the zebrafish OFT, where it is expressed as early as 72 hpf (Miao, Bruce, Bhanji, Davis, & Keeley, 2007). Knockdown of both *elna* and *elnb* reduces BA contraction, although the effect of *elnb* knockdown is more pronounced than that of *elna*, exhibiting additional hypoplasia of the BA (Moriyama et al., 2016). Ectopic cardiomyocytes are observed in the BA of *elna* and *elnb* mutants, but CPC migration patterns are not altered. Rather, *elnb* governs the differentiation of CPCs into smooth muscle cells and its absence allows differentiation into cardiomyocytes instead. Such studies improve our understanding of how individual gene deletions contribute to the overall disease etiology.

3.7 Microdeletion Syndrome 22q11.2

22q11.2 deletion syndrome (22q11.2DS) is the most common microdeletion syndrome in humans. Also known as DiGeorge syndrome or CATCH-22 syndrome (*C*ardiac abnormality, *A*bnormal facies, *T*-cell deficient due to thymic hypoplasia, *C*left palate, *H*ypercalcemia due to hypoparathyroidism resulting from **22**q11 deletion), the prevalence is 1 in 4000 births (Devriendt, Fryns, Mortier, van Thienen, & Keymolen, 1998). Nonallelic homologous recombination occurs between repeat elements, resulting in a highly reproducible series of deletions. Consequently, 90% of patients have a 3 Mb deletion which encompasses 90 genes, many of which remain poorly characterized (Emanuel, 2008). Approximately 80% of patients present with CHDs, primarily conotruncal defects, aortic arch abnormalities, and septal defects. Considering the phenotypes of 22q11.2DS are believed to arise as a result of gene dosage effects, a major advantage of using zebrafish is the ability to manipulate expression levels using morpholinos and mRNA overexpression.

Of the genes deleted in 22q11.2DS, *TBX1* appears to be most important. Deletion or knockdown of *tbx1* in zebrafish causes cardiac defects: the heart fails to jog or loop, with reduced proliferation in the FHF and decreased cell contribution from the SHF (Piotrowski et al., 2003; Zhang, Gui, Wang, Jiang, & Song, 2010). The reduced incorporation of SHF cells at the arterial pole results in impaired development of the OFT (Hami et al., 2011). TBX1 has a central role in regulating the expression of many genes and altering their expression may contribute to 22q11.2DS. For example, WNT11R has been demonstrated to be downstream of TBX1 in a linear pathway regulating heart development (Choe & Crump, 2014; Choudhry & Trede, 2013), while WNT5A is essential for SHF development (Sinha et al., 2015).

The spectrum of phenotypes observed in patients cannot be fully explained by *TBX1* haploinsufficiency, however. Many proteins deleted in 22q11.2DS are involved in mitochondrial function, potentially implicating mitochondrial dysfunction in disease development. Mitochondrial dysfunction can cause neurodevelopmental and neurodegenerative disease, reminiscent of the neurological symptoms of 22q11.2DS patients. Among the mitochondrial genes deleted, the *SLC25A1* ortholog *slc25a1a* was studied in zebrafish (Catalina-Rodriguez et al., 2012). A dose-dependent relationship was established, where decreasing levels of protein associated with increasing mitochondrial depletion and worsening developmental defects, which included reduced heart size and pericardial edema. Furthermore, the phenotypes induced by knockdown of *slc25a1a* could be suppressed by blocking autophagy, identifying a potential new clinical target (Catalina-Rodriguez et al., 2012). This study underscores the utility of zebrafish in studying the mechanism of disease development and identifying targets that might inform new therapeutic approaches.

4. CONCLUSION

Faithful cardiac development is crucial to the health and survival of vertebrate embryos. While cardiac development is well characterized in zebrafish, several important areas remain poorly understood. The details of tube extension and cardiac looping remain unclear, and the minutiae of ballooning are only recently beginning to emerge. Further quantitative research into the velocity and routes of migrating cells from the FHF, SHF, and CNC are critical to fully understand zebrafish heart development, as is understanding how their integration into the heart affects morphogenetic events. Only with a comprehensive model we will be able to fully appreciate the perturbations to the system that lead to defects.

While zebrafish hearts are simpler than their human counterparts, significant conservation exists between teleosts and humans, both in the basic cellular changes that effect heart development and in the gene mutations that disrupt this process. Each of the complex developmental syndromes discussed herein can be at least partially recapitulated by mutation, deletion, or knockdown of the same genes in zebrafish. While the links between specific human symptoms and corresponding zebrafish heart phenotypes remain complex, this may be a result of inconsistent scoring of cardiac defects within the zebrafish community. In the future, a standardized description of phenotypes in the zebrafish may prove more informative and allow a better correspondence between zebrafish and mammalian CHD phenotypes. Nevertheless, mutations that generate CHDs in humans consistently disrupt cardiac development in fish, and future work should focus on using gene editing technology to create zebrafish with mutations that are presumed to cause disease in humans.

Perhaps the most exciting aspect of modeling CHD with zebrafish is the potential for personalized medicine. Human mutations are fantastically varied and causative mutations within the same gene can have wildly differing effects. The combination of precise genetic manipulation with the ability to perform effective drug screens makes zebrafish ideal for the identification and development of new therapeutic approaches to treat CHDs. Indeed, such work has already proven informative for RASopathies and 22q11.2DS, and similar screens with new mutations will only improve our knowledge. In the future, it should be possible to identify a human CHD-associated mutation, reproduce the mutation in zebrafish to understand the molecular and cellular causes underlying the accompanying heart defect, and then screen for the most appropriate treatment regimen. Thus, zebrafish will remain at the forefront of cardiac development and CHD research.

ACKNOWLEDGMENTS

D.T.G. is funded by an award from the Founders Affiliate of the American Heart Association (#16POST31390023). R.D.B., V.L.P., and M.G.G. are supported by the National Institutes of Health Grants R01HD048584 and R01GM086537. We thank Kari Lenhart, Jon Rosen, and members of the Burdine lab for comments on the manuscript. We apologize to any colleagues whose work we failed to include or discuss more in depth due to space limitations.

REFERENCES

Abe, Y., Aoki, Y., Kuriyama, S., Kawame, H., Okamoto, N., Kurosawa, K., et al. (2012). Prevalence and clinical features of Costello syndrome and cardio-facio-cutaneous syndrome in Japan: Findings from a nationwide epidemiological survey. *American Journal of Medical Genetics. Part A*, *158A*, 1083–1094.

Ai, D., Liu, W., Ma, L., Dong, F., Lu, M. F., Wang, D., et al. (2006). Pitx2 regulates cardiac left-right asymmetry by patterning second cardiac lineage-derived myocardium. *Developmental Biology*, *296*, 437–449.

Amack, J. D. (2014). Salient features of the ciliated organ of asymmetry. *Bioarchitecture*, *4*, 6–15.

Anastasaki, C., Estep, A. L., Marais, R., Rauen, K. A., & Patton, E. E. (2009). Kinase-activating and kinase-impaired cardio-facio-cutaneous syndrome alleles have activity during zebrafish development and are sensitive to small molecule inhibitors. *Human Molecular Genetics*, *18*, 2543–2554.

Anastasaki, C., Rauen, K. A., & Patton, E. E. (2012). Continual low-level MEK inhibition ameliorates cardio-facio-cutaneous phenotypes in zebrafish. *Disease Models & Mechanisms*, *5*, 546–552.

Aoki, Y., Niihori, T., Narumi, Y., Kure, S., & Matsubara, Y. (2008). The RAS/MAPK syndromes: Novel roles of the RAS pathway in human genetic disorders. *Human Mutation*, *29*, 992–1006.

Arrington, C. B., & Yost, H. J. (2009). Extra-embryonic syndecan 2 regulates organ primordia migration and fibrillogenesis throughout the zebrafish embryo. *Development*, *136*, 3143–3152.

Auman, H. J., Coleman, H., Riley, H. E., Olale, F., Tsai, H. J., & Yelon, D. (2007). Functional modulation of cardiac form through regionally confined cell shape changes. *PLoS Biology*, *5*, e53.

Baker, K., Holtzman, N. G., & Burdine, R. D. (2008). Direct and indirect roles for Nodal signaling in two axis conversions during asymmetric morphogenesis of the zebrafish heart. *Proceedings of the National Academy of Sciences of the United States of America*, *105*, 13924–13929.

Balow, S. A., Pierce, L. X., Zentner, G. E., Conrad, P. A., Davis, S., Sabaawy, H. E., et al. (2013). Knockdown of fbxl10/kdm2bb rescues chd7 morphant phenotype in a zebrafish model of CHARGE syndrome. *Developmental Biology*, *382*, 57–69.

Bartman, T., Walsh, E. C., Wen, K. K., McKane, M., Ren, J., Alexander, J., et al. (2004). Early myocardial function affects endocardial cushion development in zebrafish. *PLoS Biology*, *2*, E129.

Basson, C. T., Cowley, G. S., Solomon, S. D., Weissman, B., Poznanski, A. K., Traill, T. A., et al. (1994). The clinical and genetic spectrum of the Holt-Oram syndrome (heart-hand syndrome). *New England Journal of Medicine*, *330*, 885–891.

Bataille, S., Demoulin, N., Devuyst, O., Audrezet, M. P., Dahan, K., Godin, M., et al. (2011). Association of PKD2 (polycystin 2) mutations with left-right laterality defects. *American Journal of Kidney Diseases: The Official Journal of the National Kidney Foundation*, *58*, 456–460.

Beis, D., Bartman, T., Jin, S. W., Scott, I. C., D'Amico, L. A., Ober, E. A., et al. (2005). Genetic and cellular analyses of zebrafish atrioventricular cushion and valve development. *Development*, *132*, 4193–4204.

Berdougo, E., Coleman, H., Lee, D. H., Stainier, D. Y., & Yelon, D. (2003). Mutation of weak atrium/atrial myosin heavy chain disrupts atrial function and influences ventricular morphogenesis in zebrafish. *Development*, *130*, 6121–6129.

Bisgrove, B. W., Snarr, B. S., Emrazian, A., & Yost, H. J. (2005). Polaris and Polycystin-2 in dorsal forerunner cells and Kupffer's vesicle are required for specification of the zebrafish left-right axis. *Developmental Biology*, *287*, 274–288.

Blake, K. D., & Prasad, C. (2006). CHARGE syndrome. *Orphanet Journal of Rare Diseases, 1*, 1–8.
Blum, M., Weber, T., Beyer, T., & Vick, P. (2009). Evolution of leftward flow. *Seminars in Cell & Developmental Biology, 20*, 464–471.
Bonetti, M., Rodriguez-Martinez, V., Paardekooper Overman, J., Overvoorde, J., van Eekelen, M., Jopling, C., et al. (2014). Distinct and overlapping functions of ptpn11 genes in Zebrafish development. *PLoS One, 9*, e94884.
Bruneau, B. G. (2008). The developmental genetics of congenital heart disease. *Nature, 451*, 943–948.
Bussmann, J., Bakkers, J., & Schulte-Merker, S. (2007). Early endocardial morphogenesis requires Scl/Tal1. *PLoS Genetics, 3*, e140.
Cai, C. L., Liang, X., Shi, Y., Chu, P. H., Pfaff, S. L., Chen, J., et al. (2003). Isl1 identifies a cardiac progenitor population that proliferates prior to differentiation and contributes a majority of cells to the heart. *Developmental Cell, 5*, 877–889.
Catalina-Rodriguez, O., Kolukula, V. K., Tomita, Y., Preet, A., Palmieri, F., Wellstein, A., et al. (2012). The mitochondrial citrate transporter, CIC, is essential for mitochondrial homeostasis. *Oncotarget, 3*, 1220–1235.
Cavanaugh, A. M., Huang, J., & Chen, J. N. (2015). Two developmentally distinct populations of neural crest cells contribute to the zebrafish heart. *Developmental Biology, 404*(2), 103–112.
Chen, P. C., Wakimoto, H., Conner, D., Araki, T., Yuan, T., Roberts, A., et al. (2010). Activation of multiple signaling pathways causes developmental defects in mice with a Noonan syndrome-associated Sos1 mutation. *The Journal of Clinical Investigation, 120*, 4353–4365.
Chi, N. C., Shaw, R. M., De Val, S., Kang, G., Jan, L. Y., Black, B. L., et al. (2008). Foxn4 directly regulates tbx2b expression and atrioventricular canal formation. *Genes & Development, 22*, 734–739.
Ching, Y. H., Ghosh, T. K., Cross, S. J., Packham, E. A., Honeyman, L., Loughna, S., et al. (2005). Mutation in myosin heavy chain 6 causes atrial septal defect. *Nature Genetics, 37*, 423–428.
Choe, C. P., & Crump, J. G. (2014). Tbx1 controls the morphogenesis of pharyngeal pouch epithelia through mesodermal Wnt11r and Fgf8a. *Development, 141*, 3583–3593.
Choudhry, P., & Trede, N. S. (2013). DiGeorge syndrome gene tbx1 functions through wnt11r to regulate heart looping and differentiation. *PLoS One, 8*, e58145.
Chryssostomidis, G., Kanakis, M., Fotiadou, V., Laskari, C., Kousi, T., Apostolidis, C., et al. (2014). Diversity of congenital cardiac defects and skeletal deformities associated with the Holt–Oram syndrome. *International Journal of Surgery Case Reports, 5*, 389–392.
Collins, R. T. (2013). Cardiovascular disease in Williams syndrome. *Circulation, 127*, 2125–2134.
de Campos-Baptista, M. I., Holtzman, N. G., Yelon, D., & Schier, A. F. (2008). Nodal signaling promotes the speed and directional movement of cardiomyocytes in zebrafish. *Developmental Dynamics: An Official Publication of the American Association of Anatomists, 237*, 3624–3633.
De Domenico, I., Vaughn, M. B., Yoon, D., Kushner, J. P., Ward, D. M., & Kaplan, J. (2007). Zebrafish as a model for defining the functional impact of mammalian ferroportin mutations. *Blood, 110*, 3780–3783.
de Pater, E., Clijsters, L., Marques, S. R., Lin, Y. F., Garavito-Aguilar, Z. V., Yelon, D., et al. (2009). Distinct phases of cardiomyocyte differentiation regulate growth of the zebrafish heart. *Development, 136*, 1633–1641.
Deardorff, M. A., Wilde, J. J., Albrecht, M., Dickinson, E., Tennstedt, S., Braunholz, D., et al. (2012). RAD21 mutations cause a human cohesinopathy. *The American Journal of Human Genetics, 90*, 1014–1027.

Devriendt, K., Fryns, J. P., Mortier, G., van Thienen, M. N., & Keymolen, K. (1998). The annual incidence of DiGeorge/velocardiofacial syndrome. *Journal of Medical Genetics, 35*, 789–790.

Emanuel, B. S. (2008). Molecular mechanisms and diagnosis of chromosome 22q11.2 rearrangements. *Developmental Disabilities Research Reviews, 14*, 11–18.

Ewart, A. K., Morris, C. A., Atkinson, D., Jin, W., Sternes, K., Spallone, P., et al. (1993). Hemizygosity at the elastin locus in a developmental disorder, Williams syndrome. *Nature Genetics, 5*, 11–16.

Fakhro, K. A., Choi, M., Ware, S. M., Belmont, J. W., Towbin, J. A., Lifton, R. P., et al. (2011). Rare copy number variations in congenital heart disease patients identify unique genes in left-right patterning. *Proceedings of the National Academy of Sciences of the United States of America, 108*, 2915–2920.

Fish, J. E., Wythe, J. D., Xiao, T., Bruneau, B. G., Stainier, D. Y., Srivastava, D., et al. (2011). A Slit/miR-218/Robo regulatory loop is required during heart tube formation in zebrafish. *Development, 138*, 1409–1419.

Forouhar, A. S., Liebling, M., Hickerson, A., Nasiraei-Moghaddam, A., Tsai, H. J., Hove, J. R., et al. (2006). The embryonic vertebrate heart tube is a dynamic suction pump. *Science, 312*, 751–753.

Fukushima, N., Ishii, I., Contos, J. J., Weiner, J. A., & Chun, J. (2001). Lysophospholipid receptors. *Annual Review of Pharmacology and Toxicology, 41*, 507–534.

Gage, P. J., Suh, H., & Camper, S. A. (1999). Dosage requirement of Pitx2 for development of multiple organs. *Development, 126*, 4643–4651.

Garavito-Aguilar, Z. V., Riley, H. E., & Yelon, D. (2010). Hand2 ensures an appropriate environment for cardiac fusion by limiting Fibronectin function. *Development, 137*, 3215–3220.

Garrity, D. M., Childs, S., & Fishman, M. C. (2002). The heartstrings mutation in zebrafish causes heart/fin Tbx5 deficiency syndrome. *Development, 129*, 4635–4645.

Gering, M., Yamada, Y., Rabbitts, T. H., & Patient, R. K. (2003). Lmo2 and Scl/Tal1 convert non-axial mesoderm into haemangioblasts which differentiate into endothelial cells in the absence of Gata1. *Development, 130*, 6187–6199.

Ghosh, T. K., Song, F. F., Packham, E. A., Buxton, S., Robinson, T. E., Ronksley, J., et al. (2009). Physical interaction between TBX5 and MEF2C is required for early heart development. *Molecular and Cellular Biology, 29*, 2205–2218.

Granados-Riveron, J. T., Ghosh, T. K., Pope, M., Bu'Lock, F., Thornborough, C., Eason, J., et al. (2010). Alpha-cardiac myosin heavy chain (MYH6) mutations affecting myofibril formation are associated with congenital heart defects. *Human Molecular Genetics, 19*, 4007–4016.

Grimes, A. C., Stadt, H. A., Shepherd, I. T., & Kirby, M. L. (2006). Solving an enigma: Arterial pole development in the zebrafish heart. *Developmental Biology, 290*, 265–276.

Guimier, A., Gabriel, G. C., Bajolle, F., Tsang, M., Liu, H., Noll, A., et al. (2015). MMP21 is mutated in human heterotaxy and is required for normal left-right asymmetry in vertebrates. *Nature Genetics, 47*, 1260–1263.

Guner-Ataman, B., Paffett-Lugassy, N., Adams, M. S., Nevis, K. R., Jahangiri, L., Obregon, P., et al. (2013). Zebrafish second heart field development relies on progenitor specification in anterior lateral plate mesoderm and nkx2.5 function. *Development, 140*, 1353–1363.

Haffter, P., Granato, M., Brand, M., Mullins, M. C., Hammerschmidt, M., Kane, D. A., et al. (1996). The identification of genes with unique and essential functions in the development of the zebrafish, Danio rerio. *Development, 123*, 1–36.

Hamada, H., & Tam, P. P. (2014). Mechanisms of left-right asymmetry and patterning: Driver, mediator and responder. *F1000prime reports, 6*, 110.

Hami, D., Grimes, A. C., Tsai, H. J., & Kirby, M. L. (2011). Zebrafish cardiac development requires a conserved secondary heart field. *Development*, *138*, 2389–2398.
Harvey, R. P., & Rosenthal, N. (1999). *Heart development*. London: Academic Press.
Hatcher, C. J., Goldstein, M. M., Mah, C. S., Delia, C. S., & Basson, C. T. (2000). Identification and localization of TBX5 transcription factor during human cardiac morphogenesis. *Developmental Dynamics: An Official Publication of the American Association of Anatomists*, *219*, 90–95.
Hinits, Y., Pan, L., Walker, C., Dowd, J., Moens, C. B., & Hughes, S. M. (2012). Zebrafish Mef2ca and Mef2cb are essential for both first and second heart field cardiomyocyte differentiation. *Developmental Biology*, *369*, 199–210.
Hisano, Y., Ota, S., Takada, S., & Kawahara, A. (2013). Functional cooperation of spns2 and fibronectin in cardiac and lower jaw development. *Biology Open*, *2*, 789–794.
Holt, M., & Oram, S. (1960). Familial heart disease with skeletal malformations. *British Heart Journal*, *22*, 236–242.
Holtzinger, A., & Evans, T. (2005). Gata4 regulates the formation of multiple organs. *Development*, *132*, 4005–4014.
Holtzinger, A., & Evans, T. (2007). Gata5 and Gata6 are functionally redundant in zebrafish for specification of cardiomyocytes. *Developmental Biology*, *312*, 613–622.
Holtzman, N. G., Schoenebeck, J. J., Tsai, H. J., & Yelon, D. (2007). Endocardium is necessary for cardiomyocyte movement during heart tube assembly. *Development*, *134*, 2379–2386.
Horne-Badovinac, S., Lin, D., Waldron, S., Schwarz, M., Mbamalu, G., Pawson, T., et al. (2001). Positional cloning of heart and soul reveals multiple roles for PKC lambda in zebrafish organogenesis. *Current Biology: CB*, *11*, 1492–1502.
Horsfield, J. A., Print, C. G., & Monnich, M. (2012). Diverse developmental disorders from the one ring: Distinct molecular pathways underlie the cohesinopathies. *Frontiers in Genetics*, *3*, 171.
Icardo, J. M., & Sanchez de Vega, M. J. (1991). Spectrum of heart malformations in mice with situs solitus, situs inversus, and associated visceral heterotaxy. *Circulation*, *84*, 2547–2558.
Ji, Y., Buel, S. M., & Amack, J. D. (2016). Mutations in zebrafish pitx2 model congenital malformations in Axenfeld-Rieger syndrome but do not disrupt left-right placement of visceral organs. *Developmental Biology*, *416*, 69–81.
Jindal, G. A., Goyal, Y., Burdine, R. D., Rauen, K. A., & Shvartsman, S. Y. (2015). RASopathies: Unraveling mechanisms with animal models. *Disease Models & Mechanisms*, *8*, 769–782.
Jindal, G. A., Goyal, Y., Yamaya, K., Futran, A. S., Kountouridis, I., Balgobin, C. A., et al. (2017). In vivo severity ranking of Ras pathway mutations associated with developmental disorders. *Proceedings of the National Academy of Sciences of the United States*. http://dx.doi.org/10.1073/pnas.1615651114, pii:201615651, (Epub ahead of print).
Jopling, C., van Geemen, D., & den Hertog, J. (2007). Shp2 knockdown and Noonan/LEOPARD mutant Shp2-induced gastrulation defects. *PLoS Genetics*, *3*, e225.
Kamura, K., Kobayashi, D., Uehara, Y., Koshida, S., Iijima, N., Kudo, A., et al. (2011). Pkd1l1 complexes with Pkd2 on motile cilia and functions to establish the left-right axis. *Development*, *138*, 1121–1129.
Kawahara, A., Nishi, T., Hisano, Y., Fukui, H., Yamaguchi, A., & Mochizuki, N. (2009). The sphingolipid transporter spns2 functions in migration of zebrafish myocardial precursors. *Science*, *323*, 524–527.
Keegan, B. R., Meyer, D., & Yelon, D. (2004). Organization of cardiac chamber progenitors in the zebrafish blastula. *Development*, *131*, 3081–3091.
Keyte, A., & Hutson, M. R. (2012). The neural crest in cardiac congenital anomalies. *Differentiation; Research in Biological Diversity*, *84*, 25–40.

Kikuchi, K., Holdway, J. E., Major, R. J., Blum, N., Dahn, R. D., Begemann, G., et al. (2011). Retinoic acid production by endocardium and epicardium is an injury response essential for zebrafish heart regeneration. *Developmental Cell, 20*, 397–404.

Kim, S.-H., Turnbull, J., & Guimond, S. (2011). Extracellular matrix and cell signalling: The dynamic cooperation of integrin, proteoglycan and growth factor receptor. *Journal of Endocrinology, 209*, 139–151.

Kodo, K., & Yamagishi, H. (2010). GATA transcription factors in congenital heart defects: A commentary on a novel GATA6 mutation in patients with tetralogy of Fallot or atrial septal defect. *Journal of Human Genetics, 55*, 637–638.

Koenighofer, M., Hung, C. Y., McCauley, J. L., Dallman, J., Back, E. J., Mihalek, I., et al. (2016). Mutations in RIT1 cause Noonan syndrome—Additional functional evidence and expanding the clinical phenotype. *Clinical Genetics, 89*, 359–366.

Kozel, B. A., Danback, J., Waxler, J., Knutsen, R. H., Fuentes, L. d. l., Reusz, G. S., et al. (2014). Williams syndrome predisposes to vascular stiffness modified by anti-hypertensive use and copy number changes in NCF1. *Hypertension, 63*, 74–79.

Krantz, I. D., McCallum, J., DeScipio, C., Kaur, M., Gillis, L. A., Yaeger, D., et al. (2004). Cornelia de Lange syndrome is caused by mutations in NIPBL, the human homolog of Drosophila melanogaster Nipped-B. *Nature Genetics, 36*, 631–635.

Kupperman, E., An, S., Osborne, N., Waldron, S., & Stainier, D. Y. (2000). A sphingosine-1-phosphate receptor regulates cell migration during vertebrate heart development. *Nature, 406*, 192–195.

LaHaye, S., Lincoln, J., & Garg, V. (2014). Genetics of valvular heart disease. *Current Cardiology Reports, 16*, 487.

Lalani, S. R., Safiullah, A. M., Molinari, L. M., Fernbach, S. D., Martin, D. M., & Belmont, J. W. (2004). SEMA3E mutation in a patient with CHARGE syndrome. *Journal of Medical Genetics, 41*, e94.

Lazic, S., & Scott, I. C. (2011). Mef2cb regulates late myocardial cell addition from a second heart field-like population of progenitors in zebrafish. *Developmental Biology, 354*, 123–133.

Lee, Y. S., Ehninger, D., Zhou, M., Oh, J. Y., Kang, M., Kwak, C., et al. (2014). Mechanism and treatment for learning and memory deficits in mouse models of Noonan syndrome. *Nature Neuroscience, 17*, 1736–1743.

Lenhart, K. F., Holtzman, N. G., Williams, J. R., & Burdine, R. D. (2013). Integration of nodal and BMP signals in the heart requires FoxH1 to create left-right differences in cell migration rates that direct cardiac asymmetry. *PLoS Genetics, 9*, e1003109.

Levin, M., Pagan, S., Roberts, D. J., Cooke, J., Kuehn, M. R., & Tabin, C. J. (1997). Left/right patterning signals and the independent regulation of different aspects of situs in the chick embryo. *Developmental Biology, 189*, 57–67.

Li, Y., Klena, N. T., Gabriel, G. C., Liu, X., Kim, A. J., Lemke, K., et al. (2015). Global genetic analysis in mice unveils central role for cilia in congenital heart disease. *Nature, 521*, 520–524.

Li, Y. X., Zdanowicz, M., Young, L., Kumiski, D., Leatherbury, L., & Kirby, M. L. (2003). Cardiac neural crest in zebrafish embryos contributes to myocardial cell lineage and early heart function. *Developmental Dynamics: An Official Publication of the American Association of Anatomists, 226*, 540–550.

Liebling, M., Forouhar, A. S., Wolleschensky, R., Zimmermann, B., Ankerhold, R., Fraser, S. E., et al. (2006). Rapid three-dimensional imaging and analysis of the beating embryonic heart reveals functional changes during development. *Developmental Dynamics: An Official Publication of the American Association of Anatomists, 235*, 2940–2948.

Lin, C. R., Kioussi, C., O'Connell, S., Briata, P., Szeto, D., Liu, F., et al. (1999). Pitx2 regulates lung asymmetry, cardiac positioning and pituitary and tooth morphogenesis. *Nature, 401*, 279–282.

Long, S., Ahmad, N., & Rebagliati, M. (2003). The zebrafish nodal-related gene southpaw is required for visceral and diencephalic left-right asymmetry. *Development*, *130*, 2303–2316.

Losada, A. (2014). Cohesin in cancer: Chromosome segregation and beyond. *Nature Reviews. Cancer*, *14*, 389–393.

Lu, M. F., Pressman, C., Dyer, R., Johnson, R. L., & Martin, J. F. (1999). Function of Rieger syndrome gene in left-right asymmetry and craniofacial development. *Nature*, *401*, 276–278.

Maceyka, M., Payne, S. G., Milstien, S., & Spiegel, S. (2002). Sphingosine kinase, sphingosine-1-phosphate, and apoptosis. *Biochimica et Biophysica Acta*, *1585*, 193–201.

Marques, S. R., Lee, Y., Poss, K. D., & Yelon, D. (2008). Reiterative roles for FGF signaling in the establishment of size and proportion of the zebrafish heart. *Developmental Biology*, *321*, 397–406.

McDermott, D. A., Bressan, M. C., He, J., Lee, J. S., Aftimos, S., Brueckner, M., et al. (2005). TBX5 genetic testing validates strict clinical criteria for Holt–Oram syndrome. *Pediatric Research*, *58*, 981–986.

Miao, M., Bruce, A., Bhanji, T., Davis, E., & Keeley, F. (2007). Differential expression of two tropoelastin genes in zebrafish. *Matrix Biology*, *26*, 115–124.

Michaelis, C., Ciosk, R., & Nasmyth, K. (1997). Cohesins: Chromosomal proteins that prevent premature separation of sister chromatids. *Cell*, *91*, 35–45.

Miura, K., Wakayama, Y., Tanino, M., Orba, Y., Sawa, H., Hatakeyama, M., et al. (2013). Involvement of EphA2-mediated tyrosine phosphorylation of Shp2 in Shp2-regulated activation of extracellular signal-regulated kinase. *Oncogene*, *32*, 5292–5301.

Miura, G. I., & Yelon, D. (2011). A guide to analysis of cardiac phenotypes in the zebrafish embryo. *Methods in Cell Biology*, *101*, 161–180.

Moriyama, Y., Ito, F., Takeda, H., Yano, T., Okabe, M., Kuraku, S., et al. (2016). Evolution of the fish heart by sub/neofunctionalization of an elastin gene. *Nature Communications*, *7*, 10397.

Mozaffarian, D., Benjamin, E. J., Go, A. S., Arnett, D. K., Blaha, M. J., Cushman, M., et al. (2016). Heart disease and stroke statistics-2016 update: A report from the American Heart Association. *Circulation*, *133*, e38–e360.

Muto, A., Calof, A. L., Lander, A. D., & Schilling, T. F. (2011). Multifactorial origins of heart and gut defects in nipbl-deficient zebrafish, a model of Cornelia de Lange Syndrome. *PLoS Biology*, *9*, e1001181.

Nakhleh, N., Francis, R., Giese, R. A., Tian, X., Li, Y., Zariwala, M. A., et al. (2012). High prevalence of respiratory ciliary dysfunction in congenital heart disease patients with heterotaxy. *Circulation*, *125*, 2232–2242.

Noel, E. S., Verhoeven, M., Lagendijk, A. K., Tessadori, F., Smith, K., Choorapoikayil, S., et al. (2013). A Nodal-independent and tissue-intrinsic mechanism controls heart-looping chirality. *Nature Communications*, *4*, 2754.

Norris, D. P., & Grimes, D. T. (2012). Developmental biology. Cilia discern left from right. *Science*, *338*, 206–207.

Osborne, N., Brand-Arzamendi, K., Ober, E. A., Jin, S. W., Verkade, H., Holtzman, N. G., et al. (2008). The spinster homolog, two of hearts, is required for sphingosine 1-phosphate signaling in zebrafish. *Current Biology: CB*, *18*, 1882–1888.

Padmanabhan, A., Lee, J. S., Ismat, F. A., Lu, M. M., Lawson, N. D., Kanki, J. P., et al. (2009). Cardiac and vascular functions of the zebrafish orthologues of the type I neurofibromatosis gene NFI. *Proceedings of the National Academy of Sciences of the United States of America*, *106*, 22305–22310.

Palencia-Desai, S., Rost, M. S., Schumacher, J. A., Ton, Q. V., Craig, M. P., Baltrunaite, K., et al. (2015). Myocardium and BMP signaling are required for endocardial differentiation. *Development*, *142*, 2304–2315.

Patten, S. A., Jacobs-McDaniels, N. L., Zaouter, C., Drapeau, P., Albertson, R. C., & Moldovan, F. (2012). Role of Chd7 in zebrafish: A model for CHARGE syndrome. *PLoS One*, *7*, e31650.

Peal, D. S., Lynch, S. N., & Milan, D. J. (2011). Patterning and development of the atrioventricular canal in zebrafish. *Journal of Cardiovascular Translational Research*, *4*, 720–726.

Perner, B., Englert, C., & Bollig, F. (2007). The Wilms tumor genes wt1a and wt1b control different steps during formation of the zebrafish pronephros. *Developmental Biology*, *309*, 87–96.

Peterson, R. T., Mably, J. D., Chen, J. N., & Fishman, M. C. (2001). Convergence of distinct pathways to heart patterning revealed by the small molecule concentramide and the mutation heart-and-soul. *Current Biology: CB*, *11*, 1481–1491.

Piotrowski, T., Ahn, D. G., Schilling, T. F., Nair, S., Ruvinsky, I., Geisler, R., et al. (2003). The zebrafish van gogh mutation disrupts tbx1, which is involved in the DiGeorge deletion syndrome in humans. *Development*, *130*, 5043–5052.

Pober, B. R. (2010). Williams–Beuren syndrome. *The New England Journal of Medicine*, *362*, 239–252.

Poon, K.-L., Liebling, M., Kondrychyn, I., Garcia-Lecea, M., & Korzh, V. (2010). Zebrafish cardiac enhancer trap lines: New tools for in vivo studies of cardiovascular development and disease. *Developmental Dynamics*, *239*, 914–926.

Ramsdell, A. F. (2005). Left-right asymmetry and congenital cardiac defects: Getting to the heart of the matter in vertebrate left-right axis determination. *Developmental Biology*, *288*, 1–20.

Rauen, K. A. (2013). The RASopathies. *Annual Review of Genomics and Human Genetics*, *14*, 355–369.

Razzaque, M. A., Komoike, Y., Nishizawa, T., Inai, K., Furutani, M., Higashinakagawa, T., et al. (2012). Characterization of a novel KRAS mutation identified in Noonan syndrome. *American Journal of Medical Genetics. Part A*, *158A*, 524–532.

Reischauer, S., Stone, O. A., Villasenor, A., Chi, N., Jin, S. W., Martin, M., et al. (2016). Cloche is a bHLH-PAS transcription factor that drives haemato-vascular specification. *Nature*, *535*, 294–298.

Richards, A. A., & Garg, V. (2010). Genetics of congenital heart disease. *Current Cardiology Reviews*, *6*, 91–97.

Roberts, A. E., Araki, T., Swanson, K. D., Montgomery, K. T., Schiripo, T. A., Joshi, V. A., et al. (2007). Germline gain-of-function mutations in SOS1 cause Noonan syndrome. *Nature Genetics*, *39*, 70–74.

Rochais, F., Mesbah, K., & Kelly, R. G. (2009). Signaling pathways controlling second heart field development. *Circulation Research*, *104*, 933–942.

Roger, G., Morisseau-Durand, M. P., Van Den Abbeele, T., Nicollas, R., Triglia, J. M., Narcy, P., et al. (1999). The CHARGE association: The role of tracheotomy. *Archives of Otolaryngology—Head & Neck Surgery*, *125*, 33–38.

Rohr, S., Bit-Avragim, N., & Abdelilah-Seyfried, S. (2006). Heart and soul/PRKCi and nagie oko/Mpp5 regulate myocardial coherence and remodeling during cardiac morphogenesis. *Development*, *133*, 107–115.

Rohr, S., Otten, C., & Abdelilah-Seyfried, S. (2008). Asymmetric involution of the myocardial field drives heart tube formation in zebrafish. *Circulation Research*, *102*, e12–e19.

Rose, B. A., Force, T., & Wang, Y. (2010). Mitogen-activated protein kinase signaling in the heart: Angels versus demons in a heart-breaking tale. *Physiological Reviews*, *90*, 1507–1546. http://dx.doi.org/10.1152/physrev.00054.02009.

Rothschild, S. C., Easley, C. A. t., Francescatto, L., Lister, J. A., Garrity, D. M., & Tombes, R. M. (2009). Tbx5-mediated expression of Ca(2+)/calmodulin-dependent

protein kinase II is necessary for zebrafish cardiac and pectoral fin morphogenesis. *Developmental Biology*, *330*, 175–184.
Ryan, A. K., Blumberg, B., Rodriguez-Esteban, C., Yonei-Tamura, S., Tamura, K., Tsukui, T., et al. (1998). Pitx2 determines left-right asymmetry of internal organs in vertebrates. *Nature*, *394*, 545–551.
Sakaguchi, T., Kikuchi, Y., Kuroiwa, A., Takeda, H., & Stainier, D. Y. (2006). The yolk syncytial layer regulates myocardial migration by influencing extracellular matrix assembly in zebrafish. *Development*, *133*, 4063–4072.
Sampath, K., Cheng, A. M., Frisch, A., & Wright, C. V. (1997). Functional differences among Xenopus nodal-related genes in left-right axis determination. *Development*, *124*, 3293–3302.
Sato, M., Tsai, H. J., & Yost, H. J. (2006). Semaphorin3D regulates invasion of cardiac neural crest cells into the primary heart field. *Developmental Biology*, *298*, 12–21.
Sato, M., & Yost, H. J. (2003). Cardiac neural crest contributes to cardiomyogenesis in zebrafish. *Developmental Biology*, *257*, 127–139.
Scherz, P. J., Huisken, J., Sahai-Hernandez, P., & Stainier, D. Y. (2008). High-speed imaging of developing heart valves reveals interplay of morphogenesis and function. *Development*, *135*, 1179–1187.
Schier, A. F., & Shen, M. M. (2000). Nodal signalling in vertebrate development. *Nature*, *403*, 385–389.
Schnetz, M. P., Bartels, C. F., Shastri, K., Balasubramanian, D., Zentner, G. E., Balaji, R., et al. (2009). Genomic distribution of CHD7 on chromatin tracks H3K4 methylation patterns. *Genome Research*, *19*, 590–601.
Schottenfeld, J., Sullivan-Brown, J., & Burdine, R. D. (2007). Zebrafish curly up encodes a Pkd2 ortholog that restricts left-side-specific expression of southpaw. *Development*, *134*, 1605–1615.
Schulz, Y., Wehner, P., Opitz, L., Salinas-Riester, G., Bongers, E. M., van Ravenswaaij-Arts, C. M., et al. (2014). CHD7, the gene mutated in CHARGE syndrome, regulates genes involved in neural crest cell guidance. *Human Genetics*, *133*, 997–1009.
Schuster, K., Leeke, B., Meier, M., Wang, Y., Newman, T., Burgess, S., et al. (2015). A neural crest origin for cohesinopathy heart defects. *Human Molecular Genetics*, *24*, 7005–7016.
Shen, L., Li, X. F., Shen, A. D., Wang, Q., Liu, C. X., Guo, Y. J., et al. (2010). Transcription factor HAND2 mutations in sporadic Chinese patients with congenital heart disease. *Chinese Medical Journal*, *123*, 1623–1627.
Shoji, W., Isogai, S., Sato-Maeda, M., Obinata, M., & Kuwada, J. Y. (2003). Semaphorin3a1 regulates angioblast migration and vascular development in zebrafish embryos. *Development*, *130*, 3227–3236.
Sinha, T., Li, D., Theveniau-Ruissy, M., Hutson, M. R., Kelly, R. G., & Wang, J. (2015). Loss of Wnt5a disrupts second heart field cell deployment and may contribute to OFT malformations in DiGeorge syndrome. *Human Molecular Genetics*, *24*, 1704–1716.
Smith, K. A., Chocron, S., von der Hardt, S., de Pater, E., Soufan, A., Bussmann, J., et al. (2008). Rotation and asymmetric development of the zebrafish heart requires directed migration of cardiac progenitor cells. *Developmental Cell*, *14*, 287–297.
Stainier, D. Y. (2001). Zebrafish genetics and vertebrate heart formation. *Nature Reviews. Genetics*, *2*, 39–48.
Stainier, D. Y., Fouquet, B., Chen, J. N., Warren, K. S., Weinstein, B. M., Meiler, S. E., et al. (1996). Mutations affecting the formation and function of the cardiovascular system in the zebrafish embryo. *Development*, *123*, 285–292.
Stainier, D. Y., Lee, R. K., & Fishman, M. C. (1993). Cardiovascular development in the zebrafish. I. Myocardial fate map and heart tube formation. *Development*, *119*, 31–40.

Stainier, D. Y., Weinstein, B. M., Detrich, H. W., 3rd, Zon, L. I., & Fishman, M. C. (1995). Cloche, an early acting zebrafish gene, is required by both the endothelial and hematopoietic lineages. *Development*, *121*, 3141–3150.

Staudt, D., & Stainier, D. (2012). Uncovering the molecular and cellular mechanisms of heart development using the zebrafish. *Annual Review of Genetics*, *46*, 397–418.

Sun, X., Zhang, R., Lin, X., & Xu, X. (2008). Wnt3a regulates the development of cardiac neural crest cells by modulating expression of cysteine-rich intestinal protein 2 in rhombomere 6. *Circulation Research*, *102*, 831–839.

Sutherland, M. J., & Ware, S. M. (2009). Disorders of left-right asymmetry: Heterotaxy and situs inversus. *American Journal of Medical Genetics. Part C, Seminars in Medical Genetics*, *151c*, 307–317.

Swaminathan, P. D., Purohit, A., Hund, T. J., & Anderson, M. E. (2012). Calmodulin-dependent protein kinase II: Linking heart failure and arrhythmias. *Circulation Research*, *110*, 1661–1677.

Szeto, D. P., Griffin, K. J. P., & Kimelman, D. (2002). hrT is required for cardiovascular development in zebrafish. *Development*, *129*, 5093–5101.

Targoff, K. L., Schell, T., & Yelon, D. (2008). Nkx genes regulate heart tube extension and exert differential effects on ventricular and atrial cell number. *Developmental Biology*, *322*, 314–321.

Tidyman, W. E., & Rauen, K. A. (2009). The RASopathies: Developmental syndromes of Ras/MAPK pathway dysregulation. *Current Opinion in Genetics & Development*, *19*, 230–236.

Tonkin, E. T., Smith, M., Eichhorn, P., Jones, S., Imamwerdi, B., Lindsay, S., et al. (2004). A giant novel gene undergoing extensive alternative splicing is severed by a Cornelia de Lange-associated translocation breakpoint at 3q26.3. *Human Genetics*, *115*, 139–148.

Totong, R., Schell, T., Lescroart, F., Ryckebusch, L., Lin, Y. F., Zygmunt, T., et al. (2011). The novel transmembrane protein Tmem2 is essential for coordination of myocardial and endocardial morphogenesis. *Development*, *138*, 4199–4205.

Trinh, L. A., & Stainier, D. Y. (2004). Fibronectin regulates epithelial organization during myocardial migration in zebrafish. *Developmental Cell*, *6*, 371–382.

Urbán, Z., Riazi, S., Seidl, T. L., Katahira, J., Smoot, L. B., Chitayat, D., et al. (2002). Connection between elastin haploinsufficiency and increased cell proliferation in patients with supravalvular aortic stenosis and Williams–Beuren syndrome. *The American Journal of Human Genetics*, *71*, 30–44.

Veerkamp, J., Rudolph, F., Cseresnyes, Z., Priller, F., Otten, C., Renz, M., et al. (2013). Unilateral dampening of Bmp activity by nodal generates cardiac left-right asymmetry. *Developmental Cell*, *24*, 660–667.

Vermot, J., Forouhar, A. S., Liebling, M., Wu, D., Plummer, D., Gharib, M., et al. (2009). Reversing blood flows act through klf2a to ensure normal valvulogenesis in the developing heart. *PLoS Biology*, 7, e1000246.

Vetrini, F., D'Alessandro, L. C., Akdemir, Z. C., Braxton, A., Azamian, M. S., Eldomery, M. K., et al. (2016). Bi-allelic mutations in PKD1L1 are associated with laterality defects in humans. *The American Journal of Human Genetics*, *99*, 886–893.

Vissers, L. E. L. M., van Ravenswaaij, C. M. A., Admiraal, R., Hurst, J. A., de Vries, B. B. A., Janssen, I. M., et al. (2004). Mutations in a new member of the chromodomain gene family cause CHARGE syndrome. *Nature Genetics*, *36*, 955–957.

Wang, Y., Kim, E., Wang, X., Novitch, B. G., Yoshikawa, K., Chang, L. S., et al. (2012). ERK inhibition rescues defects in fate specification of Nf1-deficient neural progenitors and brain abnormalities. *Cell*, *150*, 816–830.

Watrin, E., Kaiser, F. J., & Wendt, K. S. (2016). Gene regulation and chromatin organization: Relevance of cohesin mutations to human disease. *Current Opinion in Genetics & Development*, *37*, 59–66.

Wyse, R. K., al-Mahdawi, S., Burn, J., & Blake, K. (1993). Congenital heart disease in CHARGE association. *Pediatric Cardiology, 14*, 75–81.

Yashiro, K., Shiratori, H., & Hamada, H. (2007). Haemodynamics determined by a genetic programme govern asymmetric development of the aortic arch. *Nature, 450*, 285–288.

Yelon, D., Horne, S. A., & Stainier, D. Y. (1999). Restricted expression of cardiac myosin genes reveals regulated aspects of heart tube assembly in zebrafish. *Developmental Biology, 214*, 23–37.

Yelon, D., Ticho, B., Halpern, M. E., Ruvinsky, I., Ho, R. K., Silver, L. M., et al. (2000). The bHLH transcription factor hand2 plays parallel roles in zebrafish heart and pectoral fin development. *Development, 127*, 2573–2582.

Yoshiba, S., & Hamada, H. (2014). Roles of cilia, fluid flow, and Ca2+ signaling in breaking of left-right symmetry. *Trends in Genetics, 30*, 10–17.

Zariwala, M. A., Omran, H., & Ferkol, T. W. (2011). The emerging genetics of primary ciliary dyskinesia. *Proceedings of the American Thoracic Society, 8*, 430–433.

Zentner, G. E., Layman, W. S., Martin, D. M., & Scacheri, P. C. (2010). Molecular and phenotypic aspects of CHD7 mutation in CHARGE syndrome. *American Journal of Medical Genetics. Part A, 152A*, 674–686.

Zhang, L. F., Gui, Y. H., Wang, Y. X., Jiang, Q., & Song, H. Y. (2010). Effect of Tbx1 knock-down on cardiac performance in zebrafish. *Chinese Medical Journal, 123*, 1182–1189.

Zhou, Y., Cashman, T. J., Nevis, K. R., Obregon, P., Carney, S. A., Liu, Y., et al. (2011). Latent TGF-beta binding protein 3 identifies a second heart field in zebrafish. *Nature, 474*, 645–648.

CHAPTER TWO

Using Zebrafish to Study Kidney Development and Disease

Stephanie Jerman, Zhaoxia Sun[1]
Yale University School of Medicine, New Haven, CT, United States
[1]Corresponding author: e-mail address: zhaoxia.sun@yale.edu

Contents

Abstract

The kidneys are a crucial pair of organs that are responsible for filtering the blood to remove waste, maintain electrolyte and water homeostasis, and regulate blood pressure. There are a number of factors, both genetic and environmental, that can impair the function of the kidneys resulting in significant morbidity and mortality for millions

Current Topics in Developmental Biology, Volume 124
ISSN 0070-2153
http://dx.doi.org/10.1016/bs.ctdb.2016.11.008

of people affected by kidney disease worldwide. The zebrafish, *Danio rerio*, has emerged as an attractive vertebrate model in the study of kidney development and disease and has proven to be a powerful tool in the advancement of how kidney development occurs in vertebrates and how the kidney repairs itself after injury. Zebrafish share significant similarities in kidney development and composition of nephrons, the functional unit of the kidney. This makes the zebrafish a very promising model to study the mechanisms by which renal developmental defects occur. Furthermore, zebrafish are ideally suited for the study of how vertebrate kidneys respond to injury and have provided researchers with invaluable information on repair processes after kidney injury. Importantly, zebrafish have profound potential for discovering treatment modalities and, in fact, studies in zebrafish models have provided leads for therapeutics for human patients suffering from kidney disease and kidney injury. Here, we discuss the similarities and differences in zebrafish and mammalian kidney models, and highlight some of the major contributions the zebrafish has made in the understanding of kidney development and disease.

ABBREVIATIONS

ADPKD autosomal dominant polycystic kidney disease
AKI acute kidney injury
CAKUT congenital anomalies of the kidney and urinary tract
CRISPR clustered regularly interspaced short palindromic repeats
dpf day postfertilization
dpi days postinjury
ESRD end-stage renal disease
HDAC histone deacetylase
HDACi histone deacetylase inhibitor
HH Hedgehog
HNF-1β hepatocyte nuclear factor 1β
hpf hours postfertilization
ICO intraciliary calcium oscillation
IFT intraflagellar transport
KV Kupffer's vesicle
LR left-right
m4TPB methyl-4-(phenylthio) butanoate
MTZ metronidazole
MUT mutant
noi *no isthmus*
NTR *Escherichia coli* nitroreductase
Pc1 Polycystin-1
Pc2 Polycystin-2
PKD polycystic kidney disease
PTBA 4-(phenylthio) butanoic acid
TRP transient receptor potential
VPA valproic acid
VUR vesicoureteral reflux
WT wild type

1. INTRODUCTION

The kidney is a vital organ essential for removing waste and maintaining fluid homeostasis in animals. Through the basic functional unit, the nephron, it filters blood plasma through the glomerulus and reabsorbs water, ions, and other useful small molecules such as glucose through tubular and ductal epithelial cells. Defects in kidney development can lead to congenital abnormalities in the kidney structure, which may compromise kidney function. Adult renal diseases can also arise as a result of genetic defects and environmental insults. Kidney diseases affect more than 20 million Americans, and this problem is growing with the aging of the population. Better understanding of the molecular etiology of renal diseases to aid potential rational design of treatments is of profound medical importance.

With more than a million nephrons, the human kidney is a complex internal organ that poses significant challenges to renal studies. The classic mouse model has been invaluable in advancing our knowledge of kidney development, function, and disease. Additionally, in the past 20 years, the zebrafish *Danio rerio* has emerged as a powerful vertebrate model system that complements the mouse model. The zebrafish is a tropical freshwater species. It is small in size, with adults measuring about 4 cm in length. It is therefore feasible to maintain a sizable fish colony in a laboratory setting. Moreover, it has a short generation time and each pair can produce hundreds of offspring at weekly intervals. The nephron can be distinguished in embryos as young as 2 days old, and they can develop features of kidney disease that are remarkably similar to the corresponding diseases in humans before 2 days postfertilization (dpf). These features make zebrafish suitable for large-scale screens, including genetic and chemical screens which can advance kidney disease research.

The zebrafish is a relevant system for modeling mammalian kidney development and disease. Nephrons in the zebrafish pronephros are structurally similar to that of mammals, contain similar cell types, and perform similar functions (Diep et al., 2015; Drummond, 2000, 2002, 2003). A number of disease models, including congenital anomalies of the kidney and urinary tract (CAKUT) (Majumdar, Lun, Brand, & Drummond, 2000; Sun & Hopkins, 2001), polycystic kidney disease (PKD) (Coxam et al., 2014; Mangos et al., 2010; Obara et al., 2006; Paavola et al., 2013; Schottenfeld, Sullivan-Brown, & Burdine, 2007; Sun et al., 2004), and acute kidney injury (AKI) (Hellman et al., 2010; Hentschel et al., 2005; Huang et al., 2013; Johnson, Holzemer, & Wingert, 2011; McCampbell, Springer, & Wingert, 2015; McKee & Wingert, 2015; Palmyre et al., 2014; Zhou, Boucher, Bollig, Englert, & Hildebrandt, 2010; Zhou & Hildebrandt, 2012), have been established successfully in zebrafish. Zebrafish models have also been used to screen drug leads for kidney diseases (Cao et al., 2009; de Groh et al., 2010). As zebrafish kidney development has been described elsewhere (Drummond, 2000, 2002, 2003, 2005; Drummond & Davidson, 2010; Drummond & Wingert, 2016; Gerlach & Wingert, 2013; Wingert & Davidson, 2008), in this review, we highlight the similarities and differences of kidney development and disease between the zebrafish and mammals to illustrate the advantages and potential of this system and also reveal areas that can be further developed in the future.

2. CONSERVED FEATURES OF VERTEBRATE KIDNEY DEVELOPMENT: ZEBRAFISH AS A MODEL SYSTEM

Studying kidney development has led to great insight on the pathophysiology of congenital anomalies of the kidney. To fully realize the potential of zebrafish as a model of congenital kidney malformation, it is essential to understand the differences and similarities between zebrafish and mammalian kidney development. At a macroscopic level, in higher vertebrates, such as birds and mammals, three different forms of kidneys develop sequentially: the pronephros, mesonephros, and metanephros. While the pronephros is mostly vestigial, the mesonephros is functional during embryonic stages. Both the pronephros and the mesonephros are transient in mammals, and the metanephros is the permanent kidney. By contrast, the zebrafish develops pronephros and mesonephros, but not metanephros (Diep et al., 2015). Moreover, the pronephros is functional during embryonic and early larval stages. Most of the renal studies in zebrafish so far have been focused on the pronephros, as it is mature and functional by roughly 2.5 dpf, and thus is experimentally expedient and accessible. The development of the mesonephros, although it occurs later during juvenile stages, does show unique and relevant features and in recent years has attracted more attention.

The zebrafish pronephros forms from intermediate mesoderm during embryonic development (Fig. 1A). It is structurally simple and is composed of two nephrons with a pair of glomeruli fused at the midline at the level of the pectoral fin (Drummond et al., 1998; Fig. 1B and C). At the lateral side, each glomerulus is connected to a neck region, which drains filtrates through the pronephric tubule to the pronephric duct (Drummond et al., 1998;

Fig. 1B and C). The tubular region can be further divided into different segments based on differential gene expression (Wingert & Davidson, 2008).

On the cellular level, major segments and cell types of the nephron are conserved in zebrafish. Podocytes in the zebrafish glomerulus are very similar to those in mammals at the ultrastructural level (Drummond, 2000). Epithelial cells in the proximal convoluted tubule display prominent brush border, similar to the mammalian proximal tubule (Drummond et al., 1998). However, as an aquatic species, the zebrafish kidney lacks the intact

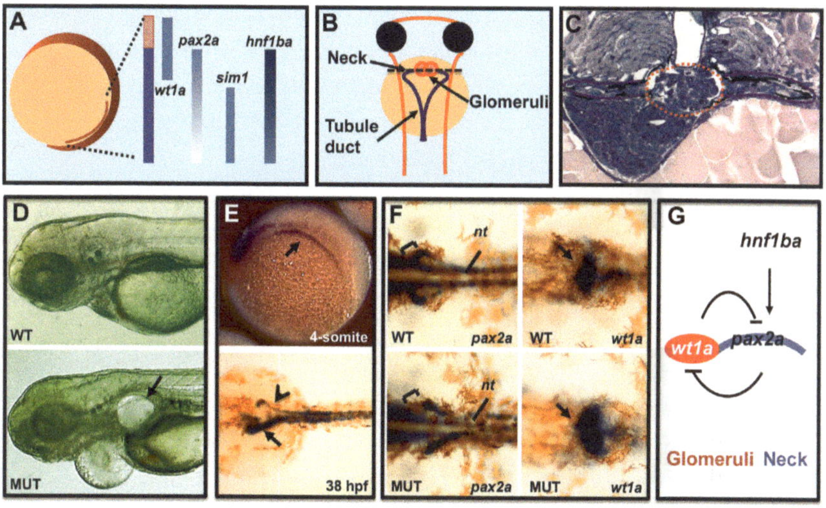

Fig. 1 Patterning of the zebrafish pronephros. (A) A diagram of a zebrafish embryo at an early-segmentation stage in a side view. A number of transcription factors are expressed in different regions of the intermediate mesoderm (*red line*), which give rise to distinct components of the pronephros. (B) Diagram of the pronephros in a zebrafish embryo at the pharyngula stage in a dorsal view. (C) A cross section through the level of the *dotted line* shown in (B) in a zebrafish embryo at 54 hour postfertilization (hfp). The fused glomeruli (*red dotted line*), and the pair of neck regions (*blue line*) can be seen. (D) Kidney cyst (*arrow*) is visible in a *hnf1ba*^{hi548} mutant (MUT) in comparison to a wild-type (WT) sibling at 4 day postfertilization (dpf). (E) *hnf1ba* expression in the developing zebrafish pronephros. At the 4-somite stage, it is expressed in the intermediate mesoderm (*arrow*). At 38 hpf, it is expressed in the neck, tubule, and duct region of the pronephros (*arrowhead*) and the gut (*arrow*). (F) Abnormal patterning of the nephron in a *hnf1ba*^{hi548} mutant (MUT) in comparison to a WT sibling at 38 hpf shown by in situ hybridization. While *pax2a* expression in the neck region (*bracket*) is absent, *wt1a* expression in the glomerular region is expanded. *nt*, neural tube. (G) A model for the function of Hnf1ba in the patterning of the zebrafish pronephros. *Panels (C–G) adapted from Sun, Z., & Hopkins, N. (2001). vhnf1, the MODY5 and familial GCKD-associated gene, regulates regional specification of the zebrafish gut, pronephros, and hindbrain.* Genes & Development, 15, *3217–3229.*

structure of the loop of Henle, which is a striking adaption by land animals to preserve water and concentrate urine through a countercurrent multiplier mechanism.

The development of the zebrafish pronephros shows both conserved and distinct features in comparison to the mammalian metanephros. Glomeruli, tubules, and ducts arise from the intermediate mesoderm to form the pronephros of zebrafish (Fig. 1A; Drummond & Davidson, 2010; Serluca & Fishman, 2001). Proper regional specification has to be achieved for the cells in the intermediate mesoderm to adopt distinct fates (Fig. 1A). Molecular mechanism for patterning different components of the kidney seems to be conserved between zebrafish and mammals. In addition, like in mammalian kidney development, mesenchymal cells in the intermediate mesoderm transform to highly polarized epithelial cells in the renal tubule and duct through mesenchymal–epithelial transition.

A significant difference between zebrafish and mammalian kidney development is the apparent lack of branching morphogenesis in zebrafish pronephros development, during which the two nephrons form de novo from the intermediate mesoderm. By contrast, in the development of more complex kidneys, branching morphogenesis is instrumental in generating the intricate arborized 3D epithelial structure.

During the transition between larval and juvenile stages, new nephrons are added around the existing pronephros to form the mesonephros in zebrafish (Diep et al., 2015). It was shown that the new nephrons arise from the nephrogenic mesenchyme, and they invade and join the pronephric tubule (Diep et al., 2015). Interestingly, new nephrons are added throughout adult stages of zebrafish, a feature which is not observed in mammals. As discussed later in more detail, the zebrafish mesonephros could be used to model kidney injury and regeneration.

3. DISEASE AND DEVELOPMENT: CAKUT IN ZEBRAFISH

The conserved features of zebrafish development discussed earlier make it a relevant model system for CAKUT. CAKUT refers to a wide spectrum of structural anomalies of the kidney and the urinary tract, including ureteropelvic junction obstruction, kidney agenesis, kidney dysplasia, and vesicoureteral reflux (VUR) (recently reviewed in Nicolaou, Renkema, Bongers, Giles, & Knoers, 2015). Consequences of these structural defects range from functionally insignificant to severely compromised kidney function. CAKUT is the leading cause of end-stage renal disease (ESRD) in children. Both environmental and genetic factors can disrupt critical steps of kidney development, leading to CAKUT.

The advancement of the next-generation sequencing technique led to the identification of a growing list of candidate genes for CAKUT (Hwang et al., 2014; Madariaga et al., 2013; Nakanishi & Yoshikawa, 2003; Nakayama et al., 2010; Negrisolo et al., 2011; Nicolaou et al., 2015). Currently mutations in *PAX2* and *HNF-1β* are the most frequent genetic cause of CAKUT. Further dissection of Pax2 and Hnf-1β function in kidney development in zebrafish will accelerate the understanding of how these genes contribute to CAKUT in patients. Consistent with the conserved patterning of different segments of the nephron in the zebrafish pronephros, zebrafish mutants of *pax2* and *hnf1β*, two genes encoding transcription factors critical for regional specification of the nephron, have provided useful insights. These are discussed later.

However, unlike these well-studied cases, a causal relationship between many of the identified genes and CAKUT remains to be established. This is particularly challenging because of the genetic heterogeneity and frequent incomplete penetrance of the genetic factors of CAKUT. Since zebrafish are genetically tractable, this is a valuable system to investigate the role of identified candidate genes in kidney development in a relatively high-throughput fashion. Understanding the genetic cause of CAKUT will aid early diagnosis and better management of CAKUT, potentially delaying the onset of ESRD.

3.1 Pax2 and CAKUT

PAX2 encodes a transcription factor of the paired-box family. Heterozygous mutation in *PAX2* was the first identified genetic cause of CAKUT (Sanyanusin et al., 1995). CAKUT patients with *PAX2* mutations show a continuum of defects ranging from renal dysplasia to VUR (Harshman & Brophy, 2012; Hwang et al., 2014; Negrisolo et al., 2011; Sanyanusin et al., 1995). In mouse, *Pax2* expression is detected in the pronephric tubule and extending nephric duct (Dressler, Deutsch, Chowdhury, Nornes, & Gruss, 1990). In the developing mouse metanephros, *Pax2* is expressed in condensing mesenchymal cells and the collecting duct. Later its expression declines with the differentiation of epithelial cells (Dressler et al., 1990). Consistent with its wide expression domain in the developing kidney, *Pax2* knockout mouse lacks both ureters and metanephric kidneys (Torres, Gomez-Pardo, Dressler, & Gruss, 1995). Interestingly, heterozygous carriers of the knockout allele or a truncating allele show renal dysplasia and VUR (Murawski, Myburgh, Favor, & Gupta, 2007; Torres et al., 1995), which bears striking resemblance to CAKUT patients carrying *PAX2* mutations. In the heterozygous carriers, the position of the ureteric bud is shifted caudally (Murawski et al., 2007). It is thought that caudal shift of the ureteric bud disrupts branching morphogenesis, causing kidney dysplasia. Moreover, caudal shift of the ureteric bud could also lead to abnormal insertion of the ureter into the bladder, resulting in VUR (Caruana & Bertram, 2015).

There are two homologs of *Pax2* in zebrafish but only one of them, *pax2a*, is expressed in the developing pronephros (Pfeffer, Gerster, Lun, Brand, & Busslinger, 1998). During early somitogenesis, *pax2a* expression is detected in the intermediate mesoderm starting at the level of somite 2 and extends to more posterior regions (Fig. 1A; Serluca & Fishman, 2001). Similar to the downregulation of *Pax2* expression as the nephron matures in the mammalian kidney, at 38 hours postfertilization (hpf), *pax2a* expression becomes restricted in the neck region and the anterior pronephric tubule in zebrafish (Krauss, Johansen, Korzh & Fjose, 1991; Majumdar et al., 2000). Complementary to *pax2a* expression, the Wilms' tumor gene *wt1a* is expressed in the future glomerular region in the intermediate mesoderm between the level of somite 1–2 and persists in mature glomeruli in zebrafish (Fig. 1A; Majumdar et al., 2000; Serluca & Fishman, 2001). This regional-specific gene expression pattern, however, is disrupted in *no isthmus* (*noi*), a zebrafish *pax2a* mutant. In this mutant, the expression of *wt1a* expands from the glomerular region to the neck region and the Na^+/K^+ ATPase signal is lost in the neck and anterior tubular regions, suggesting a patterning defect (Majumdar et al., 2000). It would be interesting to test whether a similar patterning defect is responsible for the abnormal position of the ureteric bud in *Pax2* mouse mutants and whether similar defects are seen in CAKUT patients with *PAX2* mutations.

3.2 Hnf-1β and CAKUT

Hepatocyte nuclear factor 1β (HNF-1β), also known as vHnf1, Lf-B3, or Tcf2, is a transcription factor of the homeobox family. *HNF-1β* is the second gene associated with CAKUT. A wide spectrum of kidney defects has been observed in patients with heterozygous *HNF-1β* mutations, including hypoplasia, single kidney, and multicystic dysplastic kidney (Bingham et al., 2002; Heidet et al., 2010; Hwang et al., 2014; Lindner et al., 1999; Madariaga et al., 2013; Nakayama et al., 2010; Thomas et al., 2011).

Mutations in *HNF-1β* are also associated with extrarenal diseases, most notably mature onset diabetes of the young type 5 (Horikawa et al., 1997; Nishigori et al., 1998).

Hnf-1β is expressed in multiple organs, including the kidney, liver, gut, and pancreatic primordia. In the developing mouse metanephros, its expression can be detected from early stages throughout the epithelial compartment and the ureter (Coffinier, Barra, Babinet, & Yaniv, 1999). Complete knockout is embryonically lethal by day 7.5 (Barbacci et al., 1999; Coffinier et al., 1999), precluding an analysis of kidney development.

There are two *Hnf-1β* paralogs in the zebrafish genome, *hnf1ba* and *hnf1bb* (Naylor & Davidson, 2014). In *hnf1ba*^{hi548} mutants, we found that kidney cyst is obvious in live mutant embryos under a dissecting scope (Fig. 1D; Sun & Hopkins, 2001). Specific *hnf1ba* expression domains first emerge at the end of gastrulation in the presumptive hindbrain region (Sun & Hopkins, 2001). At the 4-somite stage, *hnf1ba* expression also appears in a segment in the intermediate mesoderm (Fig. 1E; Sun & Hopkins, 2001), which is destined to become the pronephric neck, tubule, and ducts (Drummond et al., 1998; Serluca & Fishman, 2001). The *hnf1ba* expression domain in the intermediate mesoderm coincides with that of *pax2a* at this stage but persists in renal epithelial cells through the pharyngula period (Fig. 1E and F; Sun & Hopkins, 2001). *hnf1bb* is also expressed in the developing pronephros, but its expression is restricted to the proximal tubule, distal early tubule, and distal late tubule (Naylor, Przepiorski, Ren, Yu, & Davidson, 2013). In *hnf1ba*^{hi548} mutant embryos, *pax2a* expression in the neck region is absent (Fig. 1F; Sun & Hopkins, 2001). Complementarily, *wt1a* expression in the future glomerulus is expanded mediolaterally to regions corresponding to the future neck region in wild-type (WT) embryos (Fig. 1F; Sun & Hopkins, 2001). These data show that the neck region has adopted molecular attributes of the future glomeruli, suggesting that *hnf1ba* plays an important role in the specification of different fates in pronephros development (Fig. 1G).

Genetic studies of *Hnf-1β* in mouse also suggest that it is involved in regional specification. Inactivation of *Hnf-1β* in mesenchymal stem cells responsible for nephron epithelial cells results in ablation of proximal, distal, and loop of Henle segments (Heliot et al., 2013; Massa et al., 2013). More detailed analysis revealed that a specific subdomain of the S-shaped body is missing from the mutant (Heliot et al., 2013; Massa et al., 2013), suggesting a patterning defect. Interestingly, altering Hnf-1β function through tissue-specific expression of dominant-negative proteins or conditional knockout in already formed kidneys leads to renal cysts (Gresh et al., 2004; Hiesberger et al., 2004, 2005). Combined, these results suggest that *HNF-1β* is involved in multiple steps of kidney development and epithelial differentiation.

Many CAKUT genes remain unidentified; mutations in *PAX2* and *HNF-1β* combined account for only 5–15% of CAKUT cases (Madariaga et al., 2013; Thomas et al., 2011; Weber et al., 2006). The genetically accessible zebrafish system could potentially provide candidate genes for the search of novel CAKUT genes. Zebrafish could also serve as a high-throughput platform for validating candidate CAKUT genes.

4. RENAL CILIOPATHY MODELS IN ZEBRAFISH

The cilium is a rod-like cell surface organelle. It contains a microtubule axoneme in the center surrounded by a membrane that is contiguous with, yet distinct in composition from, the cell membrane. The cilium is almost ubiquitously present on vertebrate cells. Protruding from the cell surface into the environment, the ubiquitous cilium is ideally situated to function as an antenna for the cell to sense, interpret, and integrate signals. Studies in the past decade suggest that the cilium plays a critical role in signaling, particularly Hedgehog (HH) signaling (Haycraft et al., 2005; Huangfu et al., 2003). In the absence of its ligand, the HH receptor Patched is localized on the cilium (Rohatgi, Milenkovic, & Scott, 2007). There it prevents the trafficking of Smoothened into the cilium and keeps the pathway in an "off" state. Upon HH binding, Patched is removed from the cilium and Smoothened enters, leading to the activation of the pathway and in turn the expression of target genes (Rohatgi et al., 2007).

Consistent with cilia's wide distribution and critical role in signaling, cilia dysfunction and dysgenesis have been linked to a growing list of human diseases including PKD, cancer, mental retardation, and obesity, collectively referred to as ciliopathies (Hildebrandt & Zhou, 2007). Here, we discuss the use of zebrafish models to study renal ciliopathies.

4.1 Renal Ciliopathy Studies in Mice and Zebrafish: Similarities, Differences, and the Complexities of In Vivo Kidney Disease Models

Zebrafish occupies a unique niche in studying ciliopathies. Much of our earlier knowledge regarding cilia was originated in studies on the flagellum in the green algae *Chlamydomonas*. However, despite the high level of conservation in cilia biogenesis and structure, signaling mediated by cilia has diverged significantly between vertebrate and invertebrate animals. For example, in contrast to the almost ubiquitous presence of cilia on mammalian cells, most fly cells lack cilia. Despite the limited distribution of cilia on fly cells, genetic screens in this organism have been instrumental in dissecting out key steps of the HH pathway; however, the involvement of the cilium in this pathway was only revealed by genetic studies in mice (Haycraft et al., 2005; Huangfu et al., 2003). Similarly, in *Caenorhabditis elegans*, cilia presence is limited to some neuronal cells (Bae & Barr, 2008; Ingham, Nakano, & Seger, 2011). Furthermore, this organism seems to lack a functional HH pathway. This functional divergence has limited the use of these powerful classic genetic model systems to dissect cilia-mediated signaling relevant to vertebrate biology and human ciliopathies.

In contrast to worm and fly, cilia are widely distributed on zebrafish cells (Fig. 2). For example, cilia presence can be readily detected in the olfactory placode, otic vesicle, kidney, lateral line organ, and Kupffer's vesicle (KV, the zebrafish left-right (LR) organizer) (Fig. 2). Moreover, cilia-mediated signaling, particularly HH signaling, is conserved in zebrafish. Specifically, in zebrafish, Smo and Gli are targeted to cilia (Kim, Richardson, van Eeden, & Ingham, 2010), and in a maternal–zygotic mutant of *ift88*, a gene essential for cilia biogenesis, HH signaling is dampened (Huang & Schier, 2009). Importantly, a toolbox for studying cilia in zebrafish is already available, including a sizable collection of ciliary mutants and live reporters for cilia morphology and function (Table 1). Thus, zebrafish is an excellent model system for studying vertebrate-specific ciliary signaling and ciliopathies.

One of the best-studied renal ciliopathies is PKD, which is characterized by the formation of multiple epithelium-lined fluid-filled cysts in the kidney. Zebrafish is an excellent system to study kidney cyst formation. As the embryo is transparent and develops ex utero, kidney cyst formation is visible in live animals under a simple stereoscope (Fig. 1D). Large-scale genetic screens have been performed in zebrafish, and many cystic kidney mutants have been identified. The realization that many of the cystic kidney mutants isolated in fish resulted from defects in ciliary genes provided strong support for the hypothesis that the cilium is a critical organelle for PKD pathogenesis (Sun et al., 2004). These ciliary mutants in zebrafish frequently show a characteristic constellation of phenotypes, including LR asymmetry defects, ventral body curvature, and kidney cyst, providing a robust phenotypic readout of ciliary mutants.

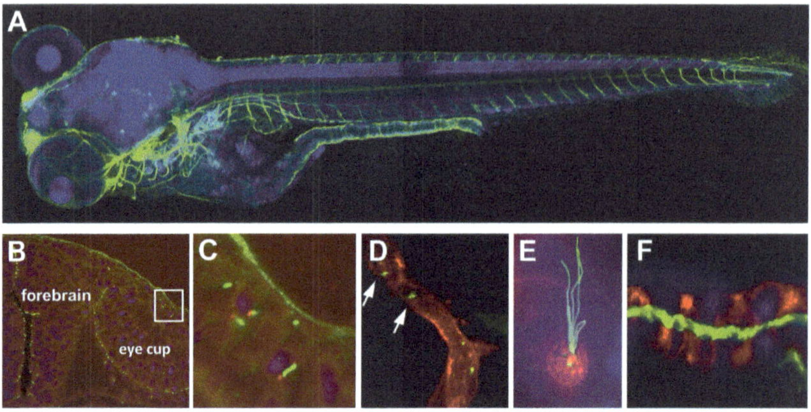

Fig. 2 Cilia are widely distributed in zebrafish. (A) A whole-mount zebrafish embryo at 2 dpf stained with antiacetylated tubulin (*green*) and TOTO3 (*blue*). (B and C) Cross section of an embryo at the 20-somite stage stained with the ciliary marker anti-Arl13b in *green*, the basal body marker anti-γ-tubulin in *red*, and DAPI in *blue*. C is an enlarged view of the *boxed area* in (B). (D) Cilia (*green*, pointed by *arrows*, labeled by Arl13b-eGFP driven by the endothelial-specific *kdrl* promoter) on endothelial cells (*red*, labeled by mRFP driven by the *kdrl* promoter) in a 24-hpf embryo. (E) Kinocilia stained with antiacetylated tubulin (*green*) in a lateral line organ in a 3-dpf embryo. F-actin is stained *red* with phalloidin. (F) Bundled cilia stained with antiacetylated tubulin (*green*) in the pronephric tubule (*red*, anti-Cdh17) in a 4-dpf embryo. *Panel (A) adapted from the cover of* JASN, 21, *1326–1333, 2010, panel (D) adapted from fig. 3B, Kallakuri, S., Yu, J. A., Li, J., Li, Y., Weinstein, B. M., Nicoli, S., et al. (2015). Endothelial cilia are essential for developmental vascular integrity in zebrafish. Journal of the American Society of Nephrology, 28, 864–875, panel (E) adapted from fig. 6A, Li, J. & Sun, Z. (2011). Qilin is essential for cilia assembly and normal kidney development in zebrafish. PLos One, 6, e27365 and panel (F) adapted from fig. S3, Zhao, L., Yuan, S., Cao, Y., Kallakuri, S., Li, Y., Kishimoto, N., et al. (2013). Reptin/Ruvbl2 is a Lrrc6/Seahorse interactor essential for cilia motility.* Proceedings of the National Academy of Sciences of the United States of America, 110, *12697–12702.*

There are notable differences between zebrafish and mammalian PKD models. While mammalian kidney epithelial cells display solitary primary cilia that bend passively in response to flow, cilia in the zebrafish pronephric duct and tubule are motile (Kramer-Zucker et al., 2005). Moreover, many of the pronephric epithelial cells are multiciliated, containing up to 16 cilia per cell (Kramer-Zucker et al., 2005). These cilia bundle together and beat vigorously and coordinate to drive fluid flow (Kramer-Zucker et al., 2005), befitting the increased need of water excretion in an aquatic species. In zebrafish, multiple cilia motility mutants develop kidney cyst (Becker-Heck et al., 2011; Panizzi et al., 2012; Zhao et al., 2013), presumably resulting from a lack of robust fluid flow and eventual defective kidney

Table 1 Markers for Analyzing Kidney Development and Function in Zebrafish

Name	Type	Utility	References
Markers for live imaging of cilia and intraciliary calcium			
Arl13b-eGFP	Plasmid template for in vitro transcription	Live imaging of cilia	Duldulao, Lee, and Sun (2009)
Tg(actin::arl13b-GFP)	Transgenic line	Live imaging of cilia in KV and CNS	Malicki, Avanesov, Li, Yuan, and Sun (2011) and Borovina, Superina, Voskas, and Ciruna (2010)
Arl13b-GCaMP6 and Arl13b-mCherry	Plasmid templates for in vitro transcription	Ratiometric imaging of intraciliary calcium	Yuan, Zhao, Brueckner, and Sun (2015)
Arl13b-Pvalb	Plasmid template for in vitro transcription	For dampening intraciliary calcium	Yuan et al. (2015)
Markers for specific cell types and segments of zebrafish kidney			
Hematoxylin and eosin	Histology stain	Distinguish renal proximal tubule from distal tubule	McCampbell et al. (2015) and McKee and Wingert (2015)
Lotus tetragonolobus	Lectin	Renal proximal tubule marker	McCampbell and Wingert (2014) and McCampbell et al. (2015)
Dolichos biflorus agglutinin	Lectin	Renal distal tubule marker	McCampbell and Wingert (2014) and McCampbell et al. (2015)
pax2a	Gene expressed in specific segments of the nephron	Renal segment marker	McCampbell et al. (2015)
scl20ala	Gene expressed in specific segments of the nephron	Marker of proximal tubules	McCampbell et al. (2015)

Continued

Table 1 Markers for Analyzing Kidney Development and Function in Zebrafish—cont'd

Name	Type	Utility	References
cdh17::GFP	Transgenic line	Labels tubules and ducts of pronephric and mesonephric kidney	Zhou et al. (2010)
Tg(pod::mCherry)	Transgenic line	Labels podocytes (marker of glomerulus)	Zhou et al. (2010)
Tg(podocin::GFP)	Transgenic line	Labels podocytes (marker of glomerulus)	He, Ebarasi, Hultenby, Tryggvason, and Betsholtz (2011)
Tg(podocin:NTR-GFP)	Transgenic line	Induces podocyte-specific ablation	Huang et al. (2013)
Tg(podocin:NTR-mCherry)	Transgenic line	Induces podocyte-specific ablation	Zhou and Hildebrandt (2012)
Tg(pod::NTR-mCherry/wt1b::GFP)	Transgenic line	Podocyte-specific ablation and a developmental marker for mesonephric nephrons	Zhou and Hildebrandt (2012)
Tg(wt1b:EGFP)	Transgenic line	Developmental marker/marker of renal progenitor cells	Zhou et al. (2010)
Tg(wt1b:mCherry)	Transgenic line	Developmental marker/marker of renal progenitor cells	Diep et al. (2011)
Tg(lhx1a:EGFP)	Transgenic line	Developmental/renal progenitor marker essential for establishing kidney field	Diep et al. (2011) and de Groh et al. (2010)

morphogenesis. As mammalian renal cilia are immotile primary cilia, cilia motility defects conceivably are not a risk factor for kidney cyst formation. Consistently, patients with primary ciliary dyskinesia, a cilia motility disease, typically show a combination of recurrent respiratory tract infection and male infertility, but not cystic kidney. Thus, cystic kidney mutants in zebrafish represent a broad collection of mutants with ciliary defects, including defective cilia biogenesis, cilia motility, and potentially cilia-mediated signaling. A subset, particularly those with specific cilia motility defects, may not be directly relevant to human PKD.

4.2 Autosomal Dominant Polycystic Kidney Disease: What We Have Learned From Zebrafish Studies

Autosomal dominant polycystic kidney disease (ADPKD) is the most common form of PKD and is one of the most common monogenic diseases (Gabow & Grantham, 1997). It affects more than 600,000 Americans and an estimated 12.5 million people worldwide. Currently, there is no directed therapy against PKD, and half of the ADPKD patients will progress to ESRD by the age of 60. ADPKD is caused by mutations in two genes, *PKD1* and *PKD2*, encoding Polycystin-1 (PC1) and -2 (PC2), respectively (Consortium, 1994; Mochizuki et al., 1996). PC2 is a six-transmembrane protein that shares homology with the transient receptor potential (TRP) channels and can complex with PC1, which contains 11-transmembrane domains, to form a voltage-sensitive, nonselective cation channel (Hanaoka et al., 2000; Koulen et al., 2002).

Among the first links between PKD and cilia were the realization that Polycystins are targeted to cilia and mutations in intraflagellar transport (IFT) genes could lead to PKD (Barr & Sternberg, 1999; Pazour et al., 2000; Yoder, Hou, & Guay-Woodford, 2002). IFT genes encode components of IFT particles. Many studies demonstrated that the formation and maintenance of the cilium depends on IFT particles (Kozminski, Johnson, Forscher, & Rosenbaum, 1993; Rosenbaum & Witman, 2002). These particles move along microtubules in the ciliary axoneme and are thought to be vehicles for transporting cargos needed for assembly, maintenance, and function of cilia.

The role of cilia in PKD pathogenesis is further highlighted by the fact that cilia-targeted trafficking of Polycystins highly correlates with their function in vivo (Cai et al., 2014; Yoshiba et al., 2012). Interestingly, although both cilia biogenesis and Polycystin mutants develop kidney cysts, complete removal of cilia ameliorated cyst progression triggered by Polycystin inactivation (Ma, Tian, Igarashi, Pazour, & Somlo, 2013), suggesting that Polycystins inhibit a novel cilia-dependent cyst-promoting pathway. The molecular basis of this pathway is currently not understood.

The zebrafish genome contains a single homolog of *PKD2*. The zebrafish *pkd2* is expressed in the KV, floor plate, and pronephric epithelial cells (Bisgrove, Snarr, Emrazian, & Yost, 2005; Schottenfeld et al., 2007). Multiple mutant alleles of *pkd2* have been isolated in zebrafish. Mutants show randomized LR development and a striking dorsal body curvature (Schottenfeld et al., 2007; Sun et al., 2004). Curiously, although Pc2 is present on pronephric cilia in zebrafish (Obara et al., 2006), the mutants do not show kidney cysts (Schottenfeld et al., 2007; Sun et al., 2004). Morpholino oligo against the translational initiation site of *pkd2*, which could block the translation of both maternally deposited and zygotic mRNA, did lead to kidney cyst formation (Schottenfeld et al., 2007; Sun et al., 2004). It is thought that maternal contribution of *pkd2* transcript may mask the early requirement of Pc2 function and hence the lack of cystic kidney phenotype. It will be important to clarify this issue by removing both maternal and zygotic *pkd2* gene products through generating maternal–zygotic mutants.

Dorsal curvature is perhaps the most visually apparent phenotype displayed by zebrafish *pkd2* mutants. Even though the underlying molecular and cellular mechanisms that generate this distinct phenotype remains unclear, it proves to be relevant and useful for ADPKD research. Using dorsal curvature as an initial readout, we performed a chemical modifier screen for $pkd2^{hi4166}$ mutants (Cao et al., 2009). Through this screen, histone deacetylase inhibitors (HDACi) were identified as suppressors of *pkd2* mutants. Moreover, valproic acid (VPA), a HDACi, ameliorated cyst progression and improved kidney function in a neonatal mouse ADPKD model based on conditional inactivation of *Pkd1*, validating that this result from zebrafish is applicable to mammalian systems and to kidney cyst caused by *PKD1* inactivation (Cao et al., 2009). This finding initiated in zebrafish has clinical implications, because VPA is an established medicine for treating seizure and bipolar disorder (Duenas-Gonzalez et al., 2008). In addition, multiple HDACi have either been approved by FDA or are in clinical trials for cancer treatment, and have known safety profiles (Mehnert & Kelly, 2007; Witt, Deubzer, Milde, & Oehme, 2009), raising the possibility of repurposing these drugs for PKD treatment.

Basic research on the LR asymmetry defect in zebrafish *pkd2* mutants also led to relevant insight regarding Pc2 function. During vertebrate development, directional fluid flow in the LR organizer breaks LR symmetry of the body plan (Nonaka, Shiratori, Saijoh, & Hamada, 2002; Nonaka et al., 1998). In both zebrafish and mouse, Pc2 is essential for proper LR development (Bisgrove et al., 2005; Pennekamp et al., 2002; Schottenfeld et al., 2007; Sun et al., 2004). By developing novel genetically encoded calcium reporters and calcium buffers that are targeted specifically to cilia and using live imaging in developing zebrafish embryos, we recently discovered novel dynamic intraciliary calcium oscillations (ICOs) at the KV (Yuan et al., 2015). Interestingly, these ICOs are left-biased and begin with the onset of cilia motility, preceding all other known molecular LR asymmetries (Yuan et al., 2015). This study further showed that Pc2 is required for ICOs and that intraciliary calcium signaling is specifically required for establishing LR asymmetry (Yuan et al., 2015). Together, these results support an important role of Pc2 in intraciliary calcium signaling in response to fluid flow.

Similar to epithelial cells lining the KV, renal epithelial cells are exposed to fluid flow. Moreover, in cultured renal epithelial cells, mechanical stress on the cilium, either by direct pipette manipulation or by application of laminar fluid flow, results in a rise in Ca^{2+} that depends on cilia and both PC1 and PC2 (Jin et al., 2014; Nauli et al., 2003; Praetorius, Frokiaer, Nielsen, & Spring, 2003). These results lead to the hypothesis that PC1 and PC2 on cilia of renal epithelial cells are involved in mechanosensory response to fluid flow and that defects in this response cause kidney cysts to form. This hypothesis, however, remains highly controversial as ciliary channel activity and kidney cyst formation in vivo seem to have distinct functional requirements for Polycystins. Data from a patch clamp study showed that there is a nonselective cation current across the ciliary membrane in multiple cell lines, including renal epithelial cells (DeCaen, Delling, Vien, & Clapham, 2013), and while PKD1L1 and PKD2L1 were found to be required for this current, both PC1 and PC2 are dispensable (DeCaen et al., 2013). By contrast, while inactivation of *Pkd1* and *Pkd2* leads to kidney cyst formation (Lu et al., 2001; Wu et al., 1998), no such defect was reported in *Pkd1l1* and *Pkd2l1* mutants (Delling, DeCaen, Doerner, Febvay, & Clapham, 2013; Field et al., 2011).

LR development also shows overlapping yet distinct requirements for Polycystins: while *Pkd2* and *Pkd1l1* are required (Field et al., 2011; Kamura et al., 2011; Pennekamp et al., 2002; Vogel et al., 2010), *Pkd1* is not (McCampbell et al., 2015) and *Pkd2l1* knockout mice have isolated intestinal malrotation with no other major organ laterality defect (DeCaen et al., 2013). As Polycystin family members have unique sequence features, it is possible that different tissues/cell lines employ different Polycystin complexes to best cope with specific functional requirements. To complicate the picture even further, a recent study failed to detect any cilia-specific calcium influxes in response to flow using transgenic mice expressing the calcium indicator GECO1.2 (Delling et al., 2016). To clarify the role of Polycystins, intraciliary calcium signaling, and kidney cyst formation, it will be critical to use calcium reporters with suitable calcium affinity and kinetics and cilia-targeted calcium buffers to analyze the function of intraciliary calcium in both mouse and zebrafish kidneys directly.

In contrast to the single *pkd2* gene, there are two zebrafish paralogs of *PKD1*, *pkd1a* and *pkd1b* (Mangos et al., 2010). While the expression of *pkd1b* seems to be restricted to the central nervous system and lateral mesoderm during embryonic stages, the expression of *pkd1a* is more widespread (Mangos et al., 2010). *lyc1*, a zebrafish truncation mutant of *pkd1a*, was isolated as a lymphangiogenesis mutant (Coxam et al., 2014). The role of Pc1 in lymphatic development was further supported by mouse mutant analysis (Coxam et al., 2014; Outeda et al., 2014). Zebrafish *lyc1* mutants, however, do not display any body curvature phenotype as is observed in *pkd2* mutants (Coxam et al., 2014). Interestingly, injection of a morpholino oligo against *pkd1b* into *lyc1* mutant embryos did lead to dorsal body curvature, suggesting a potential functional redundancy between the two orthologs (Coxam et al., 2014). This result is consistent with the previous observation that knockdown of both *pkd1a* and *pkd1b* simultaneously using morpholino oligos caused dorsal body curvature (Mangos et al., 2010). The specificity of the result is supported by the lack of any body axis phenotype in WT embryos injected with the morpholino against *pkd1b* alone (Coxam et al., 2014; Mangos et al., 2010). Genetic double mutants will further verify the role of *pkd1a* and *pkd1b* in body axis development and clarify whether complete depletion of Pc1 function in zebrafish leads to kidney cyst formation.

5. KIDNEY INJURY AND REGENERATION

The value of zebrafish as a model for kidney damage and repair has become increasingly evident, especially in the understanding of AKI. AKI is a devastating condition caused by a rapid decline in renal excretory function that occurs over a period of hours or days and results in an increase in circulatory waste products (Bellomo, Kellum, & Ronco, 2012). Incidence of AKI is increasing worldwide, with an estimated occurrence of 1 in 5 adults and 1 in 3 children that are hospitalized with acute illness and resulting in up to 60% mortality rate for patients who become critically ill (Rewa & Bagshaw, 2014). Thus, there is a critical need to identify new preventative and curative treatments for AKI.

AKI can result from a number of factors including ischemia/reperfusion, exposure to toxins, and sepsis (Lameire, Van Biesen, & Vanholder, 2005). AKI results in damage and cell death and can affect any part of the nephron including the tubular epithelia, glomerulus, interstitium, or vasculature (Basile, Anderson, & Sutton, 2012). However, the proximal tubule is especially sensitive to AKI, mainly as a consequence of their reabsorption function, which results in the direct uptake of toxins. Additionally, the proximal tubule is highly reliant on aerobic respiration; therefore, a reduction in blood supply caused by ischemic injury is particularly detrimental to this subpopulation of tubule cells. Treatment for AKI mainly consists of supportive care and patients that are sufficiently supported have the ability to recover successfully; however, it has been recognized that patients with AKI have an increased risk of progressing to chronic kidney disease and ESRD, requiring dialysis, or kidney transplantation.

Current mammalian models of AKI, including mice and rats, are limited by the fact that their kidneys are only accessible through surgery and only a very small population of renal tubules and vessels on the surface of the kidney can be imaged in vivo. This, for the most part, excludes the possibility of visual monitoring of the processes of AKI and kidney repair in real time in a living animal. Zebrafish embryos and adults have a number of attributes that make it ideally suited for the study of AKI. Over the past decade, researchers have amassed an expansive toolkit, including the advancement of genetic tools and kidney imaging methodologies (Table 1), for the study of zebrafish kidney damage and repair.

5.1 Gentamicin-Induced AKI

The aminoglycoside antibiotic gentamicin is a highly effective treatment for patients suffering from severe infections. However, the use of gentamicin is limited due to an increased risk of nephrotoxicity associated with its use (Selby, Shaw, Woodier, Fluck, & Kolhe, 2009). Understanding the mechanism of gentamicin-induced nephrotoxicity is needed in order to develop strategies to reduce AKI in patients. Zebrafish have emerged as an ideal model to study this type of AKI since gentamicin induces widespread proximal tubule damage (Fig. 3A) and renal failure that mimics what has been observed in higher vertebrates (Hentschel et al., 2005).

In a landmark study, Hentschel et al. showed that injection of gentamicin into zebrafish embryos results in flattening of the proximal tubule brush border, tubular and glomerular distention, accumulation of debris in the tubule

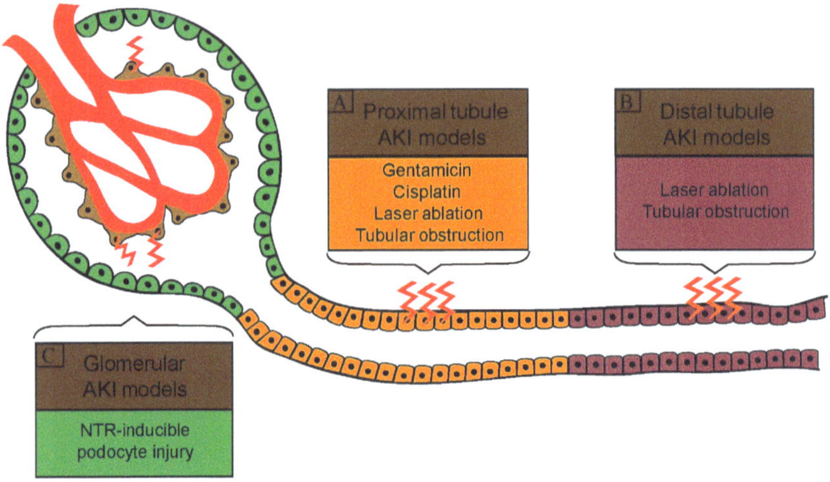

Fig. 3 Zebrafish acute kidney injury models. (A) Proximal tubule acute kidney injury zebrafish models resulting from administration of gentamicin, cisplatin, laser ablation, and mechanical tubular obstruction, (B) distal tubule acute kidney injury zebrafish models caused by laser ablation and tubular obstruction, and (C) glomerular zebrafish acute kidney injury model using transgenic zebrafish expressing the nitroreductase gene (NTR) downstream of the podocyte-specific *podocin* gene and treatment with MTZ to generate a toxic product, causing cell death.

lumen and pericardial edema, characteristics that mimic mammalian AKI (Hentschel et al., 2005). These instrumental studies revealed that zebrafish nephrotoxicity parallels that seen in higher vertebrates, including mammals, and enabled researchers a valuable model organism to study AKI. Gentamicin has since become a mainstay in the study of AKI in zebrafish. Gentamicin-induced AKI can be assessed in the transparent zebrafish embryo through the use of fluorescently labeled dextran to study kidney functionality. The 10 kDa dextran is normally endocytosed by proximal tubule cells, a process that becomes impaired in gentamicin-induced proximal tubule damage. Hentschel et al. were also among the first to recognize the power of the zebrafish for testing molecules that could be used in the recovery of AKI and demonstrated that simultaneously treating zebrafish with taurine during gentamicin-induced AKI resulted in the improved ability to clear dextran after injury (Hentschel et al., 2005). These studies were pivotal in revealing the similarities in the cellular changes in zebrafish and mammalian AKI in response to nephrotoxic agents and the potential of using zebrafish to develop treatments for AKI. However, the differences, in regard to zebrafish and mammalian kidney, are also of key interest for the understanding and development of treatments for AKI.

In mammals, the ability to make new nephrons ceases shortly after birth, whereas adult zebrafish has the ability to generate new nephrons (termed neonephrogenesis) throughout their lifetime (McCampbell et al., 2015). This remarkable capacity for life-long nephrogenesis compels researchers to understand how zebrafish retain this capacity with the aim of initiating similar processes in humans.

The similarities and differences in the complexity and function between zebrafish and mammals have made adult zebrafish an attractive model for the study of more complex AKI pathophysiology. Gentamicin-induced injury in adult zebrafish has been used extensively to study the potent regenerative powers of the zebrafish kidney. Multiple approaches and markers have been developed in zebrafish to sequence the cellular events that occur after gentamicin-induced kidney injury in zebrafish. In some cases, these approaches are similar to those used in mammals, such as hematoxylin and eosin staining to distinguish proximal tubules from distal tubules, as the proximal tubules exhibit a thick brush border on their luminal surface, which stain differentially from the distal tubule cells that lack a brush border (McCampbell et al., 2015; McKee & Wingert, 2015). Additionally, lectins, which are sugar-binding proteins, can be used to identify various tubule segments. *Lotus tetragonolobus* labels proximal tubules, whereas *Dolichos biflorus* agglutinin marks distal tubules with a high degree of specificity (McCampbell et al., 2015). The use of these and other markers (Table 1) has been instrumental in identifying how the nephron is damaged by gentamicin (McCampbell et al., 2015; McCampbell & Wingert, 2014).

After gentamicin-induced AKI, the adult zebrafish mesonephros undergoes partially coinciding waves of cell death and cell proliferation to rapidly regenerate damaged tubules (McCampbell et al., 2015). During this regeneration event, renal developmental markers such as *pax2a* are upregulated, similar to what has been observed during mammalian AKI repair (McCampbell et al., 2015). In a process distinct from mammals, gentamicin-induced injury in the adult zebrafish also results in the de novo formation of nephrons. As early as 1 day postinjury (dpi), nephrons in the adult zebrafish display extensive damage to proximal tubules as evidenced by denuded basement membranes, substantial cellular debris accumulation in the lumen, and a loss of *scl20ala+* expression (a solute transporter which marks proximal tubules) (McCampbell et al., 2015). Within 3 dpi, small basophilic *scl20ala+* clusters begin to emerge, and by 5 dpi small coiled structures appear in the damaged kidney, which were abundant by 7 dpi. Between 7 and 10 dpi, the small coiled structures elongate to form segment structures, indicative of new nephron formation, and by 14 dpi the injured zebrafish kidney displays similar *scl20ala+* expression as their WT counterparts (McCampbell et al., 2015). This careful time-course analysis provided invaluable insight into kidney regeneration in zebrafish and provides researchers with clues to understanding the process of kidney repair in humans.

5.2 Cisplatin-Induced AKI

Cisplatin (dichlorodiamino platinum) is an inorganic platinum-based chemotherapeutic agent that is used extensively to treat a variety of malignant tumors such as head and neck, lung, testis, ovarian, and bladder cancers (Shiraishi et al., 2000). However, the use of cisplatin in these therapies is limited due to the nephrotoxic effects of the drug (Shiraishi et al., 2000). Cisplatin accumulates in the kidney and causes necrosis and apoptosis of renal tubular cells, resulting in dose-dependent nephrotoxicity in up to one-third of patients (Shiraishi et al., 2000).

To improve survival of cancer patients receiving cisplatin-containing chemotherapeutic agents, it is important to understand how cisplatin causes renal injury. Zebrafish have shed light on this important area, as injection of cisplatin into zebrafish embryos resulted in destruction of proximal tubule epithelia (Fig. 3A), as evidenced by edema, cellular vacuolization, flattening and loss of brush border, distension of the tubular lumen, and a marked decrease in cell height in the proximal tubule renal epithelial cells (Hentschel et al., 2005). These are similar to the effects of cisplatin-induced nephrotoxicity in mammals. Using propidium iodine as a marker of apoptotic cells, increased cell death was reported in zebrafish nephrons 2 days after cisplatin injection, and this was proposed to lead to release of the mitochondrial protein Omi/HtrA2, a serine protease, into the cytoplasm where Omi/HtrA2 activates multiple apoptotic pathways (Hegde et al., 2002). Coinjecting zebrafish with cisplatin and Ucf-101, an inhibitor of Omi/HtrA2, resulted in decreased edema and increased survival compared to fish injected with cisplatin alone (Hentschel et al., 2005). Importantly, in a mouse model of cisplatin-induced nephrotoxicity, coadministration of Ucf-101 significantly improved renal function (Hentschel et al., 2005). These studies highlight the power of using zebrafish to discover therapeutics that can be translated in treatments for mammalian AKI.

5.3 Laser Ablation-Induced AKI

A drawback to studying drug-induced AKI in zebrafish is that the damage to nephrons is so catastrophic that it ultimately results in death of the affected fish, limiting the scope of the studies. Laser ablation of specific tubule segments in the embryonic zebrafish kidney has been a powerful alternative (Fig. 3A and B), as this technique is highly specific and has a low mortality rate, allowing for a more detailed spatiotemporal analysis of repair processes. Moreover, since the zebrafish embryonic pronephros is comprised of two nephrons, laser ablation performed on one side of the kidney leaves the contralateral side intact to serve as a control (Johnson et al., 2011). Additionally, the transparency of the zebrafish embryo allows for the visualization of fluorescently tagged dextran uptake to assess functionality of the nephron after AKI. Fluorescently tagged dextran can be intramuscularly microinjected in a trunk somite of zebrafish embryos 2 dpf, a stage when the pronephros has developed and functionality has commenced (Drummond et al., 1998; Johnson et al., 2011). After an overnight incubation, the fluorescent dextran accumulates in the proximal tubules of the pronephric duct. Following laser-induced AKI of proximal tubule segments, dextran uptake was assessed and showed the rapid regeneration of proximal tubular cells in the ablated region capable of dextran uptake. Using this technique, researchers have shown robust epithelial regeneration occurs in the zebrafish pronephros (Johnson et al., 2011).

There is compelling evidence to support the contribution of intact epithelial cells in proximity to the injured site that will migrate to the site of injury and proliferate to restore diminished epithelial cell populations (Palmyre et al., 2014). One theory suggests that these migrating cells undergo dedifferentiation to a more mesenchymal phenotype during the repair process, similar to what is observed in mammalian kidney development, which initiates with the formation of "pretubular aggregates" that undergo a mesenchymal-to-epithelial transition into renal vesicles (Dressler, 2006). A second hypothesis posits that stem and/or progenitor cells located within the kidney respond to AKI to regenerate injured tubule cells; however, this remains an area of intense debate. To address this question, laser ablation of specific pronephric proximal and distal tubule segments in zebrafish-expressing GFP was used to determine the contribution of cell proliferation and migration in kidney repair after AKI (Palmyre et al., 2014). Strikingly, after laser ablation of zebrafish nephrons, it was found that cell migration is the first response to injury, preceding cellular proliferation, and is the driving force in kidney epithelial repair after AKI (Palmyre et al., 2014). Importantly, the authors found the migratory cells that filled in the injured area did not undergo epithelial-to-mesenchymal dedifferentiation as was observed in an ischemia/reperfusion-induced injury mouse model of AKI (Bonventre, 2003). The observed differences may be a result of interspecies variation or may be due to the differences in the mode of injury, laser ablation vs ischemia/reperfusion. Although there is uncertainty of the physiological relevance of laser-induced AKI, it remains that this technique is useful to decipher kidney repair mechanisms in a highly specific injury context.

5.4 Ischemia/Reperfusion Model of AKI via Mechanical Obstruction

Mechanical obstruction of kidney tubules using fine tweezers has been used to study kidney repair in zebrafish (Hellman et al., 2010). Tubule obstruction in zebrafish embryos (Fig. 3A) at 50 hpf results in epithelial stretch-based upregulation of *foxj1* expression, a ciliary transcription factor regulating genes responsible for cilia structure and function, that was restricted to the vicinity of the injury (Hellman et al., 2010). This finding was confirmed in the adult zebrafish where surgical obstruction of the distal collecting duct (Fig. 3B) resulted in mesonephric nephron tubule distention and upregulation of *foxj1*, suggesting that *foxj1* is upregulated as a general response to kidney injury. In injured tubules, there was an increase in cilia beat rate, likely as a compensatory response to the obstruction to aid in removal of cell debris and assist in clearing the obstruction. This study highlights the importance of ciliary function not only in maintaining tubular cell homeostasis but also as a critical component of the repair process in the event of tubular injury.

5.5 Transgenic Models of Glomerular AKI to Study Zebrafish Kidney Regeneration

The genetic tractability of zebrafish has made it especially amenable to AKI research. Transgenic zebrafish harboring transgenes for both *cdh17::GFP* (which labels tubules and ducts of the pronephric and mesonephric kidney), and the podocyte-specific *pod::mCherry* (a marker of the glomerulus), made it possible to visualize the mesonephric development in zebrafish (Zhou et al., 2010; Table 1). Additionally, researchers using the transgenic *Tg(podocin:GFP)* zebrafish line confirmed that zebrafish podocytes have extended foot processes and interdigitating foot processes similar to mammalian podocytes (He et al., 2011). These pivotal studies lead to the finding that the zebrafish glomerulus and its podocytes are highly conserved with mammals and could be used to help decipher the glomerular response after AKI.

Inducible transgenic lines have also been developed which allow precise control on the timing and cell type where kidney damage occurs. Zebrafish engineered to express the *Escherichia coli* nitroreductase (NTR) gene downstream of the podocyte-specific *podocin* gene allow such NTR expression and hence kidney damage, to be restricted to the podocytes of both pronephric and mesonephric kidneys (Huang et al., 2013; Zhou & Hildebrandt, 2012). Following exposure to the prodrug metronidazole (MTZ), cells expressing the NTR protein process MTZ into a cytotoxic DNA cross-linking agent, causing cell death. This results in podocyte-specific ablation in a highly controlled context to study AKI at various developmental time points (Fig. 3C).

In *Tg(podocin:NTR-GFP)* embryos, MTZ treatment results in pericardial edema, podocyte apoptosis, a loss of nephrin and podocin expression, foot process effacement, and a leaky glomerular filtration barrier (Huang et al., 2013). Remarkably, within 7 days, after MTZ is removed, kidney repair is underway. Podocyte foot processes and glomerular barrier functionality appeared to be completely repaired, and a small group of proliferating progenitor-like cells were identified in the glomerulus during the recovery period (Huang et al., 2013). In a complimentary study in zebrafish transgenic lines expressing *Tg(podocin:NTR-mCherry)*, adult zebrafish treated with MTZ resulted in podocyte ablation as evidenced by a reduction in fluorescent reporter intensity and loss of podocyte foot processes, leakage of large proteins into the nephron tubule, and severe edema similar to human nephrotic syndrome phenotype (Zhou & Hildebrandt, 2012). Importantly, using the transgenic *Tg(pod::NTR-mCherry/wt1b::GFP)* line, it was observed that podocyte injury in zebrafish mesonephros induces *wt1b* (a marker for the glomerulus) expression within the glomerulus, suggesting that the regenerative response to podocyte injury in zebrafish involves the expression of developmental factors (Zhou & Hildebrandt, 2012). This is in agreement with a previous study using *Tg(wt1b:EGFP)* zebrafish that showed 5 days postgentamicin-induced injury, there was an increase in *wt1b:EGFP+* expressing cell clusters (Zhou et al., 2010). The *wt1b:EGFP+* cell clusters continue to increase in number until 14 dpi and are indicative of a distinct progenitor cell population. These studies provide compelling evidence for the importance of developmental programs and progenitor cells in the regenerative response to AKI in the zebrafish kidney.

6. STEM/PROGENITOR CELLS IN KIDNEY REPAIR AND REGENERATION

Stem cells have the unique capability of being able to self-renew while maintaining an undifferentiated state. By this mechanism, stem cells can produce progeny capable of differentiating into different cell types. Similar to stem cells, progenitor cells have the capability of differentiating into one or more particular cell lineages; however, unlike stem cells, progenitor cells are more specific for a particular cell lineage and cannot self-renew. As previously mentioned, zebrafish are capable of producing new nephrons throughout their lifetime, a process that could be attributed to renal stem and/or progenitor cells.

Normally, cells in the mammalian kidney have a low basal turnover rate which led to the long-held belief that the human kidney did not possess inherent regenerative powers. However, recent evidence lends supports for an important role of stem cell/progenitor cells in the repair process following mammalian AKI. A small population of resident stem cell/progenitor cells has been identified to play an important role in kidney repair after injury. Increasing these stem cell/progenitor populations in patients with AKI could be the key for regenerative medicine. Although neo-nephrogenesis does not occur in the adult mammalian kidney, studying the potent regenerative capabilities of the zebrafish mesonephros will lead to new knowledge of in vivo regulation of renal progenitor cells and may provide insights into stem cell-based therapies for kidney diseases.

Recent studies have shown that genes normally expressed during kidney development are reactivated during the recovery period after AKI. For example, the Lim1 homeobox protein, Lhx1, is essential during kidney development for establishing the kidney field and becomes reactivated after AKI (Cirio et al., 2011). Kidney organogenesis in the zebrafish closely parallels the initial steps in mammalian kidney organogenesis; therefore, zebrafish models of AKI offer a powerful tool to study how these genes become activated during the repair process and offer insights into potential therapeutic modalities in mammals to accelerate the repair process after AKI.

In a study to identify the cell population responsible for new nephron formation, adult zebrafish were subjected to gentamicin-induced AKI and observed to follow new nephron formation during kidney repair (Zhou et al., 2010). The authors reported an increase in the transcription factor *wt1b* within 2 dpi, followed by the formation of *wt1b* aggregates within 4 days, which subsequently gave rise to new nephrons. This study provided the first evidence of a *wt1b+* cell population in zebrafish adult kidney that was capable of producing new nephrons after gentamicin-induced AKI.

Closely following this discovery was a seminal study by Diep et al., which provided compelling evidence for the existence of a progenitor cell population in the zebrafish adult kidney that is capable of producing new nephrons in response to gentamicin-induced AKI. Consistent with previous reports, gentamicin-induced injury resulted in the downregulation of the proximal tubule marker *slc20a1a* and a failure to uptake 40 kDa dextran at 1 dpi followed by the emergence of proliferating, basophilic immature nephrons by 15 dpi that were capable of uptake of 40 kDa dextran (Diep et al., 2011). To test whether these new nephrons were the result of kidney progenitor cells, the authors utilized a transplantation assay in which whole-kidney marrow cells expressing either green or red fluorescent protein from donor fish were injected into adult recipients with AKI which had been irradiated to block immune rejection (Diep et al., 2011). Strikingly, this resulted in donor-derived nephrons in 100% of the recipient fish by 18 dpi, which were functional based on the ability of the new nephrons to efficiently uptake 40 kDa dextran. The finding of donor-derived nephrons in locations distant from the site of injection suggests the donor cells were capable of migration (Diep et al., 2011), and since the mature nephrons were a mosaic of red and green donor cells (Diep et al., 2011), this indicates that multiple nephron progenitors formed these new nephrons.

To distinguish stem cells from progenitors, serial transplantations were performed in which whole-kidney marrow from the recipient fish was transplanted into secondary and tertiary fish. This subsequently gave rise to donor-derived fluorescently labeled nephrons in secondary and tertiary recipients, demonstrating that nephron progenitors in the adult zebrafish kidney maintain proliferative capabilities, and suggesting these cells had self-renewing capabilities (Diep et al., 2011). In an attempt to narrow down the cell type responsible for the formation of new nephrons, the authors utilized two transgenic lines: *Tg(lhx1a:EGFP)* and *Tg(wt1b:mCherry)*, in which the expression of the fluorescent proteins are driven by the promoter of either *lhx1a* or *wt1b*, encoding transcription factors expressed in early development of nephrons. In subsequent transplantation studies, the authors found that individual *lhx1a:EGFP+* cells engrafted into a recipient fish failed to form new nephrons. However, injection of *lhx1a:EGFP+* aggregates resulted in successful engraftment in transplanted fish, demonstrating that multiple progenitors are required for new nephron formation. Additionally, injection of *lhx1a:EGFP+/wt1b:mCherry+* failed to engraft, suggesting that *lhx1a:EGFP+/wt1b+* cells may be a unique progenitor population that is already committed to a specific cell lineage (Diep et al., 2011).

Studies in zebrafish have paved the way in the study of how renal progenitor cells are activated in vivo. In a study by de Groh et al., a diverse library of small molecules was tested on zebrafish embryos to test compounds that expand renal progenitor cell populations (de Groh et al., 2010). From this screen researchers identified one compound, called 4-(phenylthio) butanoic acid (PTBA), increased expression of *lhx1a*, *pax2a*, and *pax8* (markers of renal progenitor cells). To confirm this finding, the transgenic zebrafish *Tg(lhx1a:EGFP)*^{pt303} reporter line was utilized to count the number of renal progenitor cells after treatment with PTBA. The authors observed a 2.4-fold increase in renal progenitor cells in the PTBA-treated embryos compared to controls embryos (de Groh et al., 2010). The authors show that the expansion in renal progenitor cells following PTBA treatment was dependent on proliferation of the resident renal cell population and did not occur via fate transformation of nonrenal cells. Importantly, it was shown through a structure activity study that PTBA may act as a HDACi (de Groh et al., 2010). Additional work to define the therapeutic effects of HDACi using a more active analog of PTBA, methyl-4-(phenylthio) butanoate (m4TPB), showed that mice with ischemia/reperfusion-induced AKI treated with m4TPB 24 hours postinjury (hpi) exhibited improved renal function, as well as a reduction in postinjury fibrosis compared to untreated mice (Cianciolo Cosentino et al., 2013). Importantly, treatment with m4TPB after AKI significantly decreased the proportion of epithelial cells in both G2 and M arrest phases of the cell cycle and increased the proportion of proliferating epithelial cells, suggesting that m4PTB treatment promotes cell cycle progression in regenerating epithelial cells (Cianciolo Cosentino et al., 2013). These studies are significant because after AKI in mammals, a significant proportion of proliferating epithelial cells in the injured tubule undergo G2/M arrest after injury, causing a delay in recovery and contributing to the formation of renal fibrosis, both of which could be attenuated by m4PTB treatment (Yang, Besschetnova, Brooks, Shah, & Bonventre, 2010). In summary, these preclinical studies in zebrafish hold promise for advancing drugs to treat AKI in patients.

7. USING ZEBRAFISH TO UNCOVER POTENTIAL TREATMENTS FOR AKI

To date, treatment of patients with AKI is limited and involves mainly supportive care such as dialysis, correction of fluid and electrolyte imbalance, clearance of nephrotoxins, and prevention of sepsis (Rahman, Shad, & Smith, 2012). There is a dire need to uncover new treatments for AKI. Zebrafish studies have opened new avenues for potential therapeutics for kidney repair after AKI, such as taurine treatment after gentamicin-induced AKI (Hentschel et al., 2005), Ucf-101 inhibition to reduce nephrotoxicity of cisplatin-induced AKI (Hentschel et al., 2005), and activation of renal progenitor cell after AKI via treatment with HDACi. Extending these exciting findings to the clinic will be an important future goal.

7.1 Epigenetic Targets for Potential Treatments for AKI

The finding that HDACi expand renal progenitor populations during development, and kidney repair after AKI mimics mechanisms utilized during kidney organogenesis, suggests that epigenetic factors may play a role in kidney repair and regeneration, as epigenetic factors are known to regulate key processes in kidney development.

Epigenetics is the process by which changes in gene expression are achieved via posttranslational modifications to DNA or to protein complexes associated with DNA, without changing the DNA sequence itself. Epigenetic modifications can be covalently attached to histones and DNA, enabling the modifications to be passed down to future generations. The field of epigenetic medicine is exploding, and as with many other diseases, epigenetic changes have a major impact on AKI. Epigenetics can both affect gene expression and restructure genome packaging and nuclear organization. There is a rapidly expanding list of epigenetic processes that have been linked to AKI including histone acetylation, histone methylation, DNA methylation, histone phosphorylation, histone sumoylation, miRNA, and lncRNA (Lorenzen et al., 2015; Tang & Zhuang, 2015). Using zebrafish to identify the key epigenetic modifications that occur after AKI will offer invaluable insight to identify new targets in the treatment of AKI in humans.

7.2 Retinoic Acid Signaling Promotes Kidney Repair

Retinoic acid plays a critical role in embryonic kidney development. In adult zebrafish, retinoic acid signaling becomes reactivated in limb, fin, and heart regeneration (Chiba et al., 2016). Studies have also shown an important role for retinoic acid in kidney repair after toxin-induced and ischemia/reperfusion-induced AKI (Chiba et al., 2016). Identifying the various pathways involved in the retinoic acid-dependent response to renal injury could identify potential therapeutic targets in the treatment of AKI.

The transgenic retinoic acid reporter line *Tg(12XRARE:EGFP)* has been used to identify the involvement of retinoic acid signaling in kidney repair after gentamicin-induced AKI in zebrafish embryos (Chiba et al., 2016). Expression of retinoic acid was increased throughout the kidney as soon as 6 hpi induced by gentamicin injection. Furthermore, inhibition of retinoic acid signaling using Ro41-5253, a retinoic acid receptor antagonist, reduced proximal tubule epithelial cell proliferation and decreased survival in zebrafish with gentamicin-induced AKI (Chiba et al., 2016). Taken together, these data suggest that retinoic acid signaling is crucial for kidney repair and cellular proliferation after gentamicin-induced AKI in zebrafish. The authors subsequently showed that in an ischemia/reperfusion mouse model of AKI, retinoic acid signaling becomes reactivated after kidney injury and reduces the severity of tubular injury and postinjury fibrosis after AKI (Chiba et al., 2016).

As a response to AKI, kidney epithelial cells recruit inflammatory M1 macrophages to the site of injury (Chiba et al., 2016). The inflammatory response in the injured nephron is amplified by M1 macrophages, further contributing to tissue damage after AKI (Chiba et al., 2016). The inflammatory response is eventually dampened by the recruitment of M2 macrophages that promote kidney repair. M2 macrophage activation is slow and occurs over a period of days and is dependent on a mechanism that is distinct from M1 activation. Using activators and inhibitors of retinoic acid signaling in mouse models of AKI, it was shown that retinoic acid signaling suppresses inflammatory M1 macrophages and increases activation of M2 macrophages to promote kidney repair in renal proximal tubules after AKI (Chiba et al., 2016). However, in the mouse model of AKI, reactivation of retinoic acid after AKI did not promote proliferation of tubular epithelia, suggesting that mouse kidneys may have a reduced regenerative response compared to zebrafish (Chiba et al., 2016). Deciphering the mechanism by which retinoic acid promotes tubular epithelial cell proliferation in zebrafish could provide new avenues of treatment for patients with AKI.

8. FUTURE DIRECTIONS

As an optically and genetically tractable system, the zebrafish holds great potential to be further developed and utilized to study kidney development and disease. This is becoming particularly powerful with the rapid development of imaging and gene-editing technologies in recent years.

Among the new imaging modalities, light sheet microscopy offers unprecedented opportunities for zebrafish research. By illuminating the focal plane alone and acquiring images through a separate optical path, light sheet microscopy can obtain volume images with confocal level resolution at 100 × speed of point-scanning microscopes. Combined with the optical transparency of zebrafish embryos, this technique makes high-speed longitudinal intravital imaging at subcellular resolution feasible. For example, we imaged whole-mount zebrafish embryos at 30 hpf stained with the cilia marker anti-Arl13b using a Zeiss Lightsheet Z1 microscope. A 3D image of cilia distribution throughout almost the entire embryo was obtained with ease (Fig. 4A and B). Live imaging of developing embryos will no doubt

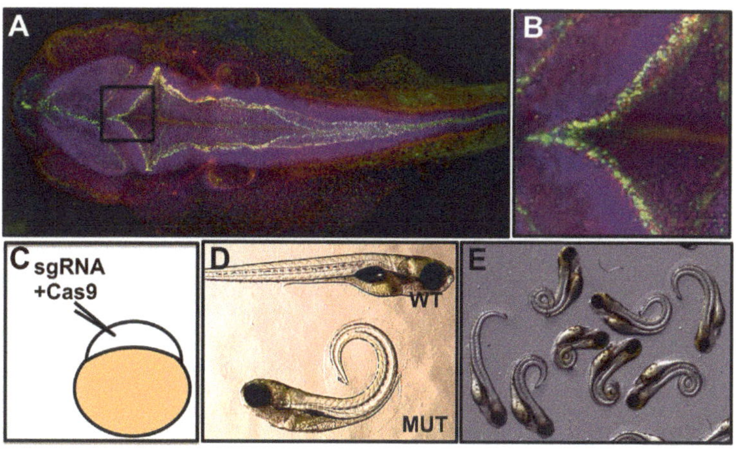

Fig. 4 Novel technologies aid the further development of the zebrafish model system. (A and B) Cilia distribution in a zebrafish embryo at the 20-somite stage revealed by light sheet microscopy. (A) Zebrafish embryo stained with anti-Arl13b (*green*) and DAPI (*blue*) in a dorsal view. (B) An enlarged view of the *boxed area* in (A). (C–E) Effective CRISPR-Cas9 in zebrafish. sgRNA can be produced through in vitro transcription and coinjected with either Cas9 mRNA or Cas9 protein into zebrafish embryos at the one-cell stage (C). (D and E) Phenotypic comparison between *pkd2*^{hi4166} mutants (D) and wild-type embryos injected with sgRNAs against *pkd2* and Cas9 (E). *MUT*, *pkd2*^{hi4166} mutant; *WT*, wild-type sibling.

provide novel insight into normal kidney development and the etiology of kidney disease. In addition, CRISPR (clustered regularly interspaced short palindromic repeats)-Cas9 is a useful and now widely used approach to create targeted mutants in zebrafish (Fig. 4C). Using this approach, we targeted *pkd2* and observed that almost 100% injected embryos display the curly tail and cystic kidney phenotypes observed in *pkd2* mutants (Fig. 4D and E). The new genome-editing technologies will allow versatile modification of the zebrafish genome, making it even more amenable to genetic analysis, particularly large-scale genetic analysis.

In summary, with the ever-increasing technical capability, the zebrafish model system is at an exciting stage to decipher the genetic and molecular pathways involved in zebrafish kidney development and disease pathogenesis. These discoveries have proven translatable to treating human kidney disorders and have the potential to greatly advance kidney therapeutics in the near future.

ACKNOWLEDGMENTS

Research in the Sun lab is supported by NIH (R01 DK092808, R01HL125885) and PKD foundation (191G14a, 213G16A). S.J. is supported by Yale Liver Center T32 Training Grant 2T32DK007356-36.

REFERENCES

Bae, Y. K., & Barr, M. M. (2008). Sensory roles of neuronal cilia: Cilia development, morphogenesis, and function in C. elegans. *Frontiers in Bioscience: A Journal and Virtual Library*, *13*, 5959–5974.

Barbacci, E., Reber, M., Ott, M. O., Breillat, C., Huetz, F., & Cereghini, S. (1999). Variant hepatocyte nuclear factor 1 is required for visceral endoderm specification. *Development*, *126*, 4795–4805.

Barr, M. M., & Sternberg, P. W. (1999). A polycystic kidney-disease gene homologue required for male mating behaviour in C. elegans. *Nature*, *401*, 386–389.

Basile, D. P., Anderson, M. D., & Sutton, T. A. (2012). Pathophysiology of acute kidney injury. *Comprehensive Physiology*, *2*, 1303–1353.

Becker-Heck, A., Zohn, I. E., Okabe, N., Pollock, A., Lenhart, K. B., Sullivan-Brown, J., et al. (2011). The coiled-coil domain containing protein CCDC40 is essential for motile cilia function and left-right axis formation. *Nature Genetics*, *43*, 79–84.

Bellomo, R., Kellum, J. A., & Ronco, C. (2012). Acute kidney injury. *Lancet*, *380*, 756–766.

Bingham, C., Ellard, S., Cole, T. R., Jones, K. E., Allen, L. I., Goodship, J. A., et al. (2002). Solitary functioning kidney and diverse genital tract malformations associated with hepatocyte nuclear factor-1beta mutations. *Kidney International*, *61*, 1243–1251.

Bisgrove, B. W., Snarr, B. S., Emrazian, A., & Yost, H. J. (2005). Polaris and Polycystin-2 in dorsal forerunner cells and Kupffer's vesicle are required for specification of the zebrafish left-right axis. *Developmental Biology*, *287*, 274–288.

Bonventre, J. V. (2003). Dedifferentiation and proliferation of surviving epithelial cells in acute renal failure. *Journal of the American Society of Nephrology: JASN*, *14*(Suppl. 1), S55–S61.

Borovina, A., Superina, S., Voskas, D., & Ciruna, B. (2010). Vangl2 directs the posterior tilting and asymmetric localization of motile primary cilia. *Nature Cell Biology*, *12*, 407–412.

Cai, Y., Fedeles, S. V., Dong, K., Anyatonwu, G., Onoe, T., Mitobe, M., et al. (2014). Altered trafficking and stability of polycystins underlie polycystic kidney disease. *The Journal of Clinical Investigation*, *124*, 5129–5144.

Cao, Y., Semanchik, N., Lee, S. H., Somlo, S., Barbano, P. E., Coifman, R., et al. (2009). Chemical modifier screen identifies HDAC inhibitors as suppressors of PKD models. *Proceedings of the National Academy of Sciences of the United States of America*, *106*, 21819–21824.

Caruana, G., & Bertram, J. F. (2015). Congenital anomalies of the kidney and urinary tract genetics in mice and men. *Nephrology*, *20*, 309–311.

Chiba, T., Skrypnyk, N. I., Skvarca, L. B., Penchev, R., Zhang, K. X., Rochon, E. R., et al. (2016). Retinoic acid signaling coordinates macrophage-dependent injury and repair after AKI. *Journal of the American Society of Nephrology: JASN*, *27*, 495–508.

Cianciolo Cosentino, C., Skrypnyk, N. I., Brilli, L. L., Chiba, T., Novitskaya, T., Woods, C., et al. (2013). Histone deacetylase inhibitor enhances recovery after AKI. *Journal of the American Society of Nephrology: JASN*, *24*, 943–953.

Cirio, M. C., Hui, Z., Haldin, C. E., Cosentino, C. C., Stuckenholz, C., Chen, X., et al. (2011). Lhx1 is required for specification of the renal progenitor cell field. *PloS One*, *6*(4). e18858. http://doi.org/10.1371/journal.pone.0018858.

Coffinier, C., Barra, J., Babinet, C., & Yaniv, M. (1999). Expression of the vHNF1/HNF1beta homeoprotein gene during mouse organogenesis. *Mechanisms of Development*, *89*, 211–213.

Consortium, T.E.P.K.D. (1994). The polycystic kidney disease 1 gene encodes a 14 kb transcript and lies within a duplicated region on chromosome 16. *Cell*, *77*, 881–894.

Coxam, B., Sabine, A., Bower, N. I., Smith, K. A., Pichol-Thievend, C., Skoczylas, R., et al. (2014). Pkd1 regulates lymphatic vascular morphogenesis during development. *Cell Reports*, *7*, 623–633.

de Groh, E. D., Swanhart, L. M., Cosentino, C. C., Jackson, R. L., Dai, W., Kitchens, C. A., et al. (2010). Inhibition of histone deacetylase expands the renal progenitor cell population. *Journal of the American Society of Nephrology: JASN*, *21*, 794–802.

DeCaen, P. G., Delling, M., Vien, T. N., & Clapham, D. E. (2013). Direct recording and molecular identification of the calcium channel of primary cilia. *Nature*, *504*, 315–318.

Delling, M., DeCaen, P. G., Doerner, J. F., Febvay, S., & Clapham, D. E. (2013). Primary cilia are specialized calcium signalling organelles. *Nature*, *504*, 311–314.

Delling, M., Indzhykulian, A. A., Liu, X., Li, Y., Xie, T., Corey, D. P., et al. (2016). Primary cilia are not calcium-responsive mechanosensors. *Nature*, *531*, 656–660.

Diep, C. Q., Ma, D., Deo, R. C., Holm, T. M., Naylor, R. W., Arora, N., et al. (2011). Identification of adult nephron progenitors capable of kidney regeneration in zebrafish. *Nature*, *470*, 95–100.

Diep, C. Q., Peng, Z., Ukah, T. K., Kelly, P. M., Daigle, R. V., & Davidson, A. J. (2015). Development of the zebrafish mesonephros. *Genesis*, *53*, 257–269.

Dressler, G. R. (2006). The cellular basis of kidney development. *Annual Review of Cell and Developmental Biology*, *22*, 509–529.

Dressler, G. R., Deutsch, U., Chowdhury, K., Nornes, H. O., & Gruss, P. (1990). Pax2, a new murine paired-box-containing gene and its expression in the developing excretory system. *Development*, *109*, 787–795.

Drummond, I. A. (2000). The zebrafish pronephros: A genetic system for studies of kidney development. *Pediatric Nephrology*, *14*, 428–435.

Drummond, I. (2002). The pronephros. *Results and Problems in Cell Differentiation*, *40*, 322–345.
Drummond, I. (2003). Making a zebrafish kidney: A tale of two tubes. *Trends in Cell Biology*, *13*, 357–365.
Drummond, I. A. (2005). Kidney development and disease in the zebrafish. *Journal of the American Society of Nephrology: JASN*, *16*, 299–304.
Drummond, I. A., & Davidson, A. J. (2010). Zebrafish kidney development. *Methods in Cell Biology*, *100*, 233–260.
Drummond, I. A., Majumdar, A., Hentschel, H., Elger, M., Solnica-Krezel, L., Schier, A. F., et al. (1998). Early development of the zebrafish pronephros and analysis of mutations affecting pronephric function. *Development*, *125*, 4655–4667.
Drummond, B. E., & Wingert, R. A. (2016). Insights into kidney stem cell development and regeneration using zebrafish. *World Journal of Stem Cells*, *8*, 22–31.
Duenas-Gonzalez, A., Candelaria, M., Perez-Plascencia, C., Perez-Cardenas, E., de la Cruz-Hernandez, E., & Herrera, L. A. (2008). Valproic acid as epigenetic cancer drug: Preclinical, clinical and transcriptional effects on solid tumors. *Cancer Treatment Reviews*, *34*, 206–222.
Duldulao, N. A., Lee, S., & Sun, Z. (2009). Cilia localization is essential for in vivo functions of the Joubert syndrome protein Arl13b/Scorpion. *Development*, *136*, 4033–4042.
Field, S., Riley, K. L., Grimes, D. T., Hilton, H., Simon, M., Powles-Glover, N., et al. (2011). Pkd1l1 establishes left-right asymmetry and physically interacts with Pkd2. *Development*, *138*, 1131–1142.
Gabow, P., & Grantham, J. (1997). Polycystic kidney disease. In R. Schrier & C. Gottschalk (Eds.), *Diseases of the kidney* (pp. 521–560). Boston: Lippincott Williams & Wilkins.
Gerlach, G. F., & Wingert, R. A. (2013). Kidney organogenesis in the zebrafish: Insights into vertebrate nephrogenesis and regeneration. *Wiley Interdisciplinary Reviews. Developmental Biology*, *2*, 559–585.
Gresh, L., Fischer, E., Reimann, A., Tanguy, M., Garbay, S., Shao, X., et al. (2004). A transcriptional network in polycystic kidney disease. *The EMBO Journal*, *23*, 1657–1668.
Hanaoka, K., Qian, F., Boletta, A., Bhunia, A. K., Piontek, K., Tsiokas, L., et al. (2000). Co-assembly of polycystin-1 and -2 produces unique cation-permeable currents. *Nature*, *408*, 990–994.
Harshman, L. A., & Brophy, P. D. (2012). PAX2 in human kidney malformations and disease. *Pediatric Nephrology*, *27*, 1265–1275.
Haycraft, C. J., Banizs, B., Aydin-Son, Y., Zhang, Q., Michaud, E. J., & Yoder, B. K. (2005). Gli2 and Gli3 localize to cilia and require the intraflagellar transport protein polaris for processing and function. *PLoS Genetics*. *1*(4), e53. http://dx.doi.org/10.1371/journal.pgen.0010053.
He, B., Ebarasi, L., Hultenby, K., Tryggvason, K., & Betsholtz, C. (2011). Podocin-green fluorescence protein allows visualization and functional analysis of podocytes. *Journal of the American Society of Nephrology: JASN*, *22*, 1019–1023.
Hegde, R., Srinivasula, S. M., Zhang, Z., Wassell, R., Mukattash, R., Cilenti, L., et al. (2002). Identification of Omi/HtrA2 as a mitochondrial apoptotic serine protease that disrupts inhibitor of apoptosis protein-caspase interaction. *The Journal of Biological Chemistry*, *277*, 432–438.
Heidet, L., Decramer, S., Pawtowski, A., Moriniere, V., Bandin, F., Knebelmann, B., et al. (2010). Spectrum of HNF1B mutations in a large cohort of patients who harbor renal diseases. *Clinical Journal of the American Society of Nephrology: CJASN*, *5*, 1079–1090.
Heliot, C., Desgrange, A., Buisson, I., Prunskaite-Hyyrylainen, R., Shan, J., Vainio, S., et al. (2013). HNF1B controls proximal-intermediate nephron segment identity in vertebrates by regulating Notch signalling components and Irx1/2. *Development*, *140*, 873–885.

Hellman, N. E., Liu, Y., Merkel, E., Austin, C., Le Corre, S., Beier, D. R., et al. (2010). The zebrafish foxj1a transcription factor regulates cilia function in response to injury and epithelial stretch. *Proceedings of the National Academy of Sciences of the United States of America, 107*, 18499–18504.

Hentschel, D. M., Park, K. M., Cilenti, L., Zervos, A. S., Drummond, I., & Bonventre, J. V. (2005). Acute renal failure in zebrafish: A novel system to study a complex disease. *American Journal of Physiology. Renal Physiology, 288*, F923–F929.

Hiesberger, T., Bai, Y., Shao, X., McNally, B. T., Sinclair, A. M., Tian, X., et al. (2004). Mutation of hepatocyte nuclear factor-1beta inhibits Pkhd1 gene expression and produces renal cysts in mice. *The Journal of Clinical Investigation, 113*, 814–825.

Hiesberger, T., Shao, X., Gourley, E., Reimann, A., Pontoglio, M., & Igarashi, P. (2005). Role of the hepatocyte nuclear factor-1beta (HNF-1beta) C-terminal domain in Pkhd1 (ARPKD) gene transcription and renal cystogenesis. *The Journal of Biological Chemistry, 280*, 10578–10586.

Hildebrandt, F., & Zhou, W. (2007). Nephronophthisis-associated ciliopathies. *Journal of the American Society of Nephrology: JASN, 18*, 1855–1871.

Horikawa, Y., Iwasaki, N., Hara, M., Furuta, H., Hinokio, Y., Cockburn, B. N., et al. (1997). Mutation in hepatocyte nuclear factor-1 beta gene (TCF2) associated with MODY. *Nature Genetics, 17*, 384–385.

Huang, J., McKee, M., Huang, H. D., Xiang, A., Davidson, A. J., & Lu, H. A. (2013). A zebrafish model of conditional targeted podocyte ablation and regeneration. *Kidney International, 83*, 1193–1200.

Huang, P., & Schier, A. F. (2009). Dampened Hedgehog signaling but normal Wnt signaling in zebrafish without cilia. *Development, 136*, 3089–3098.

Huangfu, D., Liu, A., Rakeman, A. S., Murcia, N. S., Niswander, L., & Anderson, K. V. (2003). Hedgehog signalling in the mouse requires intraflagellar transport proteins. *Nature, 426*, 83–87.

Hwang, D. Y., Dworschak, G. C., Kohl, S., Saisawat, P., Vivante, A., Hilger, A. C., et al. (2014). Mutations in 12 known dominant disease-causing genes clarify many congenital anomalies of the kidney and urinary tract. *Kidney International, 85*, 1429–1433.

Ingham, P. W., Nakano, Y., & Seger, C. (2011). Mechanisms and functions of Hedgehog signalling across the metazoa. *Nature Reviews. Genetics, 12*, 393–406.

Jin, X., Mohieldin, A. M., Muntean, B. S., Green, J. A., Shah, J. V., Mykytyn, K., et al. (2014). Cilioplasm is a cellular compartment for calcium signaling in response to mechanical and chemical stimuli. *Cellular and Molecular Life Sciences: CMLS, 71*, 2165–2178.

Johnson, C. S., Holzemer, N. F., & Wingert, R. A. (2011). Laser ablation of the zebrafish pronephros to study renal epithelial regeneration. *Journal of Visualized Experiments: JoVE, 54*, 2845. http://dx.doi.org/10.3791/2845 (Advance online publication).

Kamura, K., Kobayashi, D., Uehara, Y., Koshida, S., Iijima, N., Kudo, A., et al. (2011). Pkd1l1 complexes with Pkd2 on motile cilia and functions to establish the left-right axis. *Development, 138*, 1121–1129.

Kim, H. R., Richardson, J., van Eeden, F., & Ingham, P. W. (2010). Gli2a protein localization reveals a role for Iguana/DZIP1 in primary ciliogenesis and a dependence of Hedgehog signal transduction on primary cilia in the zebrafish. *BMC Biology, 8*, 65.

Koulen, P., Cai, Y., Geng, L., Maeda, Y., Nishimura, S., Witzgall, R., et al. (2002). Polycystin-2 is an intracellular calcium release channel. *Nature Cell Biology, 4*, 191–197.

Kozminski, K. G., Johnson, K. A., Forscher, P., & Rosenbaum, J. L. (1993). A motility in the eukaryotic flagellum unrelated to flagellar beating. *Proceedings of the National Academy of Sciences of the United States of America, 90*, 5519–5523.

Kramer-Zucker, A. G., Olale, F., Haycraft, C. J., Yoder, B. K., Schier, A. F., & Drummond, I. A. (2005). Cilia-driven fluid flow in the zebrafish pronephros, brain and Kupffer's vesicle is required for normal organogenesis. *Development*, *132*, 1907–1921.

Krauss, S., Johansen, T., Korzh, V., & Fjose, A. (1991). Expression of the zebrafish paired box gene pax[zf-b] during early neurogenesis. *Development*, *113*, 1193–1206.

Lameire, N., Van Biesen, W., & Vanholder, R. (2005). Acute renal failure. *Lancet*, *365*, 417–430.

Lindner, T. H., Njolstad, P. R., Horikawa, Y., Bostad, L., Bell, G. I., & Sovik, O. (1999). A novel syndrome of diabetes mellitus, renal dysfunction and genital malformation associated with a partial deletion of the pseudo-POU domain of hepatocyte nuclear factor-1beta. *Human Molecular Genetics*, *8*, 2001–2008.

Lorenzen, J. M., Schauerte, C., Kielstein, J. T., Hubner, A., Martino, F., Fiedler, J., et al. (2015). Circulating long noncoding RNATapSaki is a predictor of mortality in critically ill patients with acute kidney injury. *Clinical Chemistry*, *61*, 191–201.

Lu, W., Shen, X., Pavlova, A., Lakkis, M., Ward, C. J., Pritchard, L., et al. (2001). Comparison of Pkd1-targeted mutants reveals that loss of polycystin-1 causes cystogenesis and bone defects. *Human Molecular Genetics*, *10*, 2385–2396.

Ma, M., Tian, X., Igarashi, P., Pazour, G. J., & Somlo, S. (2013). Loss of cilia suppresses cyst growth in genetic models of autosomal dominant polycystic kidney disease. *Nature Genetics*, *45*, 1004–1012.

Madariaga, L., Moriniere, V., Jeanpierre, C., Bouvier, R., Loget, P., Martinovic, J., et al. (2013). Severe prenatal renal anomalies associated with mutations in HNF1B or PAX2 genes. *Clinical Journal of the American Society of Nephrology: CJASN*, *8*, 1179–1187.

Majumdar, A., Lun, K., Brand, M., & Drummond, I. A. (2000). Zebrafish no isthmus reveals a role for pax2.1 in tubule differentiation and patterning events in the pronephric primordia. *Development*, *127*, 2089–2098.

Malicki, J., Avanesov, A., Li, J., Yuan, S., & Sun, Z. (2011). Analysis of cilia structure and function in zebrafish. *Methods Cell Biology*, *101*, 39–74.

Mangos, S., Lam, P. Y., Zhao, A., Liu, Y., Mudumana, S., Vasilyev, A., et al. (2010). The ADPKD genes pkd1a/b and pkd2 regulate extracellular matrix formation. *Disease Models & Mechanisms*, *3*, 354–365.

Massa, F., Garbay, S., Bouvier, R., Sugitani, Y., Noda, T., Gubler, M. C., et al. (2013). Hepatocyte nuclear factor 1beta controls nephron tubular development. *Development*, *140*, 886–896.

McCampbell, K. K., Springer, K. N., & Wingert, R. A. (2015). Atlas of cellular dynamics during zebrafish adult kidney regeneration. *Stem Cells International*, *2015*, 19 Article ID 547636. http://dx.doi.org/10.1155/2015/547636.

McCampbell, K. K., & Wingert, R. A. (2014). New tides: Using zebrafish to study renal regeneration. *Translational Research: The Journal of Laboratory and Clinical Medicine*, *163*, 109–122.

McKee, R. A., & Wingert, R. A. (2015). Zebrafish renal pathology: Emerging models of acute kidney injury. *Current Pathobiology Reports*, *3*, 171–181.

Mehnert, J. M., & Kelly, W. K. (2007). Histone deacetylase inhibitors: Biology and mechanism of action. *Cancer Journal*, *13*, 23–29.

Mochizuki, T., Wu, G., Hayashi, T., Xenophontos, S. L., Veldhuisen, B., Saris, J. J., et al. (1996). PKD2, a gene for polycystic kidney disease that encodes an integral membrane protein. *Science*, *272*, 1339–1342.

Murawski, I. J., Myburgh, D. B., Favor, J., & Gupta, I. R. (2007). Vesico-ureteric reflux and urinary tract development in the Pax2 1Neu+/− mouse. *American Journal of Physiology. Renal Physiology*, *293*, F1736–F1745.

Nakanishi, K., & Yoshikawa, N. (2003). Genetic disorders of human congenital anomalies of the kidney and urinary tract (CAKUT). *Pediatrics International: Official Journal of the Japan Pediatric Society*, *45*, 610–616.

Nakayama, M., Nozu, K., Goto, Y., Kamei, K., Ito, S., Sato, H., et al. (2010). HNF1B alterations associated with congenital anomalies of the kidney and urinary tract. *Pediatric Nephrology, 25*, 1073–1079.

Nauli, S. M., Alenghat, F. J., Luo, Y., Williams, E., Vassilev, P., Li, X., et al. (2003). Polycystins 1 and 2 mediate mechanosensation in the primary cilium of kidney cells. *Nature Genetics, 33*, 129–137.

Naylor, R. W., & Davidson, A. J. (2014). Hnf1beta and nephron segmentation. *Pediatric Nephrology, 29*, 659–664.

Naylor, R. W., Przepiorski, A., Ren, Q., Yu, J., & Davidson, A. J. (2013). HNF1beta is essential for nephron segmentation during nephrogenesis. *Journal of the American Society of Nephrology: JASN, 24*, 77–87.

Negrisolo, S., Benetti, E., Centi, S., Della Vella, M., Ghirardo, G., Zanon, G. F., et al. (2011). PAX2 gene mutations in pediatric and young adult transplant recipients: Kidney and urinary tract malformations without ocular anomalies. *Clinical Genetics, 80*, 581–585.

Nicolaou, N., Renkema, K. Y., Bongers, E. M., Giles, R. H., & Knoers, N. V. (2015). Genetic, environmental, and epigenetic factors involved in CAKUT. *Nature Reviews. Nephrology, 11*, 720–731.

Nishigori, H., Yamada, S., Kohama, T., Tomura, H., Sho, K., Horikawa, Y., et al. (1998). Frameshift mutation, A263fsinsGG, in the hepatocyte nuclear factor-1beta gene associated with diabetes and renal dysfunction. *Diabetes, 47*, 1354–1355.

Nonaka, S., Shiratori, H., Saijoh, Y., & Hamada, H. (2002). Determination of left-right patterning of the mouse embryo by artificial nodal flow. *Nature, 418*, 96–99.

Nonaka, S., Tanaka, Y., Okada, Y., Takeda, S., Harada, A., Kanai, Y., et al. (1998). Randomization of left-right asymmetry due to loss of nodal cilia generating leftward flow of extraembryonic fluid in mice lacking KIF3B motor protein. *Cell, 95*, 829–837.

Obara, T., Mangos, S., Liu, Y., Zhao, J., Wiessner, S., Kramer-Zucker, A. G., et al. (2006). Polycystin-2 immunolocalization and function in zebrafish. *Journal of the American Society of Nephrology: JASN, 17*, 2706–2718.

Outeda, P., Huso, D. L., Fisher, S. A., Halushka, M. K., Kim, H., Qian, F., et al. (2014). Polycystin signaling is required for directed endothelial cell migration and lymphatic development. *Cell Reports, 7*, 634–644.

Paavola, J., Schliffke, S., Rossetti, S., Kuo, I. Y., Yuan, S., Sun, Z., et al. (2013). Polycystin-2 mutations lead to impaired calcium cycling in the heart and predispose to dilated cardiomyopathy. *Journal of Molecular and Cellular Cardiology, 58*, 199–208.

Palmyre, A., Lee, J., Ryklin, G., Camarata, T., Selig, M. K., Duchemin, A. L., et al. (2014). Collective epithelial migration drives kidney repair after acute injury. *PloS One. 9*. e101304. http://dx.doi.org/10.1371/journal.pone.0101304.

Panizzi, J. R., Becker-Heck, A., Castleman, V. H., Al-Mutairi, D. A., Liu, Y., Loges, N. T., et al. (2012). CCDC103 mutations cause primary ciliary dyskinesia by disrupting assembly of ciliary dynein arms. *Nature Genetics, 44*, 714–719.

Pazour, G. J., Dickert, B. L., Vucica, Y., Seeley, E. S., Rosenbaum, J. L., Witman, G. B., et al. (2000). Chlamydomonas IFT88 and its mouse homologue, polycystic kidney disease gene tg737, are required for assembly of cilia and flagella. *The Journal of Cell Biology, 151*, 709–718.

Pennekamp, P., Karcher, C., Fischer, A., Schweickert, A., Skryabin, B., Horst, J., et al. (2002). The ion channel polycystin-2 is required for left-right axis determination in mice. *Current Biology: CB, 12*, 938–943.

Pfeffer, P. L., Gerster, T., Lun, K., Brand, M., & Busslinger, M. (1998). Characterization of three novel members of the zebrafish Pax2/5/8 family: Dependency of Pax5 and Pax8 expression on the Pax2.1 (noi) function. *Development, 125*, 3063–3074.

Praetorius, H. A., Frokiaer, J., Nielsen, S., & Spring, K. R. (2003). Bending the primary cilium opens Ca^{2+}-sensitive intermediate-conductance K^{+} channels in MDCK cells. *The Journal of Membrane Biology, 191*, 193–200.

Rahman, M., Shad, F., & Smith, M. C. (2012). Acute kidney injury: A guide to diagnosis and management. *American Family Physician, 86*, 631–639.

Rewa, O., & Bagshaw, S. M. (2014). Acute kidney injury-epidemiology, outcomes and economics. *Nature Reviews. Nephrology, 10*, 193–207.

Rohatgi, R., Milenkovic, L., & Scott, M. P. (2007). Patched1 regulates hedgehog signaling at the primary cilium. *Science, 317*, 372–376.

Rosenbaum, J. L., & Witman, G. B. (2002). Intraflagellar transport. *Nature Reviews. Molecular Cell Biology, 3*, 813–825.

Sanyanusin, P., Schimmenti, L. A., McNoe, L. A., Ward, T. A., Pierpont, M. E., Sullivan, M. J., et al. (1995). Mutation of the PAX2 gene in a family with optic nerve colobomas, renal anomalies and vesicoureteral reflux. *Nature Genetics, 9*, 358–364.

Schottenfeld, J., Sullivan-Brown, J., & Burdine, R. D. (2007). Zebrafish curly up encodes a Pkd2 ortholog that restricts left-side-specific expression of southpaw. *Development, 134*, 1605–1615.

Selby, N. M., Shaw, S., Woodier, N., Fluck, R. J., & Kolhe, N. V. (2009). Gentamicin-associated acute kidney injury. *QJM: Monthly Journal of the Association of Physicians, 102*, 873–880.

Serluca, F. C., & Fishman, M. C. (2001). Pre-pattern in the pronephric kidney field of zebrafish. *Development, 128*, 2233–2241.

Shiraishi, F., Curtis, L. M., Truong, L., Poss, K., Visner, G. A., Madsen, K., et al. (2000). Heme oxygenase-1 gene ablation or expression modulates cisplatin-induced renal tubular apoptosis. *American Journal of Physiology. Renal Physiology, 278*, F726–F736.

Sun, Z., Amsterdam, A., Pazour, G. J., Cole, D. G., Miller, M. S., & Hopkins, N. (2004). A genetic screen in zebrafish identifies cilia genes as a principal cause of cystic kidney. *Development, 131*, 4085–4093.

Sun, Z., & Hopkins, N. (2001). vhnf1, the MODY5 and familial GCKD-associated gene, regulates regional specification of the zebrafish gut, pronephros, and hindbrain. *Genes & Development, 15*, 3217–3229.

Tang, J., & Zhuang, S. (2015). Epigenetics in acute kidney injury. *Current Opinion in Nephrology and Hypertension, 24*, 351–358.

Thomas, R., Sanna-Cherchi, S., Warady, B. A., Furth, S. L., Kaskel, F. J., & Gharavi, A. G. (2011). HNF1B and PAX2 mutations are a common cause of renal hypodysplasia in the CKiD cohort. *Pediatric Nephrology, 26*, 897–903.

Torres, M., Gomez-Pardo, E., Dressler, G. R., & Gruss, P. (1995). Pax-2 controls multiple steps of urogenital development. *Development, 121*, 4057–4065.

Vogel, P., Read, R., Hansen, G. M., Freay, L. C., Zambrowicz, B. P., & Sands, A. T. (2010). Situs inversus in Dpcd/Poll −/−, Nme7 −/−, and Pkd1l1 −/− mice. *Veterinary Pathology, 47*, 120–131.

Weber, S., Moriniere, V., Knuppel, T., Charbit, M., Dusek, J., Ghiggeri, G. M., et al. (2006). Prevalence of mutations in renal developmental genes in children with renal hypodysplasia: Results of the ESCAPE study. *Journal of the American Society of Nephrology: JASN, 17*, 2864–2870.

Wingert, R. A., & Davidson, A. J. (2008). The zebrafish pronephros: A model to study nephron segmentation. *Kidney International, 73*, 1120–1127.

Witt, O., Deubzer, H. E., Milde, T., & Oehme, I. (2009). HDAC family: What are the cancer relevant targets? *Cancer Letters, 277*, 8–21.

Wu, G., D'Agati, V., Cai, Y., Markowitz, G., Park, J. H., Reynolds, D. M., et al. (1998). Somatic inactivation of Pkd2 results in polycystic kidney disease. *Cell, 93*, 177–188.

Yang, L., Besschetnova, T. Y., Brooks, C. R., Shah, J. V., & Bonventre, J. V. (2010). Epithelial cell cycle arrest in G2/M mediates kidney fibrosis after injury. *Nature Medicine*, *16*, 535–543. 531 p, following 143.

Yoder, B. K., Hou, X., & Guay-Woodford, L. M. (2002). The polycystic kidney disease proteins, polycystin-1, polycystin-2, polaris, and cystin, are co-localized in renal cilia. *Journal of the American Society of Nephrology: JASN*, *13*, 2508–2516.

Yoshiba, S., Shiratori, H., Kuo, I. Y., Kawasumi, A., Shinohara, K., Nonaka, S., et al. (2012). Cilia at the node of mouse embryos sense fluid flow for left-right determination via Pkd2. *Science*, *338*, 226–231.

Yuan, S., Zhao, L., Brueckner, M., & Sun, Z. (2015). Intraciliary calcium oscillations initiate vertebrate left-right asymmetry. *Current Biology: CB*, *25*, 556–567.

Zhao, L., Yuan, S., Cao, Y., Kallakuri, S., Li, Y., Kishimoto, N., et al. (2013). Reptin/Ruvbl2 is a Lrrc6/Seahorse interactor essential for cilia motility. *Proceedings of the National Academy of Sciences of the United States of America*, *110*, 12697–12702.

Zhou, W., Boucher, R. C., Bollig, F., Englert, C., & Hildebrandt, F. (2010). Characterization of mesonephric development and regeneration using transgenic zebrafish. *American Journal of Physiology. Renal Physiology*, *299*, F1040–F1047.

Zhou, W., & Hildebrandt, F. (2012). Inducible podocyte injury and proteinuria in transgenic zebrafish. *Journal of the American Society of Nephrology: JASN*, *23*, 1039–1047.

CHAPTER THREE

Zebrafish Developmental Models of Skeletal Diseases

Lauryn N. Luderman[*,†,‡,1], **Gokhan Unlu**[*,‡,§,1], **Ela W. Knapik**[*,†,‡,§,2]
*Vanderbilt University Medical Center, Nashville, TN, United States
†Vanderbilt Brain Institute, Vanderbilt University, Nashville, TN, United States
‡Vanderbilt Genetics Institute, Vanderbilt University, Nashville, TN, United States
§Vanderbilt University, Nashville, TN, United States
[2]Corresponding author: e-mail address: ela.knapik@vanderbilt.edu

Contents

Abstract

The zebrafish skeleton shares many similarities with human and other vertebrate skeletons. Over the past years, work in zebrafish has provided an extensive understanding of

[1] These authors contributed equally.

Current Topics in Developmental Biology, Volume 124
ISSN 0070-2153
http://dx.doi.org/10.1016/bs.ctdb.2016.11.004

the basic developmental mechanisms and cellular pathways directing skeletal development and homeostasis. This review will focus on the cell biology of cartilage and bone and how the basic cellular processes within chondrocytes and osteocytes function to assemble the structural frame of a vertebrate body. We will discuss fundamental functions of skeletal cells in production and secretion of extracellular matrix and cellular activities leading to differentiation of progenitors to mature cells that make up the skeleton. We highlight important examples where findings in zebrafish provided direction for the search for genes causing human skeletal defects and also how zebrafish research has proven important for validating candidate human disease genes. The work we cover here illustrates utility of zebrafish in unraveling molecular mechanisms of cellular functions necessary to form and maintain a healthy skeleton.

1. INTRODUCTION

Over the last few decades, zebrafish (*Danio rerio*) has been used primarily to further our understanding of developmental processes and to discover genes required for development through the use of unbiased phenotype-driven forward genetic screens (Driever et al., 1996; Haffter et al., 1996; Howe et al., 2013; Knapik, 2000). In recent years, the use of zebrafish model has been expanded to address a repertoire of clinically defined disorders (Haesemeyer & Schier, 2015; Kokel et al., 2010). This trend is sustainable because zebrafish offers a compromise between physiological complexity of a mammalian vertebrate and morphological simplicity of invertebrates, along with a comprehensive toolbox of cell biological (Unlu, Levic, Melville, & Knapik, 2014; Vacaru et al., 2014) and genetic tools at a modest cost relative to mammalian models. Within the transparent body of zebrafish, morphogenetic processes and physiological activities of the skeleton are easily accessible, and the chondrocytes and osteocytes can be labeled with simple stains, allowing for rapid and high-throughput assessment of the skeleton.

As vertebrates, the zebrafish and mammalian skeletons are very similar. Although mechanics of the mammalian (terrestrial) and fish (aquatic) skeletons differ due to their habitats (Kimmel, Miller, & Moens, 2001; Miyashita, 2016; Witten & Huysseune, 2009), their similarities are strongest at the level of basic cellular functions of chondrocytes and osteocytes and their primary role of secreting extracellular matrix (ECM). The ECM is a dynamic structure that provides not only a mechanical context to cells, but it also maintains gradients of growth factors and morphogens necessary for homeostasis, proliferation, and differentiation of progenitor cells to mature tissue. The components of ECM, including bound water, proteins, and polysaccharides, modulate diffusion of signaling factors and provide structural support and scaffolding for developing tissues.

Forward genetics screens have been especially useful in identification of novel skeletal mutations in zebrafish (Andreeva et al., 2011; Neuhauss et al., 1996; Nissen, Amsterdam, & Hopkins, 2006). With forward genetic approaches and positional cloning methods, novel zebrafish models have been identified to study hereditary skeletal disorders. Availability of microsatellite-based genetic linkage maps afforded fast and reliable positional cloning to identify mutated genes in zebrafish (Fornzler et al., 1998; Knapik et al., 1996, 1998). Genetic linkage analysis is a mainstay of disease gene identification, and additional methods such as SNP-based genetic maps and whole-exome sequencing approaches have enhanced robustness of the zebrafish model (Bradley et al., 2007; Guryev et al., 2006).

Zebrafish as a model organism offers multiple routes of gene depletion strategies including classical chemical mutagenesis with *N*-ethyl-*N*-nitrosourea (ENU) (Mullins, Hammerschmidt, Haffter, & Nusslein-Volhard, 1994; Solnica-Krezel, Schier, & Driever, 1994), insertional mutagenesis (Amsterdam et al., 2004), gene knockdown via morpholino-oligonucleotide (MO) injections (Nasevicius & Ekker, 2000), and newer techniques such as zinc-finger nuclease (Doyon et al., 2008), TALEN (Huang et al., 2011), and CRISPR/Cas9 genome editing systems (Hwang et al., 2013; Jao, Wente, & Chen, 2013).

Here, we focus on the primary functions of the skeletal cells to synthesize, secrete, and modify proteins of the ECM and how disruptions of these highly conserved metabolic processes in the zebrafish model have informed our understanding of human skeletal pathology.

2. BONE AND CARTILAGE: OVERVIEW OF ZEBRAFISH SKELETAL DEVELOPMENT

The endoskeletal system is unique to vertebrates, and its components are highly conserved. Mechanisms of bone development are classified into two highly conserved processes based on the progression of mesenchymal progenitors: through either cartilage intermediate, endochondral bone formation, or directly to bone by the process of membranous ossification (Berendsen & Olsen, 2015). To begin, mesenchymal stem cells migrate to locations of future bones and condense before differentiating into either chondrocytes or osteoblasts. Endochondral ossification, on the other hand, involves intermediate step of differentiated chondrocytes secreting cartilage ECM and forming structural anlage that will ultimately be replaced by bone in areas including the skull base, axial, and appendicular skeleton. Mesenchymal stem cells that differentiate into osteoblasts will directly produce bone as part of the membranous neuro- and viscerocranium through intramembranous ossification; these bones including gill covers, shoulder girdle, and fin spine rays are often referred to as dermal bones (Miyashita, 2016).

3. SYNTHESIS OF THE CARTILAGE MATRIX COMPONENTS

Differentiating chondrocytes upregulate expression of Sox9, transcriptional activator required for the production and secretion of ECM components such as collagen type-II, -IX, -XI and aggrecan (Bell et al., 1997; Ohba, He, Hojo, & McMahon, 2015). Chondrocytes synthesize proteoglycan core proteins such as aggrecan, syndecans, and glypicans and modify them with glycosaminoglycan (GAG) chains of heparan sulfate (HSPG) and chondroitin sulfate (CSPG), the two most common GAG modifications in cartilage tissue (Holmborn et al., 2012). In addition to major fibrous components of ECM and major proteoglycans, cartilage matrix contains numerous noncollagenous proteins, including matrilins and chondromodulin.

3.1 Collagens and Fibrillar Proteins

Collagens are the major component of ECM, being expressed in tissue-specific manner (Shoulders & Raines, 2009). A larger precursor, procollagen, is synthesized and assembled in the ER before being secreted to the extracellular space, where enzymes cleave off the N- and C-propeptides to assemble mature collagen fibrils. Some types of collagen will form fibrils, while others are considered fibril-associated collagens. Fibril-associated collagens contain noncollagenous regions within the helical domains that confer flexibility to the rigid macromolecule, and they interact with fibrillar collagens to regulate collagen fiber organization and tissue mechanical properties. Since collagens make up 65–80% of ECM, mutations within genes that encode α chains can result in decreased collagen production or function, leading to developmental disorders (Arnold & Fertala, 2013).

3.1.1 Collagen Type I

Collagen type I is abundantly expressed throughout bone and skin, and its deficit has been implicated in developmental disorders. In most vertebrates, collagen I is a trimer consisting of two α1 chains and one α2 chain encoded by *COL1A1* and *COL1A2*, respectively. In contrast, the zebrafish genome contains three distinct genes for α chains of collagen I: *col1a1a*, *col1a1b*, and *col1a2* encoding for α1(I), α3(I), α2(I) chains, respectively. The zebrafish mutant *chihuahua* phenocopies *osteogenesis imperfecta (OI)*, a brittle bone disease in humans (Fisher, Jagadeeswaran, & Halpern, 2003). Mutations in collagen I genes are also associated with other human diseases including Caffey disease (MIM# 114000) (Gensure et al., 2005; Glorieux, 2005) and multiple forms of Ehlers–Danlos syndrome, EDS1 and EDS7A (MIM# 130000 & 130060) (Cabral et al., 2005; Nuytinck et al., 2000).

3.1.2 Collagen IX

Collagens play an important role in maintaining structure in skeletal elements, for example, the intervertebral disc, which confers flexibility, multiaxial spinal motion, and load transmission to the vertebral column (Erwin & Hood, 2014). ECM surrounding intervertebral discs consists of collagen-I, -II, -IX and proteoglycans/aggrecan. Collagen IX acts as a connector between other fibril-forming collagens and matrix proteins. A mutation resulting in a tryptophan substitution of Gln^{326} in *COL9A2*, a gene encoding the α2 chain of collagen IX, has been found in patients with *intervertebral disc disease* (MIM# 603932). Tryptophan, a highly hydrophobic amino acid, may disrupt the formation of the collagen fibril triple helix by preventing cross-linking by lysyl oxidase and subsequently disrupting collagen IX and collagen II interaction (Annunen et al., 1999).

Mutations in *COL9A1*, encoding for α1 chain of collagen IX, are associated with a disease known as *multiple epiphyseal dysplasia (MED)* (MIM# 614135), a pediatric disease that presents with waddling gait, mobility restriction, and painful/stiff joints. MED is polygenic and autosomal dominant, with mutations found in *COL9A1*, *COL9A2*, *COL9A3*, and genes encoding for other matrix components such as matrilin-3 and cartilage oligomeric matrix protein. Patients with mutations in *COL9A2* or *COL9A3* have similar symptoms exhibiting normal height, but epiphyseal dysplasia of knees and joints as children, and osteoarthritis as adults (Czarny-Ratajczak et al., 2001) (see Table 1).

A forward genetic screen in zebrafish identified a mutation termed *persistent plexus (prp)* that disrupts caudal fin development and regeneration. Sequencing of the *prp* mutant revealed a missense mutation (176Leu > His) in the thrombospondin repeat domain of *col9α1* gene. This is a noncollagenous

Table 1 Summary of Genes Involved in the Synthesis, Processing, and Secretion of ECM Components, Including Associated Human Diseases and Corresponding Zebrafish Mutants

Human					Zebrafish				References
Gene	Disease	OMIM	Common Mutations	Skeletal Phenotypes	Gene	Allele	Mutation/Tool	Phenotypes	
COL1A1	Osteogenesis imperfecta, type I	166200	• Missense (G94C, G178C, G43C, G901S, G12A) • Frameshift near 3' end • Early termination (G-A transition in intron 26, in-frame insertion) • Early termination (R963X)	• Normal stature • Hearing loss • Blue sclerae • Mild osteopenia • Wormian bones • Mild joint hypermobility • Thin skin	col1a1a (*chihuahua*)	dc124	Missense (G390D)	• Shortened embryos • Irregular bone density • Abnormal caudal fins • Thickened lepidotrichia • Irregular vertebrae • Breaks in rib bones	Fisher et al. (2003) and Gistelinck et al. (2016)
	Osteogenesis imperfecta, type II	166210	• Missense (G97D, G391R, G559D, G673D, G667R, G718C, G748C, G847R, G883D, G904C, G913S, G988C, G1009S, G541D, G1003S, G637V, G1006V, G973V, G256V, G802V, G415S, G565V, G355D, W94C, G586V, A1387V) • 9-bp deletion (874del9, 868del9) • 1 bp insertion (4088insT)	• Short limb dwarfism • Blue sclerae • Beaked nose • Beaded ribs • Multiple fractures • Soft calvaria • Tibial bowing • Telescoped femur • Perinatal lethal					
	Osteogenesis imperfecta, type III	259420	• Missense (G526C, G844S, G154R, G415C, G352S, G862S, G661S, G76E) • Exon 22 deletion • C-terminal mutation (T-C transition, L-P amino acid change) • 562-bp deletion (exon 34-intron 36)	• Short limb dwarfism • Facial dysmorphia (triangular face, frontal bosing, micrognathia) • Hearing loss • Blue sclerae at birth, normal with age • Severe osteoporosis • Multiple fractures at birth • Scoliosis and kyphosis • Tibial bowing					

	Osteogenesis imperfecta, type IV	166220	• Missense (G175C, G832S, G415C) • Splice site mutation (G-A transition in intron 8) • In-frame deletion of 3-bp (1964GGC) with A438del and E437D missense mutation	• Short stature • Hearing loss • Mild skeletal deformity • Multiple fractures • Scoliosis and kyphosis • Femoral bowing at birth					
COL9A1	Multiple epiphyseal dysplasia 6 (EDM6)	614135	• 1-bp insertion, T, at donor splice site of exon 8	• Multiple epiphyseal dysplasia • Early onset osteoarthritis • Endplate irregularities • Hip, knee arthralgia • Irregular epiphyses (knee)	col9α1 (*persistent plexus*)	j131	Missense (L176H)	• Small finfolds • Defects in vasculature of caudal fins • Shortened actinotrichia • Abnormal patterning of osteoblasts	Huang et al. (2009), Annunen et al. (1999), and Czarny-Ratajczak et al. (2001)
							MO		
COL9A2	Multiple epiphyseal dysplasia 2 (EDM2)	600204	• T-C transition in donor splice site of intron 3 resulting in in-frame deletion of 12 amino acids, skipping of exon 3; • G-A transition at end of exon 3, no amino acid change (proline)	• Short stature • Varus/valgus knee deformity • Knee osteoarthritis • Mildy short hands • Waddling gait and knee pain/stiffness • Intervertebral disc disease	col9α2		—	—	
MATN3	Multiple epiphyseal dysplasia 5 (EDM5)	607078	• Missense (V194D, R121W, A219D, R70H, A128P)	• Normal stature • Early onset osteoarthritis • Multiple epiphyseal dysplasia • Hip, knee arthralgia • Small femoral epiphyses • Delayed carpal, tarsal ossification	matn3		—	—	

Continued

Table 1 Summary of Genes Involved in the Synthesis, Processing, and Secretion of ECM Components, Including Associated Human Diseases and Corresponding Zebrafish Mutants—cont'd

Human					Zebrafish				References
Gene	Disease	OMIM	Common Mutations	Skeletal Phenotypes	Gene	Allele	Mutation/Tool	Phenotypes	
MATN1	Relapsing polychondritis	115437	—	• Chondrodysplasia • Joint abnormalities	matn1		MO	• Smaller bodies • Malformed head • Curly tail • Loss of cartilage tissue	Neacsu et al. (2014), Ko et al. (2005), Klatt, Becker, Neacsu, Paulsson, and Wagener (2011), and Buckner, Wu, Reife, Terato, and Eyre (2000)
UGDH	Unknown	—	—	—	ugdh (*jekyll*)	m151	Missense (I331D)	• Impaired heart valve formation • Kinked, shortened pectoral fins • Defects in cartilage differentiation • Inner ear defects	Walsh and Stainier (2001) and Neuhauss et al. (1996)
UXS1	Unknown	—	—	—	uxs1 (*man o war*)	hi954	Transgenic insertion	• Impaired chondrocyte stacking • Delayed ossification • Missing perichondrium • Dwarfism • Deformed craniofacial skeleton	Wiweger et al. (2011) and Eames et al. (2010)
						w60	Missense (R233H)	• Reduced lower jaw • Shortened body axis • Small pectoral fins • Abnormal chondrocyte intercalation	

XYLT1	Desbuquois dysplasia type 2	615777	• Missense (A115S, R481W, R598C) • Early termination • Frameshift • Splice site mutations	• Dislocations of large joints • Shortened long bones • Distinct facial features (flat, round face and low nasal bridge) • Epiphyseal dysplasia • Short metacarpals and phalanges	xylt1	b1128	Splice donor site	• Increased bone ossification • Decreased cartilage formation • Shortened upper and lower jaws • Hypoplastic midface • Bulging eyes • Craniofacial dysmorphia	Eames et al. (2011) and Schreml et al. (2014)
						b1189	Missense (S534A)		
FAM20C	Raine syndrome	259775	• Missense (G365R, R535W, D437N, I244N, G266R, P314S) • Intron 4/exon 5 spice site change • Exon 4/intron 4 splice site change	• Severe bone dysplasia • Increased ossification of skull • Distinct facial features (narrow prominent forehead, proptosis, low nasal bridge, midface hypoplasia)	fam20b	b1125	Nonsense (Q388X)	• Increased bone ossification • Decreased cartilage formation • Shortened upper and lower jaws • Hypoplastic midface • Bulging eyes • Craniofacial dysmorphia	Eames et al. (2011) and Faundes et al. (2014)
						b1127	Missense (C331R)		
B4GALT7	Ehlers–Danlos syndrome, progeroid type 1 (EDSP1)	130070	• Missense (A186D, L206P, R270C, L41P)	• Short stature • Failure to thrive • Macrocephaly and facial dysmorphia • Absent ear lobes • Joint laxity • Long, slender fingers and toes	b4galt7	—	—	—	
B3GALT6	Ehlers–Danlos syndrome, progeroid type 2 (EDSP2)	615349	• Missense (S309T, R6W) • 1-bp deletion (353A, 588G) • 9-bp deletion (NT415)	• Prominent forehead and flat face • Kyphoscoliosis • Hip dislocation and joint laxity • Elbow malalignment • Clubfeet	b3galt6	—	—	—	

Continued

Table 1 Summary of Genes Involved in the Synthesis, Processing, and Secretion of ECM Components, Including Associated Human Diseases and Corresponding Zebrafish Mutants—cont'd

Human					Zebrafish				References
Gene	**Disease**	**OMIM**	**Common Mutations**	**Skeletal Phenotypes**	**Gene**	**Allele**	**Mutation/Tool**	**Phenotypes**	
B3GAT3	Multiple joint dislocations, short stature, and craniofacial dysmorphism with or without congenital heart defects (JDSCD)	245600	• Missense (R277Q, P140L, G223S)	• Short stature • Midface hypoplasia • Low nasal bridge • Hip dislocation and joint laxity • Scoliosis • Short neck and narrow chest • Brachycephaly • Small teeth	b3gat3	hi307	Transgenic insertion	• Loss of CSPGs • Loss of HSPGs • Loss of DSPGs	
EXT1	Hereditary multiple exotoses, type I	133700	• Missense (R339L, G339D, R340C) • Early termination (1364delC, 1035ins4, 2120delT, 2077insC, 1178del8, 1664isnA)	• Short stature • Rib, scapular, pelvic exostoses • Coxa vara • Proturberances at ends of long bones • Increased risk of chondrosarcoma • Multiple cartilage-capped bone tumors • Increase in chondrocyte profileration at metaphyses of long bones	—	—	—	—	Jennes et al. (2009), Holmborn et al. (2012), Lee et al. (2004), Clement et al. (2008), and Wiweger et al. (2011)
EXT2	Hereditary multiple exotoses, type II	133701	• Missense (D227N) • Early termination (784del4, Q172X, Y222X, Q258X) • Splice variants		ext2 (*dackel*)	tw25e	Nonsense (R227X)	• Decreased chondrocyte intercalation • Shortened cartilage elements • Shortened jaw • Abnormal or loss of pectoral fin	
EXTL3	Unknown	—	—		extl3 (*boxer*)	tm70g	Missense (D831N)	Same phenotypes as ext2	
PMM2	Congenital disorder of glycosylation, type Ia (CDG-Ia)	212065	• Missense (F119L, R141H)	• Prominent forehead • Large ears • Flat nasal bridge • Osteopenia • Kyphosis • Joint contractures	pmm2	—	MO	• Craniofacial defects • Reduced motility • Decreased N-glycosylation	Chu et al. (2013), Jaeken (2013), and Leroy (2006)

TMEM-165	Congenital disorder of glycosylation, type IIK (CDG2K)	614727	• Missense (R126H, R126C, G304R); • G-A transition in intron 4 resulting in alternative splice site activation	• Psychomotor, growth retardation • Shortened stature • Osteoporosis • Epiphyseal, metaphyseal, diaphyseal dysplasia • Hepatomegaly, eye abnormalities	tmem165	—	MO	• Shortened embryos • Misshapen heads • Abnormal ventral jaw development • Decreased number of chondrocytes • Altered expression of skeletal markers	Foulquier et al. (2012) and Bammens et al. (2015)
SEC23A	Cranio-lenticulo-sutural dysplasia (CLSD)	607812	• Missense (F382L, M702V)	• Short stature • Facial dysmorphism (frontal bossing, midface hypoplasia, prominent supraorbital ridge) • Macrocephaly • Hypertelorism • Joint laxity • Ossification defects • Scoliosis • Intracellular accumulation of collagen I and II	sec23a (*crusher*)	m299	Nonsense (L402X) MO	• Malformed craniofacial skeleton • Kinked pectoral fins • Shortened body • Distended ER • Intracellular accumulations of collagen II	Lang, Lapierre, Frotscher, Goldenring, and Knapik (2006), Boyadjiev et al. (2006), and Fromme et al. (2007)
SEC23B	Congenital dyserythropoietic anemia, type II (CDAII)	224100	• Missense (E109K, R14W, R530W); • Early termination (R264X, R324X, R217X)	• Anemia • Altered erythropoeisis • Abnormal erythroblast morphology, including multinucleation	sec23b		MO	• Anemia • Craniofacial defects (neural crest) • Jaw morphology defects • Binucleated erythrocytes	Schwarz et al. (2009), Lang et al. (2006), Bianchi et al. (2009), Iolascon et al. (2010), and Russo et al. (2010)

Continued

Table 1 Summary of Genes Involved in the Synthesis, Processing, and Secretion of ECM Components, Including Associated Human Diseases and Corresponding Zebrafish Mutants—cont'd

Human					Zebrafish				References
Gene	**Disease**	**OMIM**	**Common Mutations**	**Skeletal Phenotypes**	**Gene**	**Allele**	**Mutation/Tool**	**Phenotypes**	
SEC24D	Osteogenesis imperfecta (Cole-Carpenter syndrome-2)	607186	• Missense (G205X, G978P, S1015F, Q978P)	• Fractures in clavicle and ribs • Ossification defects in the cranium • Midface hypoplasia • High palate	sec24d (*bulldog*)	m421 m494 m606 m757	Nonsense (Q811X) Frameshift (333X) Frameshift (216X) Nonsense (Q931X)	• Malformed craniofacial skeleton • Kinked pectoral fins • Shortened body • Distended ER • Intracellular accumulations of collagen II	Sarmah et al. (2010) and Yang et al. (2013)
SAR1B	Chylomicron retention disease (CMRD)	246700	• Missense (G37R, D137N, S179R, G185V) • 2-bp deletion (75delTG) • Early termination (E122X) • Splice site mutation	• Growth retardation • Failure to thrive • Fat malabsorption • Steatorrhea • Malnutrition • Neurologic deficits • Hypocholesterolemia • Defect in chylomicron secretion	sar1b		MO	• Dietary lipid malabsorption • Reduction in body length • Reduced digestive organ size • Disruption of neural development • Shortened head • Kinked pectoral fins	Levic et al. (2015) and Charcosset et al. (2008)
SEC13	Unknown	—	—	—	sec13	sq198	Splicing and frameshift	• Abnormal craniofacial cartilage • Short pectoral fins • Small eyes	Townley et al. (2008) and Niu et al. (2012)
							MO		
CREB3L2	Unknown	—	—	—	creb3L2 (*feelgood*)	m662	Missense (N301K)	• Malformed head skeleton • Short body length • Accumulations of collagen II and IV	Melville et al. (2011)

MO, morpholino oligonucleotide.

domain within α-helix of collagen IX (Bork, 1992) that likely interacts with thrombospondin, a matrix glycoprotein. Experiments in zebrafish confirmed that *prp* mutant failed to develop normal fin folds and collagen-rich structural matrices. Disruption of lepidotrichia resulted in abnormal patterning of osteoblast anlage and blood vessels. These results indicate that collagen IX acts as a linker for collagenous matrices and is essential for development and vascularization of peripheral appendages and skeleton (Huang et al., 2009).

3.2 Proteoglycans

Proteoglycans (PG) are ubiquitously expressed throughout animal cells and tissues, consisting of protein cores decorated with long GAG side chains (Prydz & Dalen, 2000). PG and their degree of sulfation play important roles in cell signaling for growth and development through interactions with other ECM components and growth factors. Addition of GAG chains to a protein core is equally necessary for trafficking of the proteoglycan as it is for its function. Synthesis of PG begins with initiation of the common linker chain of tetrasaccharide on the Ser/Thr residue of a core protein (Fig. 1), which is further processed by the addition of amino sugars and glucuronic acid moieties in alternating pattern. The tetrasaccharide linker, repeating disaccharide composition, and sulfation patterns can differentiate distinct proteoglycans: heparan sulfate (HSPG), CSPG, keratin sulfate, and dermatan sulfate. Mutations in genes of the PG synthesis pathway lead to decreased or aberrant production of PGs and consequently alter composition of ECM leading to skeletal malformations (Brown & Eames, 2016) (Fig. 1).

3.2.1 UDP-Glucose Dehydrogenase and UDP-Glucuronic Acid Decarboxylase

To date, no human syndromes were linked to deficiency in the enzymes that carry out the first steps in PG synthesis, and thus the zebrafish mutants provided the first insight into their requirement for craniofacial development. Zebrafish mutants in *ugdh/jekyll*^{m151} locus have diminished UDP-glucose dehydrogenase (Ugdh) activity and present with defective craniofacial cartilage and cardiac cushion development (Neuhauss et al., 1996; Walsh & Stainier, 2001). UDP-glucuronic acid decarboxylase (UGD, also known as *UDP-xylose synthase*) is essential for production of UDP-xylose, the substrate molecule in the production of the tetrasaccharide linker on core proteins. The *mow*^{w60} mutation in zebrafish causes an amino acid substitution R233H (1283G > A) in Uxs1 protein (Eames et al., 2010) and is allelic to

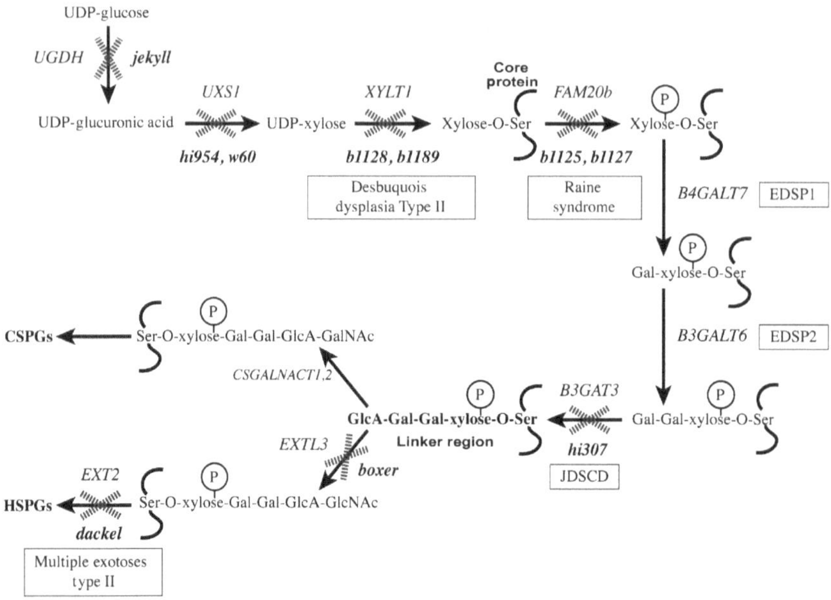

Fig. 1 O-Glycosylation pathway of proteoglycans. Schematic representation of O-glycosylation pathway ending with proteoglycan (CSPGs, HSPGs) synthesis. *Crossed arrows* indicate mutations of human genes (*capitalized italics*) and corresponding zebrafish mutant alleles (*lower case italics*). Associated human diseases are boxed.

the viral insertion mutant, *hi3357* (Golling et al., 2002; Nissen et al., 2006). While the missense *uxs1/mow*^{w60} disrupts protein dimerization interface, the *hi3357* insertion disrupts sequences directly upstream of translation initiation, likely resulting in a null allele, similarly to the *hi954* allele (Wiweger et al., 2011). UGD enzyme activity assays confirmed both missense and presumptive null *uxs1* alleles lack enzymatic activity. Immunofluorescence showed that *uxs1* is essential for production and organization of many components of ECM, including both PG and collagens. Confocal microscopy of GFP-labeled cranial neural crest cells revealed that *uxs1* is essential for the normal development of chondrocytes, perichondrium, and bone, and in situ hybridization experiments confirmed that loss of Uxs1 leads to upregulation of early chondrogenic markers *sox9a*, *sox9b*, and *runx2b* (Eames et al., 2010). This provides a mechanistic explanation for the premature endochondral ossification of craniofacial skeleton.

3.2.2 Xylosyltransferase 1

Addition of UDP-xylose onto a serine residue of a core protein is the subsequent step in PG synthesis and is conducted by one of the two xylosyltransferases (Gotting, Kuhn, Zahn, Brinkmann, & Kleesiek, 2000; Stoolmiller, Horwitz, & Dorfman, 1972). XT1 and XT2 are encoded by distinct genes (*XYLT1* and *2*) that are differentially expressed: XT2 is expressed in kidney, lung, and liver, and XT1 in cartilage and neural tissue (Hurtado, Podinin, Oudega, & Grimpe, 2008). Schreml and colleagues identified a family with two individuals carrying a missense mutation in a conserved arginine (R481W) of xylosyltransferase 1 (*XYLT1*) gene. They present short stature and distinct facial features including deep nasal ridges, long philtrum, unibrow (synophrys), relative macrocephaly, and mild kyphoscoliosis (Schreml et al., 2014). This mutation disrupts the putative catalytic domain resulting in decreased enzymatic activity. Function and subcellular localization of ECM-bound cargos examined by immunofluorescence and Western blot analysis in patients' fibroblasts showed decreased glycosylation and mislocalization of target proteins.

Prior to identification of a human mutation in *XYLT1*, chemical mutagenesis screen followed by positional cloning uncovered a zebrafish mutant in the *xylotransferase 1* (*xylt1*) characterized by hypoplastic midface and bulging eyes. Biochemical, cellular, and developmental analyses revealed decreased levels of CSPG, reduced cartilage matrix, and premature initiation of perichondral bone formation. During chondral bone formation in vertebrates, differentiated chondrocytes secrete CSPG to enrich cartilage ECM. As development progresses, maturation factors such as Indian hedgehog (Ihh) and collagen type X are secreted, inducing surrounding perichondrium cells to differentiate into osteoblasts, producing perichondral bone by endochondral ossification. The mechanism responsible for *xylt1*-dependent skeletal dysmorphology discovered in the zebrafish mutant was shown to stem from precocious maturation of chondrocytes in craniofacial skeleton (Eames et al., 2011).

3.2.3 Glycosaminoglycan Xylosyl Kinase (fam20b)

fam20b is involved in PG synthesis by carrying phosphorylation of the newly added xylose onto the core protein (Fig. 1). The skeletal phenotypes of zebrafish *fam20b* mutant are essentially identical to those described above for *xylt1* mutants. Homozygous *fam20b* mutants exhibited decreased Alcian blue staining in cartilage and increased Alizarin red staining, indicative of excessive ossification. Immunohistochemistry and HPLC analysis of GAG disaccharide levels revealed CSPGs were reduced in both *xylt1* and *fam20b* mutants, whereas HSPGs were less affected. Interestingly, HSPGs around chondrocytes appear to be more strongly affected in *fam20b* than in *xylt1* mutants (Eames et al., 2011).

Results from zebrafish experiments support the hypothesis that not all PG are affected by mutations in early enzymes of the PG synthesis pathway, and cartilage-specific PG are particularly sensitive to deficits in polysaccharide linker production. Additionally, cartilage specificity of *fam20b* and *xylt1* mutants is likely derived from their high level of expression in cartilage as compared to other tissues. The zebrafish *fam20b* and *xylt1* mutants are homozygous viable, providing the only in vivo vertebrate models to study their functions during skeletal development and regeneration.

Genomic deletions of human *FAM20C* have been linked to *Raine syndrome*, an early onset skeletal dysmorphia with characteristic facial features and brain abnormalities (Faundes et al., 2014) (MIM# 611061). Typical features include macrocephaly, hypoplastic nasal bridge, gingival nodules, and early postnatal death (Raine, Winter, Davey, & Tucker, 1989). *fam20b* in zebrafish is paralogous human FAM20C and is involved in phosphorylation of xylose on the core protein, although other functions were also reported (Ishikawa, Xu, Ogura, Manning, & Irvine, 2012). Mechanistic and phenotypic analyses of *fam20b* zebrafish mutants make human *FAM20B* a strong candidate disease gene for a yet uncharacterized skeletal disorder.

3.2.4 Glucuronosyltransferase 1

Glucuronosyltransferase 1 (b3gat3) adds the fourth sugar to the tetrasaccharide linker and is the final enzyme before the divergence to HSPG- and CSPG-specific synthesis pathways (Fig. 1). Zebrafish mutations in *b3gat3* led to decreased chondrocyte intercalation and short, thick cartilage elements particularly evident in the developing jaw (Wiweger et al., 2011). Reverse phase ion-pair HPLC (RPIP-HPLC) total PG profile analysis in zebrafish revealed that *b3gat3* mutant fish had reduced HSPG and CSPG content correlating well with the PG profile in patients with mutations in *B3GAT3* gene (Holmborn et al., 2012). Missense mutations of human *B3GAT3* lead to the *JDSCD* syndrome characterized by multiple joint dislocations, short stature, craniofacial dysmorphism, and congenital heart defects (MIM# 245600)

3.2.5 Exostosin-1 and Exostosin-2

In patients, mutations in Exostosin-1 (*EXT1*) and Exostosin-2 (*EXT2*) can result in *hereditary multiple exostoses (HME)*, a disorder of multiple cartilage-capped bone tumors (Townley et al., 2008). Both EXT1 and EXT2 are transmembrane glycosyltransferases involved in polymerization of HS chains during the formation of proteoglycans (Fig. 1). Prior to transport from the ER en route to Golgi, the two proteins interact to form a hetero-oligomeric complex that elongates HS-chains onto growing GAG chains. HME is an inherited autosomal dominant disorder caused by excessive chondrocyte proliferation at the metaphyses of long bones (femur, tibia, fibula, and humerus) that need to be surgically corrected, often in children. These tumors are benign but can lead to complications associated with tumor burden including compression of nerves, tendons, blood vessels, other skeletal deformities, and short stature. HSPG are known to direct growth and proliferation of progenitors by facilitating diffusion of signaling molecules such as Ihh, and altered production or secretion of HSPG results in excessive proliferation of chondrocytes and tumor formation (Jennes et al., 2009) (Table 1).

EXT genes are highly conserved across multiple species, which enabled generation of animal models through depletion of functional homologs. Mutations in *EXT1* and *EXT2* orthologs in *Drosophila* have been shown to reduce HSPG synthesis, affecting gradients of signaling molecules during development (Han, Belenkaya, Khodoun, Tauchi, & Lin, 2004). However, since *Drosophila* lack neural crest cartilage, this is an impractical model to investigate human skeletal disorders. In contrast, mouse *Ext1* and *Ext2* genes are highly homologous to the human proteins, but homozygous mutants are embryonic lethal before cartilage formation can be examined. Although heterozygotes exhibit decreased HSPG synthesis, they do not develop tumors similar to patients (Stickens, Zak, Rougier, Esko, & Werb, 2005; Zak et al., 2011). Homologs of human *EXT1*, *EXT2*, and *EXTL3* exist in zebrafish, a vertebrate with skeletal biology similar to that of tetrapods making it a tractable animal model.

Exostosin 2 (ext2) and exostosin-like 3 (extl3) are glycotransferases processing two consecutive steps of HSPG production (Lind, Tufaro, McCormick, Lindahl, & Lidholt, 1998; Van Hul et al., 1998). Zebrafish missense and nonsense mutations, *dackel/ext2* and *boxer/extl3*, are characterized by decreased HSPG production, impaired chondrocyte stacking, and reduced mineralization, similar to patients (Clement et al., 2008; Lee et al., 2004). A long-standing question about pathophysiology of HME is the mechanism of *Osteochondroma* formation. Zebrafish studies led to a hypothesis that loss of heterozygosity (LOH) at EXT loci in skeletal cells might result in unregulated growth and clonal expansion. Reported data show that hypertrophic differentiation of chondrocytes and subsequent bone formation are lost in mutant larvae (Clement et al., 2008). Importantly, although most homozygous mutant clones in the replacement experiments were rescued, some *dackel*$^{-/-}$ chondrocytes behaved autonomously lending credence to the LOH model. The glycan profile analysis showed low content of HSPG and increased CSPG in *extl3*-deficient samples (Holmborn et al., 2012). Possible explanations for overproduction of CSPG in *extl3* mutants are changes in expression of CSPG core proteins or a change in sulfation patterns.

3.2.6 Glypicans

Glypicans (Gpc) are family of core proteins anchored to the cell membrane and modified by addition of multiple HS chains to their extracellular region. Glypicans regulate development by modulating availability of ligands such as Wnts, Bmps, and Fgfs. HS-modified glypicans can capture and present ligands in the matrix to cell surface receptors. Human GLYPICAN 6 (GPC6) mutations are linked to *omodysplasia type I* (MIM# 258315), characterized by limb shortening and facial dysmorphism (Campos-Xavier et al., 2009). Glypican 4 (*gpc4*) is a close homolog of *gpc6* and was found to be the mutated in the *knypek* zebrafish line (Caneparo et al., 2007; Topczewski et al., 2001). By mRNA rescue, investigators promoted *knypek* mutant survival to adults, which revealed dermal bone defects in the head skeleton presenting small head size, domed skull, and short jaw (LeClair, Mui, Huang, Topczewska, & Topczewski, 2009). A more recent study revealed cell-autonomous roles of *gpc4* in chondrocyte stacking and organization (Sisson, Dale, Mui, Topczewska, & Topczewski, 2015). Human *GPC6* and zebrafish *gpc4* studies linked glypicans to skeletal development; however, it is conceivable that other members of the glypican family might contribute to skeletal phenotypes and other defects. For example, mutations in human GPC3 lead to congenital heart defects (Simpson–Golabi–Behmel syndrome) (Agatep et al., 2014). Future studies of *GPC6* mutations may benefit from available zebrafish tools to unravel pathophysiology of these debilitating human diseases.

3.3 Noncollagenous Proteins

Another important component of the ECM are proteins that act as adaptors, binding to fibrillar collagens, PG, and other noncollagenous macromolecules to assist in ECM assembly and structure. We review the genes encoding these proteins that have been identified as disease genes in humans and have been studied for their role in chondrocyte development in zebrafish.

3.3.1 Matrilins

Matrilins are family of secreted noncollagenous proteins (Klatt et al., 2009). Vertebrate genomes contain four genes: matrilin-1 (matn1), matrilin-2 (matn2), matrilin-3 (matn3), and matrilin-4 (matn4). Although zebrafish lack matrilin-2, it was postulated that splice variants of matrilin-4 might functionally substitute for matrilin-2 activity (Ko, Kobbe, Paulsson, & Wagener, 2005).

Characterizing mechanisms of human skeletal disorders caused by Matrilin deficiency is difficult in knockout mouse models because mice lack traceable phenotypes (Aszodi et al., 1999; Mates et al., 2004). However, expression of zebrafish *matn1* and *matn3a/b* in the developing skeleton and close sequence similarity to human and mouse genes make zebrafish an efficient model to characterize function of matrilin-1 during skeletal development.

Both matrilin-1 and matrilin-3 are coexpressed in cartilage ECM and were shown to directly interact with collagen II and XI in in vitro biochemical assays. Matrilin-1 (previously known as *cartilage matrix protein*) was first identified as an aggrecan-associated protein through its interactions with CSPG (Ko et al., 2005). Mutations in *MATRILIN-1* (*MATN1*) are associated with *relapsing polychondritis*, an autoimmune disease involving chronic inflammation and degradation of cartilage tissues. Patients with relapsing polychondritis produce autoantibodies against MATN1, serving as a biomarker of the disease (Buckner et al., 2000; Hansson, Heinegård, Piette, Burkhardt, & Holmdahl, 2001). GWAS also linked *MATN1* mutations to osteoarthritis and idiopathic scoliosis (IS) (Montanaro et al., 2006), in select ethnic populations (see Table 1).

Zebrafish *matn1* is expressed strictly in the skeleton, but prior to 24 hpf, it is mainly detected in mesenchymal precursors in the head, notochord, and distal tail. Morpholino knockdown of matrilin-1 resulted in short body, malformed head, curly tail, disrupted notochord integrity, and misshapen somites. Morphants' cartilaginous head skeleton was replaced with a connective tissue that lacked collagens -II and -IX, and GAGs in the ECM. This zebrafish model revealed that matrilin-1 functions primarily in skeletal cells, and in its absence, proteins accumulate in distended ER, while chondrocyte progenitors fail to stack and secrete ECM as observed by TEM. However, protein backlog in the ER did not trigger activation of the *xbp1* gene, suggesting that the ER stress response was not fully activated. Additionally, *matn1* morphants contained fewer chondrocytes with increased TUNEL-positive cells in jaw, signifying elevated chondrocyte death (Neacsu et al., 2014). Together, these results indicate a possible role for matrilin-1 in the development of skeletal tissues and establish a model for further mechanistic studies.

3.3.2 Chondromodulin

Although cartilage has been characterized as an avascular tissue, it still expresses angiogenic factors such as FGF and TGF-β (Grimaud, Heymann, & Redini, 2002; Krejci, Krakow, Mekikian, & Wilcox, 2007). To maintain cartilage antiangiogenic properties, chondrocytes secrete chondromodulin (ChM-I), a type II transmembrane glycoprotein (Shukunami & Hiraki, 2007). Similar activity was identified in tenocytes by a highly homologous protein, tenomodulin (TeM). Although zebrafish cartilage and bone are avascular, both proteins are expressed in zebrafish suggesting a highly conserved function in cartilage, bone, and tendon development (Sachdev et al., 2001). Currently, there are no zebrafish models for *chm1* or *tnmd*, but the discovery of human orthologs and their expression patterns in relevant tissues indicate zebrafish may be an efficient model for studying avascular tissues such as cartilage and tendons.

4. PROCESSING OF ECM MACROMOLECULES

Molecules destined for the ECM are modified in the ER and Golgi complex. Modification and processing of ECM components involve glycosylation, sulfation, and phosphorylation among others. The unique pattern of posttranslational modifications a protein or lipid receives contributes to its form and function within the ECM.

4.1 Procollagen-Lysine, 2-Oxoglutarate 5-Dioxygenases

Collagen biosynthesis requires several posttranslational modifications including hydroxylation of lysine residues within the triple-helical region and N- and C-terminal telopeptides (Yamauchi & Sricholpech, 2012). Lysine hydroxylation, catalyzed by lysyl hydroxylases (LH, also known as procollagen-lysine, 2-oxoglutarate 5-dioxygenases (PLODs)), is critical for collagen triple-helix formation and cross-linking that conveys structural stability. Human LHs are encoded by three genes: *PLOD1*, *PLOD2*, and *PLOD3*, each with distinct expression pattern and substrate specificity (Hautala et al., 1992; Valtavaara et al., 1997; Valtavaara, Szpirer, Szpirer, & Myllyla, 1998). For instance, *PLOD1* and *PLOD3* are expressed in undifferentiated osteoprogenitors and throughout differentiation, while *PLOD2* expression is low in undifferentiated state and during early phases of differentiation, but increases at the onset of matrix mineralization (Uzawa et al., 1999). Expressions of *plod1* and *plod2* were reported in neurocranium, mandible, and pectoral fins (Schneider & Granato, 2007). LHs process preferred substrates, e.g., PLOD1 hydroxylates lysine residues within the regions of triple helix, whereas PLOD2 carries out hydroxylation of procollagen I telopeptides. Function and expression of PLOD genes can be studied using zebrafish models due to conservation of collagen posttranslational modifications and skeletal development pathways between humans and zebrafish.

Patients with mutations in PLOD genes develop a range of skeletal anomalies. For instance, mutations in PLOD1 gene cause *Ehlers–Danlos syndrome type VIA*, which is characterized by hyperextendable skin, joint hypermobility, and kyphoscoliosis (MIM# 225400) (Yeowell & Walker, 2000), and *plod1* zebrafish morphants showed significantly reduced fin-fold area around caudal fin. Although bone defects have not yet been analyzed, knockdown tools are in place to study other skeletal dysmorphologies of *plod1* deficiency in zebrafish (Duran, Mari-Beffa, Santamaria, Becerra, & Santos-Ruiz, 2011).

4.2 The Importance of N-Linked Glycosylation Pathway in Bone Development

The complex cartilage and bone ECM is made up primarily of glycoproteins. Numerous enzymes are required for proper glycosylation of matrix proteins, and mutations in these enzymes lead to Congenital Disorders of Glycosylation (CDGs). Without proper glycosylation, secreted proteins destined for ECM may be subject to misfolding, decreased stability, and mislocalization (Freeze, 2006). Loss of proper posttranslational modifications can affect connective tissue ECM by altering the signals directing development of skeletal elements (Bateman, Boot-Handford, & Lamande, 2009). Here, we review four such examples where zebrafish made significant contributions to CDG field, more examples of CDGs can be found in expert reviews (Jaeken & Matthijs, 2001; Moremen, Tiemeyer, & Nairn, 2012)

N-linked glycosylation refers to addition of the glycan to an asparagine residue of a nascent protein in the ER. Glycan synthesis starts with the formation of a lipid-linked oligosaccharide (LLO) anchored in the ER membrane and the 14-residue carbohydrate chain is transferred to the nascent protein. Glycoproteins exit the ER en route to Golgi where they undergo further processing to form complex or hybrid glycan chains.

4.3 Zebrafish Models of CDG

CDG patients commonly carry mutations in genes coding for processing enzymes, and they present with a range of features, including failure to thrive, metabolic defects such as hypoglycemia and hypocholesterolemia, hypotonia, dysplasia, dwarfism, skeletal malformations, facial dysmorphia, and macrocephaly (Marquardt & Denecke, 2003). The skeletal and craniofacial features of CDG have been attributed to a defect in glycosylation of ECM proteins. Identifying the defective glycoprotein(s) that contribute to CDG pathology has long been sought but remains elusive. Animal models provide a platform for these studies, but null mutations in genes such as *phosphomannomutase 2* (*Pmm2*) and *mannose phosphate isomerase (Mpi)* that cause CDG are early embryonic lethal in mouse (DeRossi et al., 2006; Thiel, Lubke, Matthijs, von Figura, & Korner, 2006). Recent mechanistic studies of genes in glycosylation pathways in zebrafish provided essential knowledge to advance understanding of the pathophysiology of CDG (Clement et al., 2008; Eames et al., 2010, 2011).

PMM2 and MPI are essential enzymes for mannose metabolism, which are required precursors for building the oligosaccharide used for N-glycosylation. Zebrafish *mpi* morpholino injection reduced enzyme levels to those found in *MPI-CDG* patients (Chu et al., 2013). This model illustrates the utility of MO for studying an inborn metabolic error where null mutation causes lethality, as the morpholino is tunable to achieve the desired degree of enzyme depletion. *mpi* morphants developed multiple defects, including reduced body length and dysmorphic jaws (Table 1; Chu et al., 2013). MPI-CDG patients are treated with oral mannose, as this bypasses the enzymatic requirement for MPI in generating mannose-6-phosphate. The zebrafish MPI-CDG model is supported by the finding that mannose supplementation completely rescues *mpi* morphant phenotypes (Chu et al., 2013). This model will be important in studies of the MPI-CDG disease mechanism and mannose treatment.

PMM2-CDG is the most common form of CDG. *pmm2* morpholino injection into zebrafish modestly reduced Pmm2 enzymatic activity and minimized *pmm2* deficiency (Cline et al., 2012). *pmm2* morphants presented with a mild phenotype consisting of craniofacial cartilage defects similar to PMM2-CDG patients (MIM# 212065). Both *mpi* and *pmm2* morphants showed decreased LLO levels and global decrease in N-glycosylation of proteins. The Pmm2 substrate, *mannose-6-phosphate* (*M6P*), was elevated in *pmm2* morphants, and this was corrected by reducing M6P production by coinjection of *pmm2* and *mpi* MOs. This improved chondrocyte morphology (Cline et al., 2012), suggesting that the accumulation of M6P is, in part, responsible for the PMM2-CDG phenotype, reminiscent of the "honeybee effect" where M6P is toxic to honeybees (Sols, Cadenas, & Alvarado, 1960).

CDG-IIk (MIM# 614727) is a CDG subtype caused by mutation of *transmembrane protein 165 (TMEM165)* (Nabavi, Pustylnik, & Harrison, 2012). Patients with *TMEM165* mutations present with dwarfism, facial hypoplasia and dysmorphia, bone and cartilage dysplasia (Foulquier et al., 2012). Variants in *TMEM165* result in different clinical presentations depending on the type of mutation (Table 1). However, little is known about the function of this protein, or how mutations result in the phenotypes seen in CDG-IIk patients.

CDG-IIk was modeled in zebrafish using a morpholino by Bammens et al. *tmem165* morphants were shorter than wild-type controls and exhibited craniofacial defects. Disruption of osteoblast maturation resulted in deficient bone formation similar to TMEM165-CDG patients. This was attributed to a decrease in the total number of chondrocytes in morphants, reduced expression of aggrecan, marker for an intermediate stage of chondrocyte differentiation, and an increase in sox9, a modulator of chondrocyte differentiation (Bammens et al., 2015). Glycome analysis of *tmem165* morphants revealed deficits in N-glycosylation determined by a decrease in total glycoprotein, and decreased glycan modification. Altered bone differentiation was indicated by decreased expression of the osteoblast marker *col10a1,* and *osterix*, an osteoblast maturation factor. Importantly, while injection of wild-type *tmem165* mRNA restored cartilage development in *tmem165* morphants, mRNA with mutations found in CDG-IIk patients did not, both confirming target specificity of the MO and establishing that patient-derived *tmem165* mutations are pathogenic.

A putative novel CDG was identified by the zebrafish mutant, *trappc11/ foie gras*, which is characterized by fatty liver and jaw defects (Cinaroglu, Gao, Imrie, & Sadler, 2011; Sadler, Amsterdam, Soroka, Boyer, & Hopkins, 2005). Patients with *TRAPPC11* mutations present with fatty liver and muscular dystrophy (Liang et al., 2015). *trappc11* zebrafish mutants have defects in protein glycosylation similar to those found in patients (DeRossi et al., 2016). Future studies will be needed to detect potential skeletal deficits in patients.

4.4 Secretion of ECM Macromolecules

The ECM plays an important role in cell migration, differentiation, and survival. ECM components are synthesized by specialized cells and are secreted to the extracellular milieu by exocytic machinery in an organized, regulated fashion. Specialized coat proteins supporting vesicular carriers are critical to secretion of ECM, especially large structural components such as collagens. Secreted, extracellular cargos are synthesized in the ER and are carried forward for central processing and sorting in the Golgi complex, from which the post-Golgi system of specialized coats direct them to their final destinations (Schekman & Orci, 1996). COPII (coat protein II complex) mediates the first leg of the secretory pathway leading to secretion of ECM components. The importance of COPII proteins in the skeleton has been highlighted by studies in patients with skeletal dysmorphology syndromes and zebrafish with mutations in COPII genes.

4.5 ER Egress of COPII Carriers

Export from ER is facilitated by COPII-dependent translocation of vesicles to the ER-to-Golgi intermediate compartment (EGRIC) and then to the Golgi complex. Details on this process were recently reviewed (Brandizzi & Barlowe, 2013; Melville & Knapik, 2011; Szul & Sztul, 2011; Unlu et al., 2014). Here, we focus on how studies in zebrafish have illustrated the importance of COPII complex genes in skeletal formation.

COPII complex assembly begins when Sar1, a cytoplasmic GTPase, is activated through GTP binding (Barlowe, d'Enfert, & Schekman, 1993). Sar1 recruits heterodimers, Sec23/Sec24, to form the "inner coat" (Bi, Corpina, & Goldberg, 2002), and Sec12, a guanine nucleotide exchange factor for Sar1 cooperates with Sec16, a scaffold protein, to assist with vesicular carrier formation. Sec24 acts as cargo loading adaptor (Barlowe & Schekman, 1993). When the vesicle is ready to dissociate from ER membrane, Sec23, a GTPase activating protein, deactivates Sar1 (Yoshihisa, Barlowe, & Schekman, 1993). Following assembly of the inner coat, Sec13/Sec31 heterotetramer assists in the formation of an outer coat that stabilizes the entire complex as it travels from the ER to the EGRIC (Copic, Latham, Horlbeck, D'Arcangelo, & Miller, 2012).

4.5.1 Sec24 Paralogs

Four genes encode for Sec24 paralogs in all vertebrates (Tang, Kausalya, Low, Lock, & Hong, 1999), and each paralog may exhibit selective binding of diverse cargos (Wendeler, Paccaud, & Hauri, 2007). The distinct skeletal phenotypes observed in Sec24C and Sec24D mutants suggested distinct use of cargo adaptors in secretion. For instance, the zebrafish *sec24d* mutant, *bulldog* presents with disrupted secretion of collagen II and matrilin in chondrocytes, fibroblasts, and notochord sheath cells (Sarmah et al., 2010). Mutant chondrocytes exhibit distended rough ER with accumulation of collagen II (Fig. 2A, B, E), while other trafficked ECM and transmembrane proteins are unaffected. This suggests that Sec24d is essential for loading of collagen II into COPII vesicles. Although Sec24c and Sec24d recognize similar binding motifs in cargo proteins (Miller, Antonny, Hamamoto, & Schekman, 2002), depletion of *sec24c* in zebrafish does not result in the craniofacial skeleton dysmorphology seen in *sec24d* mutants. However, a double knockdown of *sec24c/sec24d* results in a more severe craniofacial and notochord phenotypes indicating that Sec24 paralogs function redundantly in the secretion of matrix proteins, but both are required for formation of basement membrane surrounding the notochord. A mouse model has not been as successful in determining Sec24D role in secretion of ECM-bound proteins because gene knockout models result in preimplantation lethality, and haploinsufficient animals present with no visible phenotype (Baines, Adams, Zhang, & Ginsburg, 2013). It is likely that zebrafish models of secretory protein deficiencies develop normally through the critical early gastrulation and morphogenesis stages because of maternally deposited Sec24D protein and mRNA. Zebrafish and medaka models of *sec24d* deficiency were helpful guides in the identification of *SEC24D* mutations in a syndromic form of OI by a whole-exome sequencing of the genome of a 7-year-old boy with severe ossification defects, and two fetuses with impaired skull formation and fractures throughout the skeleton (Garbes et al., 2015). Of the five candidate genes, *SEC24D* was prioritized due to the phenotypes reported for zebrafish and medaka *sec24d* mutants (Ohisa, Inohaya, Takano, & Kudo, 2010; Sarmah et al., 2010). The short stature, craniofacial dysmorphology, micrognathia, and nonfusing sutures in the skull were highly correlated to zebrafish *sec24d/bulldog* mutants. Consequently, the identification of compound heterozygotes for *SEC24D* mutations in these patients and backlog of collagen in the ER supported the causal effect of this gene in OI. The molecular phenotypes of zebrafish *sec24d* mutant chondrocytes were also found in patient cells, underscoring the power of zebrafish to model human skeletal disorders.

4.5.2 Sec23 Paralogs

Sec23a and Sec23b are the two vertebrate paralogs (Paccaud et al., 1996). A zebrafish *sec23a* mutant, *crusher*, displays malformed craniofacial skeleton, kinked pectoral fins, and shortened body length (Lang et al., 2006). *crusher* carries a nonsense mutation (1287T > A) in the Sec23a protein producing an early stop codon and predicted truncated protein. TEM and immunofluorescence analysis of *crusher* chondrocytes revealed a distended rough ER with

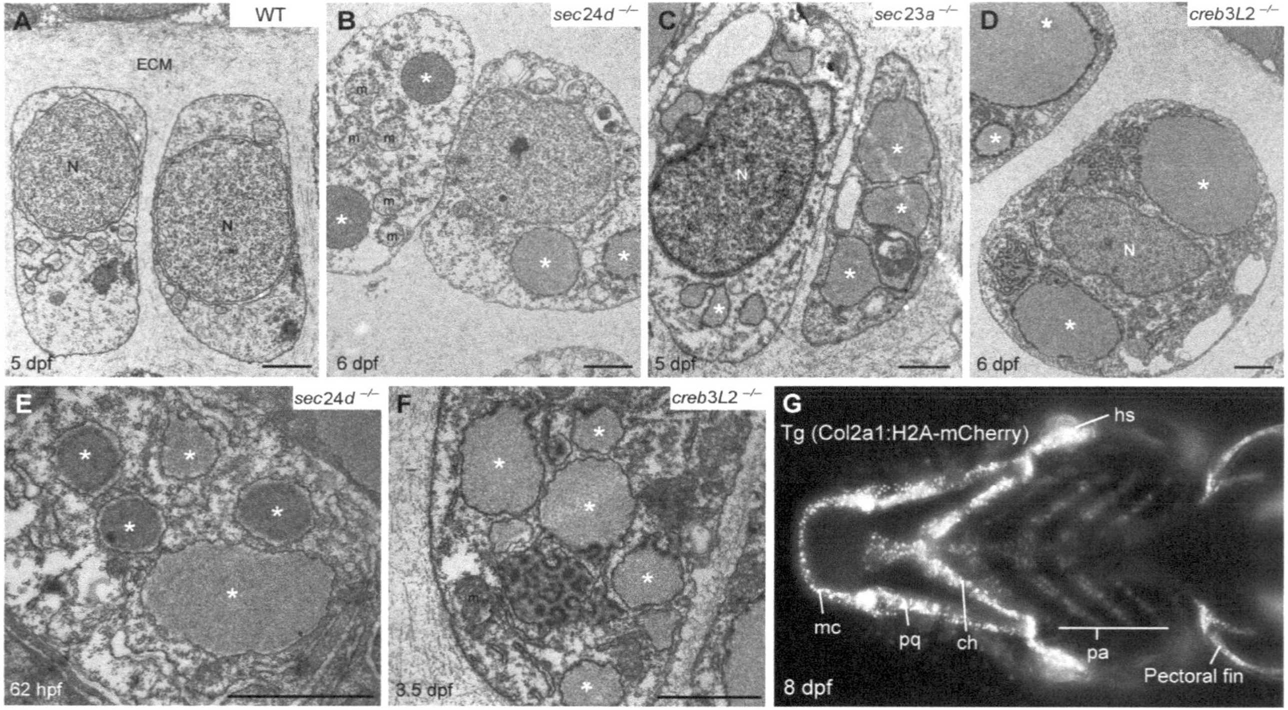

Fig. 2 Zebrafish craniofacial chondrocytes are a useful model to study collagen secretion, expression, and cartilage differentiation. Transmission electron micrographs (TEM) of zebrafish craniofacial chondrocytes of (A) wild type (WT), (B,E) *sec24d*$^{-/-}$ (*bulldog*^{m757}), (C) *sec23a*$^{-/-}$ (*crusher*^{m299}), (D,F) *creb3L2*$^{-/-}$ (*feelgood*^{m662}). N, nucleus; m, mitochondrion; * marks distended rough ER. Scale bar: 2 μm. (G) Ventral view of eight dpf zebrafish skeleton marked by chondrocyte-specific expression of histone H2A-mCherry nuclear marker driven under 1.7 kb Col2a1a promoter (Dale & Topczewski, 2011). p5E Tol2kit vector containing 1.7 kb Col2a1a promoter sequence was gifted by Topczewski laboratory. Using Tol2kit cloning strategy, middle entry element, "pME-H2A-mCherry" was recombined with "p5E-1.7 kb Col2a1a promoter" and "p3E-polyA" into destination vector "pDest-pA2." Destination vector was injected into one-cell stage zebrafish embryos along with transposase mRNA and grown into adulthood. Verified founders are maintained in AB background. *Tg(Col2a1:H2A-mCherry)* transgenics displays nuclear mCherry expression in craniofacial chondrocytes, notochord, floor plate, hypochord, and fins, recapitulating *col2a1a* expression pattern. *ch*, ceratohyal; *hs*, hyosymplectic cartilage; *mc*, meckel's cartilage; *pq*, palatoquadrate; *pa*, pharyngeal arches.

intracellular accumulations of collagen II deposits resulting in reduced collagen and matrilin presence in cartilage ECM (Fig. 2C). A genetrap *Sec23a* deficiency has been identified in mice, although these are embryonic lethal (Zhu et al., 2015), phenotypes observed prior to death included developmental defects of extraembryonic membranes and neural tube closure. Collagen secretion by skin fibroblasts was affected by *Sec23a* deficiency and resulted in intracellular accumulation similar to the zebrafish mutant model. These results indicate an important role for Sec23A in secretion of large cargos such as collagen across multiple vertebrate species.

Human mutations in *SEC23A* were reported to cause cranio-lenticulo-sutural-dysplasia (CLSD) (MIM# 607812). Patient phenotypes for this autosomal recessive disorder include facial dysmorphism and axial skeletal defects (Boyadjiev et al., 2006). As in zebrafish *crusher*, CLSD patient fibroblasts have distended ER with collagen accumulation. CLSD patient mutations include two missense mutations that alter the SEC31-binding site on the SEC23A protein (Table 1). Alteration of the SEC13–SEC31 outer coat produces a malformed vesicle or causes premature dissociation from the membrane (Fromme et al., 2007).

Mutations in the second *SEC23* paralog, *SEC23B*, reportedly involved in erythrocyte differentiation, have been associated with congenital dyserythropoietic anemia type II (CDAII) (MIM# 224100) (Table 1; Bianchi et al., 2009; Schwarz et al., 2009). Phenotypes in patients include altered erythropoiesis and hemolysis as well as multinucleated erythroblasts present in bone marrow. The ECM contained hypoglycosylated proteins suggesting that a defect in trafficking of glycosylated proteins to the ECM could account for the phenotype caused by *sec23* deficiency. CDAII patients with anemic phenotypes typically have one of over 50 identified missense mutations found throughout *SEC23B* (Russo et al., 2010). In mice, *Sec23b* deficiency did not present an anemic phenotype but caused embryonic lethality (Tao et al., 2012), whereas a knockdown of zebrafish *sec23b* presented with anemia and hemolysis and craniofacial defects (Unlu et al., 2014). Thus, in the case of CDAII, the zebrafish model may more faithfully recapitulate the human syndrome than the mouse model.

4.5.3 Sar1 Paralogs

The two Sar1 paralogs found in vertebrates: Sar1a and Sar1b differ by only 20 amino acids. Sequence identity between mammalian and zebrafish homologs is over 90% (Levic et al., 2015). Sar1a and Sar1b functions differentially affect COPII coat vesicle formation: Sar1A has a higher affinity for the outer coat complex resulting in smaller vesicles, whereas Sar1B, which hydrolyzes GTP more slowly than Sar1A, may allow a larger vesicle to form which could then contain larger cargo, such as procollagen (Fromme, Orci, & Schekman, 2008).

In addition, Sar1b participates in trafficking of other large cargos, such as chylomicrons. This is exemplified by the clinical presentation of patients with *SAR1B* mutation who develop chylomicron retention disease (CMRD) (MIM#246700), a disorder involving lipid malabsorption characterized by defects in intestinal lipid uptake and hypocholesterolemia (Peretti et al., 2010; Treepongkaruna et al., 2009). The pathophysiology of Sar1b deficiency was first described in a MO model in zebrafish. *sar1b* morphants exhibited accumulation of dietary lipids as well as defects in organogenesis (Levic et al., 2015). Furthermore, decreased bone mineral density observed in patients correlated well with collagen secretion defects and craniofacial cartilage dysmorphology in zebrafish. This model of CMRD recapitulates symptoms of human patients and offers potential for additional studies to investigate functions of Sar1B in trafficking of large vesicular carriers.

4.5.4 Sec13–Sec31 Complex

The outer coat of COPII vesicles consists of Sec13–Sec31 heterotetramers and provides vesicular carriers with a more rigid structure. There have been no identified human patients with mutations in either *SEC13* or *SEC31*, suggesting deficits in the function of these genes could cause in utero lethality or other genes may replace their activity. Furthermore, a mouse *Sec13* knockout is embryonic lethal (Moreira et al., 2015). While search for a human condition needs to be continued, zebrafish provide a robust in vivo model to examine COPII outer coat loss-of-function phenotypes.

Knockdown zebrafish model for *Sec13* resulted in phenotypes including abnormal craniofacial cartilage elements, short pectoral fins, small eyes, and cardiac edema (Townley et al., 2008). Organogenesis of digestive system organs was also affected in *sec13* genetic mutants leading to hypoplastic intestine, liver, and exocrine pancreas (Niu et al., 2012). Sec13 is a dual function protein involved in COPII-mediated traffic and nucleocytoplasmic traffic through the nuclear pore complex (NPC). Elegant studies in the zebrafish model discerned that retinal lamination and developmental deficits were inflicted by the loss of function within the NPC and not COPII. Zebrafish with *sec13* deficiency provides evidence on the importance of COPII outer coat formation in the development of highly secretory organs and tissues such as cartilage, intestine, pancreas, and liver.

4.6 Creb3L2, Transcriptional Regulation of the Secretory Pathway

Secretory machinery is a cellular service system charged with delivery of specific cargos to match the functional demands of cells. Cells of the skeletal system produce large amounts of ECM cargos that change in volume and composition during development, repair, and homeostasis. To match supply to demand, cells regulate availability of their transit system. One way to regulate the secretory pathway is through transcriptional control of secretory machinery. Evidence of this form of regulation has been established in multiple model systems including mouse, zebrafish, and fly (Fox, Hanlon, & Andrew, 2010; Melville et al., 2011; Saito et al., 2009; Unlu et al., 2014).

Creb3L2 (cAMP responsive element-binding protein-3 like 2), an ER synthesized transmembrane transcription factor highly expressed in chondrocytes, has been associated with secretion of collagen in vertebrates. Following translation in the ER, Creb3L2 is exported via COPII vesicles to the Golgi complex for processing by *site-1-protease (S1P)* before being translocated to the nucleus where it dimerizes and activates expression of target genes. Mouse knockout model of *Creb3L2* results in chondrodysplasia with collagen II accumulation and distended ER (Saito et al., 2009). In zebrafish *feelgood* mutants, a missense mutation in the DNA-binding domain of zebrafish *creb3L2* was identified. *Sec23a* promoter has previously been shown to be under *Creb3L2* transcriptional control, and *feelgood* mutants have significant reduction in *creb3L2* transcriptional activity and decreased expression of *sec23a* (Melville et al., 2011), leading to defective secretion of collagen II and IV from chondrocytes and notochord sheath cells, respectively. TEM analysis showed distended ER structures with intracellular protein accumulation and a loss of cartilage matrix surrounding chondrocytes similar to observations in other COPII mutants (see Fig. 2D and F). These results indicate Creb3L2 functions to regulate trafficking machinery necessary for secretion of large ECM cargo to keep up with developmental demands of the skeleton. Multiple potential targets of Creb3L2 were identified by qPCR and microarray approaches in mouse, frog, and fish (Melville et al., 2011; Saito et al., 2009; Tanegashima, Zhao, Rebbert, & Dawid, 2009), and further analysis will be required to determine the functional targets and their physiological relevance. Novel regulatory pathways to be discovered in the future and acting similarly to the Creb3L2–Sec24D–Sec23A axis (Melville & Knapik, 2011) may play a role in yet to be uncovered human skeletal disorders.

5. ZEBRAFISH MODELS OF AXIAL SKELETON DEFECTS

Scoliosis is a class of axioskeletal defects that are defined by lateral spine curvatures >10 degree. Congenital vertebral malformations (CVMs) such as wedge-shaped vertebrae, fusions, and hemivertebrae can lead to congenital scoliosis (CS), one of the most prevalent skeletal malformations (0.013–0.05% in live newborns) apparent at birth (Giampietro, 2012). Spinal deformities more frequently occur with no detectable structural defects in the vertebrae and are collectively called idiopathic scoliosis (IS), the most common pediatric skeletal disorder (Miller, 1999). Historically, rodents were used to study scoliosis by generating vertebral deformities through surgical procedures (Janssen, de Wilde, Kouwenhoven, & Castelein, 2011). However, these models failed to recapitulate effects of genetic mutations and are not suitable to investigate progression of IS. The ease of axial skeleton analysis, the biological similarity of vertebral phenotypes, and the fact that spinal defects can be observed as early as 72 hpf makes zebrafish a robust and efficient genetic tool to model scoliosis. Here, we review discoveries of pathophysiology and of scoliosis genes in zebrafish models.

5.1 GPR126

A GWAS conducted in Japanese, Han Chinese, and European populations revealed that a novel variant of an orphan *G-protein coupled receptor, GPR126*, is associated with adolescent IS (Kou et al., 2013). This showed that GPR126 is expressed at a higher level in human cartilage than other tissues tested. MO knockdown of *gpr126* in zebrafish resulted in shorter embryonic body length and delayed ossification of vertebrae, supporting causative effects of GPR126 variants in scoliosis development.

5.2 Centriolar Protein Gene 5 (POC5)

Patten et al. identified three variants within the coding region of a *centriolar protein gene, POC5,* conducting genetic linkage analyses and exome sequencing in multimember families with IS (Patten et al., 2015). The variants A429V, A446T, and A455P were predicted to be missense mutations. Overexpression of WT mRNA, as control, had no effect on axial development, whereas overexpressing mutant POC5 mRNAs in zebrafish led to specific spinal deformities in 3-day old larvae, confirming that all variants are functional contributing to IS pathophysiology.

5.3 COL27A1

Zebrafish notochord serves as axial skeleton at early embryonic stages maintaining posture. Notochord extension, structure, and stiffness are heavily dependent upon ECM deposition by notochord sheath cells to the basement membrane. Col27a1a and Col27a1b are essential components of axial skeleton. Christiansen et al. showed depletion of *col27a1a* alone or in conjunction with *col27a1b* result in embryonic vertebral curvatures reminiscent of scoliosis (Christiansen, Lang, Pace, & Parichy, 2009). Later in development, dysmorphic structures remained throughout vertebral column of *col27a1*-depleted zebrafish. This observation in zebrafish models paved the way for identification of missense variant "G697R" in COL27A1 in probands suffering from Steel syndrome (Gonzaga-Jauregui et al., 2015), a type of osteochondrodysplasia with main clinical features including bilateral hip dysplasia, short stature, and scoliosis (Steel, Piston, Clancy, & Betz, 1993).

5.4 PAX1 Enhancer

PAX1 is a key transcription factor regulating vertebral development (Monsoro-Burq, 2005). Disease-causing nucleotide variants can be tested in zebrafish by genetic tools such as *Tol2kit* genome integration approach enabling mapping enhancers and their activity in vivo. This tool was instrumental in discovery of IS susceptibility locus distal to PAX1 gene, where 10 SNPs associated with a form of IS were evaluated using the Tol2 enhancer assay and narrowed down to a single IS-associated SNP "rs16931," as disruptive to PAX1 expression (Sharma et al., 2015).

5.5 Protein Tyrosine Kinase 7

Protein Tyrosine Kinase 7 (Ptk7) has been implicated in planar cell polarity signaling and regulation of cell movements, thereby shaping body axis (Keller, 2002). Hayes et al. generated zygotic and maternal zygotic mutants of *ptk7* and showed these are functional models of IS and CS, respectively (Hayes et al., 2014). Zygotic *ptk7* mutants displayed minor vertebral malformations at 28 days (dpf), and by 40 dpf, all had axial deformities that increased in severity by ~3 months. These findings led to sequencing of the *PTK7* exons from multiple cohorts of IS patients, and identification of a heterozygous missense mutation, P545A. While WT *ptk7* mRNA overexpression could restore axial extension in a maternal zygotic mutant, mRNA carrying P545A variant failed to rescue, further confirming pathogenicity of P545A. CVMs, such as those observed in maternal zygotic *ptk7* mutants, are primary symptoms of CS. This was a landmark, as the first congenital model of scoliosis in zebrafish (Hayes, Naito, Daulat, Angers, & Ciruna, 2013). This group further described that scoliosis phenotype observed in *ptk7* loss of function is linked to its role in development of ependymal ciliary cells, which direct cerebrospinal fluid (CSF) flow. Transgenic reintroduction of Ptk7 in motile-ciliated cells was able to restore reduced CSF flow to normal levels in *ptk7* mutant fish and ultimately preventing spinal curves (Grimes et al., 2016). Through explaining contributions of motile cilia and CSF flow in scoliosis progression, this study presents new opportunities for treatment of spinal deformities by targeting CSF flow and its downstream effectors. Availability of several zebrafish reporter lines (Dale & Topczewski, 2011; Ellis, Bagwell, & Bagnat, 2013; Vacaru et al., 2014) to track skeletal cells could facilitate such efforts (Fig. 2G).

5.6 COL8A1

The knowledge of mechanisms and causes of CVMs are limited due to their low prevalence, hence difficulty in conducting genetic association studies, and ineffective animal models. ENU-based, unbiased forward genetic screens carried out in zebrafish provided an entry point into understanding CMVs and CS progression. Gray et al. characterized mutant alleles of *leviathan* that result in folded embryonic notochord with progression into adult vertebral malformations (Gray et al., 2014). Mutant fish lacking functional *col8a1a* gene displayed kinks in the notochord at 1 dpf and lived to adulthood presenting with fused vertebral elements and kinked structures throughout vertebrae, i.e., scoliosis phenotypes, as analyzed by micro-CT imaging and shorter body length. To specifically test how *col8a1a* impacts vertebral body formation, they utilized splice blocking antisense MO against *col8a1a* (Gansner & Gitlin, 2008). Since MOs degrade and dilute over the course of ~5 days, they were able to test the effect of early *col81a* depletion on later developmental stages. MO-injected zebrafish recapitulated both embryonic notochord folding phenotypes and vertebral fusions and shorter body length in adult. There has not been a reported case of *COL8A1* mutation in CS patients yet; however, unbiased animal studies will help geneticists prioritize their results.

6. CONCLUDING REMARKS

Zebrafish has successfully transitioned from an embryology-focused organism to a workhorse model system in service of human disease gene

Table 2 Commonly Used Tools and Markers to Study Zebrafish Skeletal Development

Marker Name	Marker Type	Cell/Tissue Type	Labeled Structure	References
Tg[-1.7col2a1a:caax-EGFP]	Transgenic	Craniofacial cartilage, ear, notochord, floor plate, hypochord, and fins	Plasma membrane with EGFP	Dale and Topczewski (2011)
Tg[-1.7col2a1a:H2A-mCherry]	Transgenic	Craniofacial cartilage, ear, notochord, floor plate, hypochord, and fins	Nuclei with mCherry	This publication
Tg[osx:nuGFP]	Transgenic	Early osteoblasts	Nuclei with EGFP	Vanoevelen et al. (2011)
Tg[osc:GFP]	Transgenic	Mature osteoblasts	Cytoplasm with GFP	Vanoevelen et al. (2011)
Tg[sp7:EGFP]	Transgenic	Otic vesicle, skeletal structures (embryos and adult)	Cytoplasm with GFP	DeLaurier et al. (2010)
Anti-Collagen2α1	Antibody	Chondrocytes/cartilage	Cartilage matrix	Lang et al. (2006)
Wheat Germ Agglutinin	Lectin	Cartilage matrix	GAGs	Sarmah et al. (2010)
Anti-Fibronectin	Antibody	Mesenchymal condensations in cartilage primordia	Condensation Matrix	Sarmah et al. (2010)
Peanut Agglutinin	Lectin	Mesenchymal condensations in cartilage primordia	Glycosylated proteins in matrix	Sarmah et al. (2010)
Anti-Laminin	Antibody	Notochord basement membrane	Notochord sheath	Melville et al. (2011)
Anti-Collagen type IV	Antibody	Notochord basement membrane	Notochord sheath	Melville et al. (2011)
Alcian Blue	Dye	Cartilage	Mucopolysaccharides in matrix	Walker and Kimmel (2007)
Alizarin Red	Dye	Bone	Mineralized matrix	Walker and Kimmel (2007)

discovery and drug development. The combination of vertebrate biology, in vivo imaging techniques, genetic, and transient tools to manipulate gene expression levels and study ensuing phenotypes positioned zebrafish model in a "sweet spot" between simplicity and complexity of experimental analysis (Table 2). The least mentioned, but perhaps some of the most important, factors that have driven the model to its current status are rapid turnaround of experimental outcomes at a cost that is affordable relative to rodents.

As we have highlighted through the current review, the approaches are bidirectional, from zebrafish models to human disease, and back from human disease to zebrafish modeling. Examples of "from fish to man knowledge flow" are initial discoveries of the zebrafish mutant phenotypes including Sec24D/*bulldog* that facilitated prioritization of GWAS candidate genes for OI patients and sequencing of *SEC24D* gene (Garbes et al., 2015; Sarmah et al., 2010), or a well-described scoliosis phenotype of zebrafish Col27a1 mutant that lead to sequencing of Steel syndrome patients for mutations in *COL27A1* (Christiansen et al., 2009; Gonzaga-Jauregui et al., 2015).

The reverse direction from human to zebrafish is driven by the necessity to validate disease causing gene variants. A good example would be testing of the idiopathic scoliosis variant P545A in *PTK7* gene for its potential to rescue the notochord phenotype in the corresponding zebrafish mutant (Hayes et al., 2014). Whole-exome sequencing of probands from human pedigrees segregating a specific skeletal syndrome began to generate novel disease genes whose pathology is often unknown.

Although rapidly growing lists of new Mendelian disease genes are continuously supplied by WES efforts around the globe, there is still a sizable group of genes for which there are no associated diseases. A good example is CREB3L2, a transcription factor that regulates collagen secretory pathway. COPII genes, *SEC23A* and *SEC24D*, were implicated in CLSD and OI, while *CREB3L1*, a close paralog of *CREB3L2*, has also been identified as an OI gene (Symoens et al., 2013). The zebrafish cellular phenotypes of Creb3L2–Sec24D–Sec23A axis are nearly identical (Fig. 2), suggesting that yet an undiagnosed skeletal syndrome is caused by mutations in *CREB3L2*. For these reasons, zebrafish are poised to advance skeletal biology, disease gene discovery, and the search for new therapeutics.

FUNDING AND ACKNOWLEDGMENTS

The authors were supported in part by the Zebrafish Initiative of the Vanderbilt University Academic Venture Capital Fund, the NIH R01 Grants DE018477, HL92217, EY012018 (E.W.K.), T32HD007502 Training Program in Developmental Biology (L.N.L.), the

Vanderbilt International Scholar Program (VISP), and AHA Predoctoral Fellowship 15PRE22940041 (G.U.). The authors thank Jeff Davidson and Brian Eames for insightful discussions. Authors apologize to all colleagues whose work could not be cited due to space limitations.

Competing Interests: The authors declare that they do not have any competing or financial interests.

REFERENCES

Agatep, R., Shuman, C., Steele, L., Parkinson, N., Weksberg, R., & Stockley, T. L. (2014). Paternal germline mosaicism for a GPC3 deletion in X-linked Simpson-Golabi-Behmel syndrome. *American Journal of Medical Genetics. Part A, 164A*, 2682–2684.

Amsterdam, A., Nissen, R. M., Sun, Z., Swindell, E. C., Farrington, S., & Hopkins, N. (2004). Identification of 315 genes essential for early zebrafish development. *Proceedings of the National Academy of Sciences of the United States of America, 101*, 12792–12797.

Andreeva, V., Connolly, M. H., Stewart-Swift, C., Fraher, D., Burt, J., Cardarelli, J., et al. (2011). Identification of adult mineralized tissue zebrafish mutants. *Genesis, 49*, 360–366.

Annunen, S., Paassilta, P., Lohiniva, J., Perala, M., Pihlajamaa, T., Karppinen, J., et al. (1999). An allele of COL9A2 associated with intervertebral disc disease. *Science, 285*, 409–412.

Arnold, W. V., & Fertala, A. (2013). Skeletal diseases caused by mutations that affect collagen structure and function. *The International Journal of Biochemistry & Cell Biology, 45*, 1556–1567.

Aszodi, A., Bateman, J. F., Hirsch, E., Baranyi, M., Hunziker, E. B., Hauser, N., et al. (1999). Normal skeletal development of mice lacking matrilin 1: Redundant function of matrilins in cartilage? *Molecular and Cellular Biology, 19*, 7841–7845.

Baines, A. C., Adams, E. J., Zhang, B., & Ginsburg, D. (2013). Disruption of the sec24d gene results in early embryonic lethality in the mouse. *PloS One, 8*e61114.

Bammens, R., Mehta, N., Race, V., Foulquier, F., Jaeken, J., Tiemeyer, M., et al. (2015). Abnormal cartilage development and altered N-glycosylation in Tmem165-deficient zebrafish mirrors the phenotypes associated with TMEM165-CDG. *Glycobiology, 25*, 669–682.

Barlowe, C., d'Enfert, C., & Schekman, R. (1993). Purification and characterization of SAR1p, a small GTP-binding protein required for transport vesicle formation from the endoplasmic reticulum. *The Journal of Biological Chemistry, 268*, 873–879.

Barlowe, C., & Schekman, R. (1993). SEC12 encodes a guanine-nucleotide-exchange factor essential for transport vesicle budding from the ER. *Nature, 365*, 347–349.

Bateman, J. F., Boot-Handford, R. P., & Lamande, S. R. (2009). Genetic diseases of connective tissues: Cellular and extracellular effects of ECM mutations. *Nature Reviews. Genetics, 10*, 173–183.

Bell, D. M., Leung, K. K., Wheatley, S. C., Ng, L. J., Zhou, S., Ling, K. W., et al. (1997). SOX9 directly regulates the type-II collagen gene. *Nature Genetics, 16*, 174–178.

Berendsen, A. D., & Olsen, B. R. (2015). Bone development. *Bone, 80*, 14–18.

Bi, X., Corpina, R. A., & Goldberg, J. (2002). Structure of the Sec23/24-Sar1 pre-budding complex of the COPII vesicle coat. *Nature, 419*, 271–277.

Bianchi, P., Fermo, E., Vercellati, C., Boschetti, C., Barcellini, W., Iurlo, A., et al. (2009). Congenital dyserythropoietic anemia type II (CDAII) is caused by mutations in the SEC23B gene. *Human Mutation, 30*, 1292–1298.

Bork, P. (1992). The modular architecture of vertebrate collagens. *FEBS Letters, 307*, 49–54.

Boyadjiev, S. A., Fromme, J. C., Ben, J., Chong, S. S., Nauta, C., Hur, D. J., et al. (2006). Cranio-lenticulo-sutural dysplasia is caused by a SEC23A mutation leading to abnormal endoplasmic-reticulum-to-Golgi trafficking. *Nature Genetics, 38*, 1192–1197.

Bradley, K. M., Elmore, J. B., Breyer, J. P., Yaspan, B. L., Jessen, J. R., Knapik, E. W., et al. (2007). A major zebrafish polymorphism resource for genetic mapping. *Genome Biology*, *8*, R55.

Brandizzi, F., & Barlowe, C. (2013). Organization of the ER-Golgi interface for membrane traffic control. *Nature Reviews. Molecular Cell Biology*, *14*, 382–392.

Brown, D. S., & Eames, B. F. (2016). Emerging tools to study proteoglycan function during skeletal development. *Methods in Cell Biology*, *134*, 485–530.

Buckner, J. H., Wu, J. J., Reife, R. A., Terato, K., & Eyre, D. R. (2000). Autoreactivity against matrilin-1 in a patient with relapsing polychondritis. *Arthritis & Rheumatism*, *43*(4), 939.

Cabral, W. A., Makareeva, E., Colige, A., Letocha, A. D., Ty, J. M., Yeowell, H. N., et al. (2005). Mutations near amino end of alpha1(I) collagen cause combined osteogenesis imperfecta/Ehlers-Danlos syndrome by interference with N-propeptide processing. *The Journal of Biological Chemistry*, *280*, 19259–19269.

Campos-Xavier, A. B., Martinet, D., Bateman, J., Belluoccio, D., Rowley, L., Tan, T. Y., et al. (2009). Mutations in the heparan-sulfate proteoglycan glypican 6 (GPC6) impair endochondral ossification and cause recessive omodysplasia. *American Journal of Human Genetics*, *84*, 760–770.

Caneparo, L., Huang, Y. L., Staudt, N., Tada, M., Ahrendt, R., Kazanskaya, O., et al. (2007). Dickkopf-1 regulates gastrulation movements by coordinated modulation of Wnt/beta catenin and Wnt/PCP activities, through interaction with the Dally-like homolog Knypek. *Genes & Development*, *21*, 465–480.

Charcosset, M., Sassolas, A., Peretti, N., Roy, C. C., Deslandres, C., Sinnett, D., et al. (2008). Anderson or chylomicron retention disease: Molecular impact of five mutations in the SAR1B gene on the structure and the functionality of Sar1b protein. *Molecular Genetics and Metabolism*, *93*, 74–84.

Christiansen, H. E., Lang, M. R., Pace, J. M., & Parichy, D. M. (2009). Critical early roles for col27a1a and col27a1b in zebrafish notochord morphogenesis, vertebral mineralization and post-embryonic axial growth. *PloS One*, *4*.

Chu, J., Mir, A., Gao, N., Rosa, S., Monson, C., Sharma, V., et al. (2013). A zebrafish model of congenital disorders of glycosylation with phosphomannose isomerase deficiency reveals an early opportunity for corrective mannose supplementation. *Disease Models & Mechanisms*, *6*, 95–105.

Cinaroglu, A., Gao, C., Imrie, D., & Sadler, K. C. (2011). Activating transcription factor 6 plays protective and pathological roles in steatosis due to endoplasmic reticulum stress in zebrafish. *Hepatology*, *54*, 495–508.

Clement, A., Wiweger, M., von der Hardt, S., Rusch, M. A., Selleck, S. B., Chien, C. B., et al. (2008). Regulation of zebrafish skeletogenesis by ext2/dackel and papst1/pinscher. *PLoS Genetics*, *4*, e1000136.

Cline, A., Gao, N., Flanagan-Steet, H., Sharma, V., Rosa, S., Sonon, R., et al. (2012). A zebrafish model of PMM2-CDG reveals altered neurogenesis and a substrate-accumulation mechanism for N-linked glycosylation deficiency. *Molecular Biology of the Cell*, *23*, 4175–4187.

Copic, A., Latham, C. F., Horlbeck, M. A., D'Arcangelo, J. G., & Miller, E. A. (2012). ER cargo properties specify a requirement for COPII coat rigidity mediated by Sec13p. *Science*, *335*, 1359–1362.

Czarny-Ratajczak, M., Lohiniva, J., Rogala, P., Kozlowski, K., Perala, M., Carter, L., et al. (2001). A mutation in COL9A1 causes multiple epiphyseal dysplasia: Further evidence for locus heterogeneity. *American Journal of Human Genetics*, *69*, 969–980.

Dale, R. M., & Topczewski, J. (2011). Identification of an evolutionarily conserved regulatory element of the zebrafish col2a1a gene. *Developmental Biology*, *357*, 518–531.

DeLaurier, A., Eames, B. F., Blanco-Sanchez, B., Peng, G., He, X., Swartz, M. E., et al. (2010). Zebrafish sp7:EGFP: A transgenic for studying otic vesicle formation, skeletogenesis, and bone regeneration. *Genesis*, *48*, 505–511.

DeRossi, C., Bode, L., Eklund, E. A., Zhang, F., Davis, J. A., Westphal, V., et al. (2006). Ablation of mouse phosphomannose isomerase (Mpi) causes mannose 6-phosphate accumulation, toxicity, and embryonic lethality. *The Journal of Biological Chemistry*, *281*, 5916–5927.

DeRossi, C., Vacaru, A., Rafiq, R., Cinaroglu, A., Imrie, D., Nayar, S., et al. (2016). trappc11 is required for protein glycosylation in zebrafish and humans. *Molecular Biology of the Cell*, *27*, 1220–1234.

Doyon, Y., McCammon, J. M., Miller, J. C., Faraji, F., Ngo, C., Katibah, G. E., et al. (2008). Heritable targeted gene disruption in zebrafish using designed zinc finger nucleases. *Nature Biotechnology*, *26*, 702–708.

Driever, W., Solnica-Krezel, L., Schier, A. F., Neuhauss, S. C., Malicki, J., Stemple, D. L., et al. (1996). A genetic screen for mutations affecting embryogenesis in zebrafish. *Development*, *123*, 37–46.

Duran, I., Mari-Beffa, M., Santamaria, J. A., Becerra, J., & Santos-Ruiz, L. (2011). Actinotrichia collagens and their role in fin formation. *Developmental Biology*, *354*, 160–172.

Eames, B. F., Singer, A., Smith, G. A., Wood, Z. A., Yan, Y. L., He, X., et al. (2010). UDP xylose synthase 1 is required for morphogenesis and histogenesis of the craniofacial skeleton. *Developmental Biology*, *341*, 400–415.

Eames, B. F., Yan, Y. L., Swartz, M. E., Levic, D. S., Knapik, E. W., Postlethwait, J. H., et al. (2011). Mutations in fam20b and xylt1 reveal that cartilage matrix controls timing of endochondral ossification by inhibiting chondrocyte maturation. *PLoS Genetics*, 7. e1002246.

Ellis, K., Bagwell, J., & Bagnat, M. (2013). Notochord vacuoles are lysosome-related organelles that function in axis and spine morphogenesis. *The Journal of Cell Biology*, *200*, 667–679.

Erwin, W. M., & Hood, K. E. (2014). The cellular and molecular biology of the intervertebral disc: A clinician's primer. *Journal of the Canadian Chiropractic Association*, *58*, 246–257.

Faundes, V., Castillo-Taucher, S., Gonzalez-Hormazabal, P., Chandler, K., Crosby, A., & Chioza, B. (2014). Raine syndrome: An overview. *European Journal of Medical Genetics*, *57*, 536–542.

Fisher, S., Jagadeeswaran, P., & Halpern, M. E. (2003). Radiographic analysis of zebrafish skeletal defects. *Developmental Biology*, *264*, 64–76.

Fornzler, D., Her, H., Knapik, E. W., Clark, M., Lehrach, H., Postlethwait, J. H., et al. (1998). Gene mapping in zebrafish using single-strand conformation polymorphism analysis. *Genomics*, *51*, 216–222.

Foulquier, F., Amyere, M., Jaeken, J., Zeevaert, R., Schollen, E., Race, V., et al. (2012). TMEM165 deficiency causes a congenital disorder of glycosylation. *American Journal of Human Genetics*, *91*, 15–26.

Fox, R. M., Hanlon, C. D., & Andrew, D. J. (2010). The CrebA/Creb3-like transcription factors are major and direct regulators of secretory capacity. *The Journal of Cell Biology*, *191*, 479–492.

Freeze, H. H. (2006). Genetic defects in the human glycome. *Nature Reviews. Genetics*, 7, 537–551.

Fromme, J. C., Orci, L., & Schekman, R. (2008). Coordination of COPII vesicle trafficking by Sec23. *Trends in Cell Biology*, *18*, 330–336.

Fromme, J. C., Ravazzola, M., Hamamoto, S., Al-Balwi, M., Eyaid, W., Boyadjiev, S. A., et al. (2007). The genetic basis of a craniofacial disease provides insight into COPII coat assembly. *Developmental Cell*, *13*, 623–634.

Gansner, J. M., & Gitlin, J. D. (2008). Essential role for the alpha 1 chain of type VIII collagen in zebrafish notochord formation. *Developmental Dynamics*, *237*, 3715–3726.

Garbes, L., Kim, K., Riess, A., Hoyer-Kuhn, H., Beleggia, F., Bevot, A., et al. (2015). Mutations in SEC24D, encoding a component of the COPII machinery, cause a syndromic form of osteogenesis imperfecta. *American Journal of Human Genetics*, *96*, 432–439.

Gensure, R. C., Mäkitie, O., Barclay, C., Chan, C., DePalma, S. R., Bastepe, M., et al. (2005). A novel COL1A1 mutation in infantile cortical hyperostosis (Caffey disease) expands the spectrum of collagen-related disorders. *The Journal of Clinical Investigation*, *115*, 1250–1257.

Giampietro, P. F. (2012). Genetic aspects of congenital and idiopathic scoliosis. *Scientifica (Cairo)*, *2012*, 1–15.

Gistelinck, C., Witten, P. E., Huysseune, A., Symoens, S., Malfait, F., Larionova, D., et al. (2016). Loss of type I collagen telopeptide lysyl hydroxylation causes musculoskeletal abnormalities in a zebrafish model of Bruck syndrome. *Journal of Bone and Mineral Research*, *31*(11), 1930–1942.

Glorieux, F. H. (2005). Caffey disease: An unlikely collagenopathy. *The Journal of Clinical Investigation*, *115*, 1142–1144.

Golling, G., Amsterdam, A., Sun, Z., Antonelli, M., Maldonado, E., Chen, W., et al. (2002). Insertional mutagenesis in zebrafish rapidly identifies genes essential for early vertebrate development. *Nature Genetics*, *31*, 135–140.

Gonzaga-Jauregui, C., Gamble, C. N., Yuan, B., Penney, S., Jhangiani, S., Muzny, D. M., et al. (2015). Mutations in COL27A1 cause Steel syndrome and suggest a founder mutation effect in the Puerto Rican population. *European Journal of Human Genetics*, *23*, 342–346.

Gotting, C., Kuhn, J., Zahn, R., Brinkmann, T., & Kleesiek, K. (2000). Molecular cloning and expression of human UDP-d-Xylose:proteoglycan core protein beta-d-xylosyltransferase and its first isoform XT-II. *Journal of Molecular Cell Biology*, *304*, 517–528.

Gray, R. S., Wilm, T. P., Smith, J., Bagnat, M., Dale, R. M., Topczewski, J., et al. (2014). Loss of col8a1a function during zebrafish embryogenesis results in congenital vertebral malformations. *Developmental Biology*, *386*, 72–85.

Grimaud, E., Heymann, D., & Redini, F. (2002). Recent advances in TGF-beta effects on chondrocyte metabolism. Potential therapeutic roles of TGF-beta in cartilage disorders. *Cytokine & Growth Factor Reviews*, *13*, 241–257.

Grimes, D. T., Boswell, C. W., Morante, N. F., Henkelman, R. M., Burdine, R. D., & Ciruna, B. (2016). Zebrafish models of idiopathic scoliosis link cerebrospinal fluid flow defects to spine curvature. *Science*, *352*, 1341–1344.

Guryev, V., Koudijs, M. J., Berezikov, E., Johnson, S. L., Plasterk, R. H., van Eeden, F. J., et al. (2006). Genetic variation in the zebrafish. *Genome Research*, *16*, 491–497.

Haesemeyer, M., & Schier, A. F. (2015). The study of psychiatric disease genes and drugs in zebrafish. *Current Opinion in Neurobiology*, *30*, 122–130.

Haffter, P., Granato, M., Brand, M., Mullins, M. C., Hammerschmidt, M., Kane, D. A., et al. (1996). The identification of genes with unique and essential functions in the development of the zebrafish, Danio rerio. *Development*, *123*, 1–36.

Han, C., Belenkaya, T. Y., Khodoun, M., Tauchi, M., & Lin, X. (2004). Distinct and collaborative roles of Drosophila EXT family proteins in morphogen signalling and gradient formation. *Development*, *131*, 1563–1575.

Hansson, A. S., Heinegård, D., Piette, J. C., Burkhardt, H., & Holmdahl, R. (2001). The occurrence of autoantibodies to matrilin 1 reflects a tissue-specific response to cartilage of the respiratory tract in patients with relapsing polychondritis. *Arthritis & Rheumatism*, *44*(10), 2402–2412.

Hautala, T., Byers, M. G., Eddy, R. L., Shows, T. B., Kivirikko, K. I., & Myllyla, R. (1992). Cloning of human lysyl hydroxylase: Complete cDNA-derived amino acid sequence and assignment of the gene (PLOD) to chromosome 1p36.3—p36.2. *Genomics*, *13*, 62–69.

Hayes, M., Gao, X., Yu, L. X., Paria, N., Henkelman, R. M., Wise, C. A., et al. (2014). ptk7 mutant zebrafish models of congenital and idiopathic scoliosis implicate dysregulated Wnt signalling in disease. *Nature Communications*, *5*.

Hayes, M., Naito, M., Daulat, A., Angers, S., & Ciruna, B. (2013). Ptk7 promotes non-canonical Wnt/PCP-mediated morphogenesis and inhibits Wnt/beta-catenin-dependent cell fate decisions during vertebrate development. *Development*, *140*, 1807–1818.

Holmborn, K., Habicher, J., Kasza, Z., Eriksson, A. S., Filipek-Gorniok, B., Gopal, S., et al. (2012). On the roles and regulation of chondroitin sulfate and heparan sulfate in zebrafish pharyngeal cartilage morphogenesis. *The Journal of Biological Chemistry*, *287*, 33905–33916.

Howe, D. G., Bradford, Y. M., Conlin, T., Eagle, A. E., Fashena, D., Frazer, K., et al. (2013). ZFIN, the zebrafish model organism database: Increased support for mutants and transgenics. *Nucleic Acids Research*, *41*, D854–D860.

Huang, C., Wang, T. C., Lin, B. H., Wang, Y. W., Johnson, S. L., & Yu, J. (2009). Collagen IX is required for the integrity of collagen II fibrils and the regulation of vascular plexus formation in zebrafish caudal fins. *Developmental Biology*, *332*, 360–370.

Huang, P., Xiao, A., Zhou, M., Zhu, Z., Lin, S., & Zhang, B. (2011). Heritable gene targeting in zebrafish using customized TALENs. *Nature Biotechnology*, *29*, 699–700.

Hurtado, A., Podinin, H., Oudega, M., & Grimpe, B. (2008). Deoxyribozyme-mediated knockdown of xylosyltransferase-1 mRNA promotes axon growth in the adult rat spinal cord. *Brain*, *131*, 2596–2605.

Hwang, W. Y., Fu, Y., Reyon, D., Maeder, M. L., Tsai, S. Q., Sander, J. D., et al. (2013). Efficient genome editing in zebrafish using a CRISPR-Cas system. *Nature Biotechnology*, *31*, 227–229.

Iolascon, A., Russo, R., Esposito, M. R., Asci, R., Piscopo, C., Perrotta, S., et al. (2010). Molecular analysis of 42 patients with congenital dyserythropoietic anemia type II: New mutations in the SEC23B gene and a search for a genotype-phenotype relationship. *Haematologica*, *95*, 708–715.

Ishikawa, H. O., Xu, A., Ogura, E., Manning, G., & Irvine, K. D. (2012). The raine syndrome protein FAM20C is a Golgi kinase that phosphorylates bio-mineralization proteins. *PloS One*, *7*, e42988.

Jaeken, J. (2013). Congenital disorders of glycosylation. *Handbook of Clinical Neurology*, *113*, 1737–1743.

Jaeken, J., & Matthijs, G. (2001). Congenital disorders of glycosylation. *Annual Review of Genomics and Human Genetics*, *2*, 129–151.

Janssen, M. M., de Wilde, R. F., Kouwenhoven, J. W., & Castelein, R. M. (2011). Experimental animal models in scoliosis research: A review of the literature. *The Spine Journal*, *11*, 347–358.

Jao, L. E., Wente, S. R., & Chen, W. (2013). Efficient multiplex biallelic zebrafish genome editing using a CRISPR nuclease system. *Proceedings of the National Academy of Sciences of the United States of America*, *110*, 13904–13909.

Jennes, I., Pedrini, E., Zuntini, M., Mordenti, M., Balkassmi, S., Asteggiano, C. G., et al. (2009). Multiple osteochondromas: Mutation update and description of the multiple osteochondromas mutation database (MOdb). *Human Mutation*, *30*, 1620–1627.

Keller, R. (2002). Shaping the vertebrate body plan by polarized embryonic cell movements. *Science*, *298*, 1950–1954.

Kimmel, C. B., Miller, C. T., & Moens, C. B. (2001). Specification and morphogenesis of the zebrafish larval head skeleton. *Developmental Biology*, *233*, 239–257.

Klatt, A. R., Becker, A. K. A., Neacsu, C. D., Paulsson, M., & Wagener, R. (2011). The matrilins: Modulators of extracellular matrix assembly. *The International Journal of Biochemistry & Cell Biology, 43*(3), 320–330.
Klatt, A. R., Klinger, G., Paul-Klausch, B., Kuhn, G., Renno, J. H., Wagener, R., et al. (2009). Matrilin-3 activates the expression of osteoarthritis-associated genes in primary human chondrocytes. *FEBS Letters, 583*, 3611–3617.
Knapik, E. W. (2000). ENU mutagenesis in zebrafish—From genes to complex diseases. *Mammalian Genome, 11*, 511–519.
Knapik, E. W., Goodman, A., Atkinson, O. S., Roberts, C. T., Shiozawa, M., Sim, C. U., et al. (1996). A reference cross DNA panel for zebrafish (Danio rerio) anchored with simple sequence length polymorphisms. *Development, 123*, 451–460.
Knapik, E. W., Goodman, A., Ekker, M., Chevrette, M., Delgado, J., Neuhauss, S., et al. (1998). A microsatellite genetic linkage map for zebrafish (Danio rerio). *Nature Genetics, 18*, 338–343.
Ko, Y. P., Kobbe, B., Paulsson, M., & Wagener, R. (2005). Zebrafish (Danio rerio) matrilins: Shared and divergent characteristics with their mammalian counterparts. *The Biochemical Journal, 386*, 367–379.
Kokel, D., Bryan, J., Laggner, C., White, R., Cheung, C. Y., Mateus, R., et al. (2010). Rapid behavior-based identification of neuroactive small molecules in the zebrafish. *Nature Chemical Biology, 6*, 231–237.
Kou, I., Takahashi, Y., Johnson, T. A., Takahashi, A., Guo, L., Dai, J., et al. (2013). Genetic variants in GPR126 are associated with adolescent idiopathic scoliosis. *Nature Genetics, 45*, 676–679.
Krejci, P., Krakow, D., Mekikian, P. B., & Wilcox, W. R. (2007). Fibroblast growth factors 1, 2, 17, and 19 are the predominant FGF ligands expressed in human fetal growth plate cartilage. *Pediatric Research, 61*, 267–272.
Lang, M. R., Lapierre, L. A., Frotscher, M., Goldenring, J. R., & Knapik, E. W. (2006). Secretory COPII coat component Sec23a is essential for craniofacial chondrocyte maturation. *Nature Genetics, 38*, 1198–1203.
LeClair, E. E., Mui, S. R., Huang, A., Topczewska, J. M., & Topczewski, J. (2009). Craniofacial skeletal defects of adult zebrafish glypican 4 (knypek) mutants. *Developmental Dynamics, 238*, 2550–2563.
Lee, J. S., von der Hardt, S., Rusch, M. A., Stringer, S. E., Stickney, H. L., Talbot, W. S., et al. (2004). Axon sorting in the optic tract requires HSPG synthesis by ext2 (dackel) and extl3 (boxer). *Neuron, 44*, 947–960.
Leroy, J. G. (2006). Congenital disorders of N-glycosylation including diseases associated with O- as well as N-glycosylation defects. *Pediatric Research, 60*, 643–656.
Levic, D. S., Minkel, J. R., Wang, W. D., Rybski, W. M., Melville, D. B., & Knapik, E. W. (2015). Animal model of Sar1b deficiency presents lipid absorption deficits similar to Anderson disease. *Journal of Molecular Medicine (Berlin, Germany), 93*, 165–176.
Liang, W. C., Zhu, W., Mitsuhashi, S., Noguchi, S., Sacher, M., Ogawa, M., et al. (2015). Congenital muscular dystrophy with fatty liver and infantile-onset cataract caused by TRAPPC11 mutations: Broadening of the phenotype. *Skeletal Muscle, 5*, 29.
Lind, T., Tufaro, F., McCormick, C., Lindahl, U., & Lidholt, K. (1998). The putative tumor suppressors EXT1 and EXT2 are glycosyltransferases required for the biosynthesis of heparan sulfate. *The Journal of Biological Chemistry, 273*, 26265–26268.
Marquardt, T., & Denecke, J. (2003). Congenital disorders of glycosylation: Review of their molecular bases, clinical presentations and specific therapies. *European Journal of Pediatrics, 162*, 359–379.
Mates, L., Nicolae, C., Morgelin, M., Deak, F., Kiss, I., & Aszodi, A. (2004). Mice lacking the extracellular matrix adaptor protein matrilin-2 develop without obvious abnormalities. *Matrix Biology, 23*, 195–204.

Melville, D. B., & Knapik, E. W. (2011). Traffic jams in fish bones: ER-to-Golgi protein transport during zebrafish development. *Cell Adhesion & Migration*, *5*, 114–118.

Melville, D. B., Montero-Balaguer, M., Levic, D. S., Bradley, K., Smith, J. R., Hatzopoulos, A. K., et al. (2011). The feelgood mutation in zebrafish dysregulates COPII-dependent secretion of select extracellular matrix proteins in skeletal morphogenesis. *Disease Models & Mechanisms*, *4*, 763–776.

Miller, N. H. (1999). Cause and natural history of adolescent idiopathic scoliosis. *The Orthopedic Clinics of North America*, *30*, 343–352. vii.

Miller, E., Antonny, B., Hamamoto, S., & Schekman, R. (2002). Cargo selection into COPII vesicles is driven by the Sec24p subunit. *The EMBO Journal*, *21*, 6105–6113.

Miyashita, T. (2016). Fishing for jaws in early vertebrate evolution: A new hypothesis of mandibular confinement. *Biological Reviews of the Cambridge Philosophical Society*, *91*, 611–657.

Monsoro-Burq, A. H. (2005). Sclerotome development and morphogenesis: When experimental embryology meets genetics. *The International Journal of Developmental Biology*, *49*, 301–308.

Montanaro, L., Parisini, P., Greggi, T., Di Silvestre, M., Campoccia, D., Rizzi, S., et al. (2006). Evidence of a linkage between matrilin-1 gene (MATN1) and idiopathic scoliosis. *Scoliosis*, *1*, 21.

Moreira, T. G., Zhang, L., Shaulov, L., Harel, A., Kuss, S. K., Williams, J., et al. (2015). Sec13 regulates expression of specific immune factors involved in inflammation in vivo. *Scientific Reports*, *5*.

Moremen, K. W., Tiemeyer, M., & Nairn, A. V. (2012). Vertebrate protein glycosylation: Diversity, synthesis and function. *Nature Reviews. Molecular Cell Biology*, *13*, 448–462.

Mullins, M. C., Hammerschmidt, M., Haffter, P., & Nusslein-Volhard, C. (1994). Large-scale mutagenesis in the zebrafish: In search of genes controlling development in a vertebrate. *Current Biology*, *4*, 189–202.

Nabavi, N., Pustylnik, S., & Harrison, R. E. (2012). Rab GTPase mediated procollagen trafficking in ascorbic acid stimulated osteoblasts. *PloS One*, *7*, e46265.

Nasevicius, A., & Ekker, S. C. (2000). Effective targeted gene 'knockdown' in zebrafish. *Nature Genetics*, *26*, 216–220.

Neacsu, C. D., Ko, Y. P., Tagariello, A., Rokenes Karlsen, K., Neiss, W. F., Paulsson, M., et al. (2014). Matrilin-1 is essential for zebrafish development by facilitating collagen II secretion. *The Journal of Biological Chemistry*, *289*, 1505–1518.

Neuhauss, S. C., Solnica-Krezel, L., Schier, A. F., Zwartkruis, F., Stemple, D. L., Malicki, J., et al. (1996). Mutations affecting craniofacial development in zebrafish. *Development*, *123*, 357–367.

Nissen, R. M., Amsterdam, A., & Hopkins, N. (2006). A zebrafish screen for craniofacial mutants identifies wdr68 as a highly conserved gene required for endothelin-1 expression. *BMC Developmental Biology*, *6*, 28.

Niu, X., Gao, C., Jan Lo, L., Luo, Y., Meng, C., Hong, J., et al. (2012). Sec13 safeguards the integrity of the endoplasmic reticulum and organogenesis of the digestive system in zebrafish. *Developmental Biology*, *367*, 197–207.

Nuytinck, L., Freund, M., Lagae, L., Pierard, G. E., Hermanns-Le, T., & De Paepe, A. (2000). Classical Ehlers-Danlos syndrome caused by a mutation in type I collagen. *American Journal of Human Genetics*, *66*, 1398–1402.

Ohba, S., He, X., Hojo, H., & McMahon, A. P. (2015). Distinct transcriptional programs underlie Sox9 regulation of the mammalian chondrocyte. *Cell Reports*, *12*, 229–243.

Ohisa, S., Inohaya, K., Takano, Y., & Kudo, A. (2010). sec24d encoding a component of COPII is essential for vertebra formation, revealed by the analysis of the medaka mutant, vbi. *Developmental Biology*, *342*, 85–95.

Paccaud, J. P., Reith, W., Carpentier, J. L., Ravazzola, M., Amherdt, M., Schekman, R., et al. (1996). Cloning and functional characterization of mammalian homologues of the COPII component Sec23. *Molecular Biology of the Cell*, *7*, 1535–1546.

Patten, S. A., Margaritte-Jeannin, P., Bernard, J. C., Alix, E., Labalme, A., Besson, A., et al. (2015). Functional variants of POC5 identified in patients with idiopathic scoliosis. *The Journal of Clinical Investigation*, *125*, 1124–1128.

Peretti, N., Sassolas, A., Roy, C. C., Deslandres, C., Charcosset, M., Castagnetti, J., et al. (2010). Guidelines for the diagnosis and management of chylomicron retention disease based on a review of the literature and the experience of two centers. *Orphanet Journal of Rare Diseases*, *5*, 24.

Prydz, K., & Dalen, K. T. (2000). Synthesis and sorting of proteoglycans. *Journal of Cell Science*, *113*(Pt. 2), 193–205.

Raine, J., Winter, R. M., Davey, A., & Tucker, S. M. (1989). Unknown syndrome: Microcephaly, hypoplastic nose, exophthalmos, gum hyperplasia, cleft palate, low set ears, and osteosclerosis. *Journal of Medical Genetics*, *26*, 786–788.

Russo, R., Esposito, M. R., Asci, R., Gambale, A., Perrotta, S., Ramenghi, U., et al. (2010). Mutational spectrum in congenital dyserythropoietic anemia type II: Identification of 19 novel variants in SEC23B gene. *American Journal of Hematology*, *85*, 915–920.

Sachdev, S. W., Dietz, U. H., Oshima, Y., Lang, M. R., Knapik, E. W., Hiraki, Y., et al. (2001). Sequence analysis of zebrafish chondromodulin-1 and expression profile in the notochord and chondrogenic regions during cartilage morphogenesis. *Mechanisms of Development*, *105*, 157–162.

Sadler, K. C., Amsterdam, A., Soroka, C., Boyer, J., & Hopkins, N. (2005). A genetic screen in zebrafish identifies the mutants vps18, nf2 and foie gras as models of liver disease. *Development*, *132*, 3561–3572.

Saito, A., Hino, S., Murakami, T., Kanemoto, S., Kondo, S., Saitoh, M., et al. (2009). Regulation of endoplasmic reticulum stress response by a BBF2H7-mediated Sec23a pathway is essential for chondrogenesis. *Nature Cell Biology*, *11*, 1197–1204.

Sarmah, S., Barrallo-Gimeno, A., Melville, D. B., Topczewski, J., Solnica-Krezel, L., & Knapik, E. W. (2010). Sec24D-dependent transport of extracellular matrix proteins is required for zebrafish skeletal morphogenesis. *PloS One*, *5*, e10367.

Schekman, R., & Orci, L. (1996). Coat proteins and vesicle budding. *Science*, *271*, 1526–1533.

Schneider, V. A., & Granato, M. (2007). Genomic structure and embryonic expression of zebrafish lysyl hydroxylase 1 and lysyl hydroxylase 2. *Matrix Biology*, *26*, 12–19.

Schreml, J., Durmaz, B., Cogulu, O., Keupp, K., Beleggia, F., Pohl, E., et al. (2014). The missing "link": An autosomal recessive short stature syndrome caused by a hypofunctional XYLT1 mutation. *Human Genetics*, *133*, 29–39.

Schwarz, K., Iolascon, A., Verissimo, F., Trede, N. S., Horsley, W., Chen, W., et al. (2009). Mutations affecting the secretory COPII coat component SEC23B cause congenital dyserythropoietic anemia type II. *Nature Genetics*, *41*, 936–940.

Sharma, S., Londono, D., Eckalbar, W. L., Gao, X., Zhang, D., Mauldin, K., et al. (2015). A PAX1 enhancer locus is associated with susceptibility to idiopathic scoliosis in females. *Nature Communications*, *6*, 6452.

Shoulders, M. D., & Raines, R. T. (2009). Collagen structure and stability. *Annual Review of Biochemistry*, *78*, 929–958.

Shukunami, C., & Hiraki, Y. (2007). Chondromodulin-I and tenomodulin: The negative control of angiogenesis in connective tissue. *Current Pharmaceutical Design*, *13*, 2101–2112.

Sisson, B. E., Dale, R. M., Mui, S. R., Topczewska, J. M., & Topczewski, J. (2015). A role of glypican4 and wnt5b in chondrocyte stacking underlying craniofacial cartilage morphogenesis. *Mechanisms of Development*, *138*(Pt. 3), 279–290.

Solnica-Krezel, L., Schier, A. F., & Driever, W. (1994). Efficient recovery of Enu-induced mutations from the zebrafish germline. *Genetics*, *136*, 1401–1420.

Sols, A., Cadenas, E., & Alvarado, F. (1960). Enzymatic basis of mannose toxicity in honey bees. *Science, 131*, 297–298.
Steel, H. H., Piston, R. W., Clancy, M., & Betz, R. R. (1993). A syndrome of dislocated hips and radial heads, carpal coalition, and short stature in Puerto Rican children. *The Journal of Bone and Joint Surgery. American Volume, 75*, 259–264.
Stickens, D., Zak, B. M., Rougier, N., Esko, J. D., & Werb, Z. (2005). Mice deficient in Ext2 lack heparan sulfate and develop exostoses. *Development, 132*, 5055–5068.
Stoolmiller, A. C., Horwitz, A. L., & Dorfman, A. (1972). Biosynthesis of the chondroitin sulfate proteoglycan. Purification and properties of xylosyltransferase. *Journal of Biological Chemistry, 247*, 3525–3532.
Symoens, S., Malfait, F., D'hondt, S., Callewaert, B., Dheedene, A., Steyaert, W., et al. (2013). Deficiency for the ER-stress transducer OASIS causes severe recessive osteogenesis imperfecta in humans. *Orphanet Journal of Rare Diseases, 8*, 154.
Szul, T., & Sztul, E. (2011). COPII and COPI traffic at the ER-Golgi interface. *Physiology (Bethesda), 26*, 348–364.
Tanegashima, K., Zhao, H., Rebbert, M. L., & Dawid, I. B. (2009). Coordinated activation of the secretory pathway during notochord formation in the Xenopus embryo. *Development, 136*, 3543–3548.
Tang, B. L., Kausalya, J., Low, D. Y., Lock, M. L., & Hong, W. (1999). A family of mammalian proteins homologous to yeast Sec24p. *Biochemical and Biophysical Research Communications, 258*, 679–684.
Tao, J., Zhu, M., Wang, H., Afelik, S., Vasievich, M. P., Chen, X. W., et al. (2012). SEC23B is required for the maintenance of murine professional secretory tissues. *Proceedings of the National Academy of Sciences of the United States of America, 109*, E2001–E2009.
Thiel, C., Lubke, T., Matthijs, G., von Figura, K., & Korner, C. (2006). Targeted disruption of the mouse phosphomannomutase 2 gene causes early embryonic lethality. *Molecular and Cellular Biology, 26*, 5615–5620.
Topczewski, J., Sepich, D. S., Myers, D. C., Walker, C., Amores, A., Lele, Z., et al. (2001). The zebrafish glypican knypek controls cell polarity during gastrulation movements of convergent extension. *Developmental Cell, 1*, 251–264.
Townley, A. K., Feng, Y., Schmidt, K., Carter, D. A., Porter, R., Verkade, P., et al. (2008). Efficient coupling of Sec23-Sec24 to Sec13-Sec31 drives COPII-dependent collagen secretion and is essential for normal craniofacial development. *Journal of Cell Science, 121*, 3025–3034.
Treepongkaruna, S., Chongviriyaphan, N., Suthutvoravut, U., Charoenpipop, D., Choubtum, L., & Wattanasirichaigoon, D. (2009). Novel missense mutations of SAR1B gene in an infant with chylomicron retention disease. *Journal of Pediatric Gastroenterology and Nutrition, 48*, 370–373.
Unlu, G., Levic, D. S., Melville, D. B., & Knapik, E. W. (2014). Trafficking mechanisms of extracellular matrix macromolecules: Insights from vertebrate development and human diseases. *The International Journal of Biochemistry & Cell Biology, 47*, 57–67.
Uzawa, K., Grzesik, W. J., Nishiura, T., Kuznetsov, S. A., Robey, P. G., Brenner, D. A., et al. (1999). Differential expression of human lysyl hydroxylase genes, lysine hydroxylation, and cross-linking of type I collagen during osteoblastic differentiation in vitro. *Journal of Bone and Mineral Research, 14*, 1272–1280.
Vacaru, A. M., Unlu, G., Spitzner, M., Mione, M., Knapik, E. W., & Sadler, K. C. (2014). In vivo cell biology in zebrafish—Providing insights into vertebrate development and disease. *Journal of Cell Science, 127*, 485–495.
Valtavaara, M., Papponen, H., Pirttila, A. M., Hiltunen, K., Helander, H., & Myllyla, R. (1997). Cloning and characterization of a novel human lysyl hydroxylase isoform highly expressed in pancreas and muscle. *The Journal of Biological Chemistry, 272*, 6831–6834.

Valtavaara, M., Szpirer, C., Szpirer, J., & Myllyla, R. (1998). Primary structure, tissue distribution, and chromosomal localization of a novel isoform of lysyl hydroxylase (lysyl hydroxylase 3). *The Journal of Biological Chemistry*, *273*, 12881–12886.

Van Hul, W., Wuyts, W., Hendrickx, J., Speleman, F., Wauters, J., De Boulle, K., et al. (1998). Identification of a third EXT-like gene (EXTL3) belonging to the EXT gene family. *Genomics*, *47*, 230–237.

Vanoevelen, J., Janssens, A., Huitema, L. F., Hammond, C. L., Metz, J. R., Flik, G., et al. (2011). Trpv5/6 is vital for epithelial calcium uptake and bone formation. *FASEB Journal*, *25*, 3197–3207.

Walker, M. B., & Kimmel, C. B. (2007). A two-color acid-free cartilage and bone stain for zebrafish larvae. *Biotechnic & Histochemistry*, *82*, 23–28.

Walsh, E. C., & Stainier, D. Y. (2001). UDP-glucose dehydrogenase required for cardiac valve formation in zebrafish. *Science*, *293*(5535), 1670–1673.

Wendeler, M. W., Paccaud, J. P., & Hauri, H. P. (2007). Role of Sec24 isoforms in selective export of membrane proteins from the endoplasmic reticulum. *EMBO Reports*, *8*, 258–264.

Witten, P. E., & Huysseune, A. (2009). A comparative view on mechanisms and functions of skeletal remodelling in teleost fish, with special emphasis on osteoclasts and their function. *Biological Reviews of the Cambridge Philosophical Society*, *84*, 315–346.

Wiweger, M. I., Avramut, C. M., de Andrea, C. E., Prins, F. A., Koster, A. J., Ravelli, R. B., et al. (2011). Cartilage ultrastructure in proteoglycan-deficient zebrafish mutants brings to light new candidate genes for human skeletal disorders. *The Journal of Pathology*, *223*, 531–542.

Yamauchi, M., & Sricholpech, M. (2012). Lysine post-translational modifications of collagen. *Essays in Biochemistry*, *52*, 113–133.

Yang, X. Y., Zhou, X. Y., Wang, Q. Q., Li, H., Chen, Y., Lei, Y. P., et al. (2013). Mutations in the COPII vesicle component gene SEC24B are associated with human neural tube defects. *Human Mutation*.

Yeowell, H. N., & Walker, L. C. (2000). Mutations in the lysyl hydroxylase 1 gene that result in enzyme deficiency and the clinical phenotype of Ehlers-Danlos syndrome type VI. *Molecular Genetics and Metabolism*, *71*, 212–224.

Yoshihisa, T., Barlowe, C., & Schekman, R. (1993). Requirement for a GTPase-activating protein in vesicle budding from the endoplasmic reticulum. *Science*, *259*, 1466–1468.

Zak, B. M., Schuksz, M., Koyama, E., Mundy, C., Wells, D. E., Yamaguchi, Y., et al. (2011). Compound heterozygous loss of Ext1 and Ext2 is sufficient for formation of multiple exostoses in mouse ribs and long bones. *Bone*, *48*, 979–987.

Zhu, M., Tao, J., Vasievich, M. P., Wei, W., Zhu, G., Khoriaty, R. N., et al. (2015). Neural tube opening and abnormal extraembryonic membrane development in SEC23A deficient mice. *Scientific Reports*, *5*, 15471.

CHAPTER FOUR

Netting Novel Regulators of Hematopoiesis and Hematologic Malignancies in Zebrafish

Wanda Kwan*, Trista E. North*,†,1

*Beth Israel Deaconess Medical Center, Harvard Medical School, Boston, MA, United States
†Harvard Stem Cell Institute, Cambridge, MA, United States
[1]Corresponding author: e-mail address: tnorth@bidmc.harvard.edu

Contents

Abstract

Zebrafish are one of the preeminent model systems for the study of blood development (hematopoiesis), hematopoietic stem and progenitor cell (HSPC) biology, and hematopathology. The zebrafish hematopoietic system shares strong similarities in functional populations, genetic regulators, and niche interactions with its mammalian counterparts. These evolutionarily conserved characteristics, together with emerging

Current Topics in Developmental Biology, Volume 124
ISSN 0070-2153
http://dx.doi.org/10.1016/bs.ctdb.2016.11.005

technologies in live imaging, compound screening, and genetic manipulation, have been employed to successfully identify and interrogate novel regulatory mechanisms and molecular pathways that guide hematopoiesis. Significantly, perturbations in many of the key developmental signals controlling hematopoiesis are associated with hematological disorders and disease, including anemia, bone marrow failure syndromes, and leukemia. Thus, understanding the regulatory pathways controlling HSPC production and function has important clinical implications. In this review, we describe how the blood system forms and is maintained in zebrafish, with particular focus on new insights into vertebrate hematological regulation gained using this model. The interplay of factors controlling development and disease in the hematopoietic system combined with the unique attributes of the zebrafish make this a powerful platform to discover novel targets for the treatment of hematological disease.

1. INTRODUCTION

Zebrafish (*Danio rerio*) were first used in hematology research in the early 1960s (Colle-Vandevelde, 1963); however, over the last 20 years, they have become instrumental in driving progress in the field. External fertilization and transparent embryonic stages allow real-time spatiotemporal tracking of the birth, movements, and cellular interactions of hematopoietic stem and progenitor cells (HSPCs) and their differentiated progeny during development. Further, high fecundity and rapid growth enable high-throughput compound screening and genetic analysis. A range of powerful genetic approaches, including rapid gene knockdown and epistasis studies, allow for both discovery and thorough evaluation of conserved pathways and/or signaling mechanisms relevant to mammalian biology (Carroll & North, 2014). Large-scale forward genetic screens in zebrafish have yielded many models of human hematologic diseases; in particular, advances in the study of anemia and leukemia have resulted from zebrafish research and will be reviewed in detail here. Additionally, the large repository of hematology-related fluorescent reporter lines (Table 1) has strengthened our ability to visualize molecular signals and cellular interactions important for vertebrate hematopoiesis. Together, these advantages make zebrafish an important tool for hematology research which has already had significant impact on the field.

Key genes and/or signaling pathways regulating blood development, as well as the structure and function of various committed blood cell types in mammals, are highly conserved in zebrafish (de Jong & Zon, 2005). One prominent focus of zebrafish hematology research has been geared toward modeling hematological disorders by disrupting critical genetic factors via

Table 1 Markers for Cell Types in the Zebrafish Hematopoietic System

Cell Type	Transgenic Marker	Functional Assay
Erythrocytes	*gata1:gfp* *gata1:dsred*	*o*-Dianisidine staining
Endothelial cells	*flk1:dsred* *flk1:gfp*	—
Prehematopoietic vasculature; lymphatics	*fli1a:egfp* *fli1a:dsred*	—
Prehematopoietic vasculature; erythrocytes	*lmo2:gfp* *lmo2:mcherry* *lmo2:dsred*	—
Hemogenic endothelium	*scl:gfp*	—
Erythromyeloid progenitors	*runx1p1:gfp*	—
Hematopoietic stem cells	*itga2b:gfp*^{lo} *flk1:dsred; cmyb:gfp* *runx1p2:gfp* *runx1+23:gfp* *cmyb:gfp*	In vitro stromal cell culture; in vitro methylcellulose culture; HSCT
Thrombocytes	*itga2b:gfp*^{hi}	—
Macrophages	*mpeg1:egfp* *mpeg1:mcherry* *mpeg1:gal4*	(see chapter "Modeling Infection Diseases in the Context of a Developing Immune System" by Masud et al.)
Neutrophils	*lyz:dsRed* *lyz:gfp* *mpx:egfp*	(see chapter "Modeling Infection Diseases in the Context of a Developing Immune System" by Masud et al.)
B cells; dendritic cells	*mhcII:gfp*	—
Lymphocytes (immature B and T-cells)	*rag2:gfp* *rag2:dsred*	—
Mature T-cells	*lck:egfp*	—
Pan-leukocyte	*ptprc:dsred*	
Kidney	*cdh17:gfp* *cdh17:mcherry* *gata3:amcyan*	—
Thymus	*foxn1:mcherry* *ikzf1:egfp*	—

(1) large-scale ENU-based modeling of the disease and (2) gene targeting. Unbiased screens have successfully identified a number of anemic mutants that exhibit cellular dysfunction similar to that of human disorders. These provide in vivo functional models to interrogate the significance of gene disruptions impacting primitive or definitive erythropoiesis (Ransom et al., 1996). Importantly, in select instances, zebrafish mutants revealed the causative foundation of human red blood cell disease (Carroll & North, 2014). Further, compound screens have identified novel regulatory pathways and/or pharmacological modifiers with potential application for increasing the success of hematopoietic stem cell (HSC) transplantation and/or targeting leukemia progression (North et al., 2007). Finally, engineering zebrafish to express genes that bear mutations which are causative in human disease has facilitated research on leukemia and bone marrow failure syndromes. In this review, we will focus on the advances in our understanding of key developmental stages and regulatory pathways controlling HSPC production and homeostasis in the zebrafish model, as well as their utility for deciphering disease regulatory networks or novel therapeutics to alleviate human hematological disorders.

2. DEVELOPMENT OF THE HEMATOPOIETIC SYSTEM IN ZEBRAFISH

HSCs give rise to each of the blood lineages found throughout the lifetime of the organism (see also Fig. 1 of chapter "Modeling Infectious Diseases in the Context of a Developing Immune System" by Masud et al.). The genetic programs regulating HSC development are highly conserved between vertebrate species. There are two waves of blood cell formation that occur throughout vertebrate development: (1) the primitive wave gives rise primarily to erythrocytes and committed myeloid cells and (2) the definitive wave is responsible for multipotent progenitor emergence. The transient primitive wave is presumed to ensure the embryo has adequate oxygen supply for development. In zebrafish, the primitive wave occurs in the intermediate cell mass (ICM), which is the functional equivalent of the early blood islands in the mammalian yolk sac (murine embryonic days E7.5–9.5) (Carroll & North, 2014). Significantly, mutations affecting primitive hematopoiesis are embryonic lethal in mammals, but are often viable in zebrafish due to their aqueous development.

The definitive wave of hematopoiesis initiates with the spatial–temporal emergence of definitive erythromyeloid precursors (EMPs) that arise independently of HSCs. EMPs originate from the posterior blood island (PBI) as early as 24 h postfertilization (hpf) in zebrafish, or in the yolk sac at E8.5 in mice. These cells are believed to be nonself-renewing progenitors with limited proliferative and multilineage differentiation potential (Bertrand, Kim, Teng, & Traver, 2008); however, recent reports suggest EMPs may contribute to tissue-specific immune populations, such as the microglia (Kierdorf et al., 2013). Subsequent to EMP emergence, long-term repopulating HSCs that possess both life-long self-renewal and multilineage differentiation potential are produced.

In zebrafish, definitive HSCs arise de novo from the specialized endothelium in the ventral wall of the dorsal aorta (VDA) between 28 and 36 hpf (Bertrand et al., 2010). These HSCs are functionally analogous to those arising from "hemogenic" endothelium in the mammalian aorta-gonad mesonephros (AGM) region. Once formed, these newly derived HSPCs migrate to the vascular plexus within the posterior region of the zebrafish tail, termed the caudal hematopoietic tissue (CHT). This site is presumed to act similarly to the mammalian fetal liver, supporting both HSC expansion and lineage differentiation. HSCs and committed lymphoid progenitors seed the adult sites of hematopoiesis, the thymus and kidney marrow (KM), by 3 and 4 days postfertilization (dpf), respectively (Bertrand et al., 2008). As there is high conservation of HSPC biology from hemogenic endothelium specification to secondary sites of expansion and maintenance (Carroll & North, 2014), zebrafish are an excellent model for discovering and characterizing signaling networks regulating hematopoiesis, including alterations leading to hematological disease.

2.1 Primitive Hematopoiesis

Understanding the key signals controlling primitive and/or definitive hematopoiesis can elucidate pathways contributing to human hematopoietic disorders or novel therapeutic targets. During the onset of primitive hematopoiesis, cells in both the anterior and posterior lateral plate mesoderm (ALM and PLM), which will form the ICM, begin to express hematovascular markers, including *gata1*, *lmo2*, *fli1*, and *scl* (*stem cell leukemia*). Cells coexpressing markers of each lineage are historically termed hemangioblasts, the theoretical common progenitors for both hematopoietic cells and endothelial cells. Importantly, gene expression and function studies across vertebrates have indicated these factors act as so-called master regulators of hematopoiesis (Carroll & North, 2014).

Primitive hematopoiesis is controlled by the coordinated regulatory actions of *gata1* and *pu.1*, promoting erythroid and myeloid differentiation, respectively. Gata1+ erythroid progenitors first form bilaterally in the PLM and then migrate toward the midline to form the ICM, the main site of primitive erythropoiesis. These cells differentiate into proerythroblasts, which enter circulation around 24 hpf with the initiation of heartbeat and undergo terminal maturation. As erythroblasts mature, they express erythroid-specific genes for hemoglobin synthesis, including α and β embryonic globin chains, *alas2*, *fch*, and *urod* (de Jong & Zon, 2005), necessary for heme biosynthesis and oxygen transport. Mutations in these genes result in anemia in zebrafish and humans.

Pu.1 is a central regulator of myeloid fate (de Jong & Zon, 2005). Expression of early myeloid genes, *pu.1* and *l-plastin,* occurs in both the ALM and PLM (Chen & Zon, 2009), with committed myeloid cell types including macrophages, granulocytes, and neutrophils observed in the PBI by 22–24 hpf. In zebrafish, *l-plastin* is expressed in the anterior yolk region at 18 hpf, and within the posterior ICM by 28 hpf, which suggests zebrafish macrophages originate from two different sites, with the latter population potentially being EMP-derived (Chen & Zon, 2009; de Jong & Zon, 2005). During the primitive wave, production of erythroid and myeloid cells is precisely regulated to maintain homeostatic balance sufficient to support continued embryonic growth. This is achieved by mutual antagonistic regulation by the *gata1* and *pu.1* transcription factors themselves in the progenitor pool (Galloway, Wingert, Thisse, Thisse, & Zon, 2005). Disruption of this equilibrium during normal development results in lineage bias that can be potentially fatal to the developing embryo.

2.2 Definitive Hematopoiesis

The definitive wave of hematopoiesis produces multipotent progenitors, including HSCs that give rise to all hematopoietic lineages for the lifetime of the organism. In mammals, HSCs are produced de novo from hemogenic endothelium in VDA of the AGM region during early mid-gestation (Medvinsky & Dzierzak, 1996). This conserved process of HSPC specification, budding, and egress is now called endothelial-to-hematopoietic transition (EHT) and was first visualized in vivo in real time in the zebrafish model (Bertrand et al., 2010; Herbomel, Thisse, & Thisse, 2001). Zebrafish definitive hematopoiesis begins between 24 and 28 hpf with specification of the first hemogenic endothelial cells, and hematopoietic maturation marked by the sequential expression of the highly conserved regulators of HSC production and function: *runx1*, *myb* (formerly *cmyb*), and *integrin 2b* (*itga2b*, formerly *cd41*) (Carroll & North, 2014).

By 44 hpf, Myb+ cells appear in the CHT (Bertrand et al., 2008), which is thought to be the functional equivalent of the mammalian fetal liver, supporting HSC expansion. Recently, high-resolution imaging has revealed that upon arrival in the CHT, zebrafish HSCs extravasate and are then surrounded by endothelial cells, which are presumed to further direct cell fate (Tamplin et al., 2015). Between 48 and 96 hpf, HSPCs begin to seed the pronephros, where they continue to be maintained within the KM for the lifetime of the adult zebrafish to sustain multilineage hematopoiesis (Bertrand et al., 2008). The zebrafish thymus is colonized by T lymphocyte progenitors by 72 hpf, with mature T-cells emerging by 7 dpf. B lymphoid progenitors are formed in the KM, and do not differentiate into mature B-cells until 28 dpf (Fig. 1; see also Fig. 1 in chapter "Modeling Infectious Diseases in the Context of a Developing Immune System" by Masud et al.).

2.2.1 Essential HSC Regulators: Runx1

RUNX1 [Runt-related transcription factor 1; also known as acute myeloid leukemia 1 (AML)] is an essential transcription factor for definitive hematopoiesis (Kalev-Zylinska et al., 2002). Translocation of *RUNX1* occurs in a significant percentage of acute myelogenous and lymphoblastic leukemias (Goessling & North, 2014). In the mouse, *Runx1* is expressed in hemogenic endothelium and hematopoietic clusters in the AGM at E10.5 (North et al., 1999) and is required for HSC formation (North et al., 2002), where it controls EHT (Carroll & North, 2014). *runx1* is similarly present in hemogenic endothelium and presumptive HSCs in the VDA of zebrafish (Burns et al., 2002). Similar to mice, *runx1* knockdown in zebrafish does not significantly alter primitive erythropoiesis, but blocks HSC formation and multilineage differentiation (Burns et al., 2002; Kalev-Zylinska et al., 2002). Characterization of a zebrafish *runx1* truncation mutant (*runx1*^{w84x}) revealed some homozygotes overcome this initial bloodless phase, with Itga2b:GFP+ cells emerging and seeding the KM at a later time point (Sood et al., 2010).

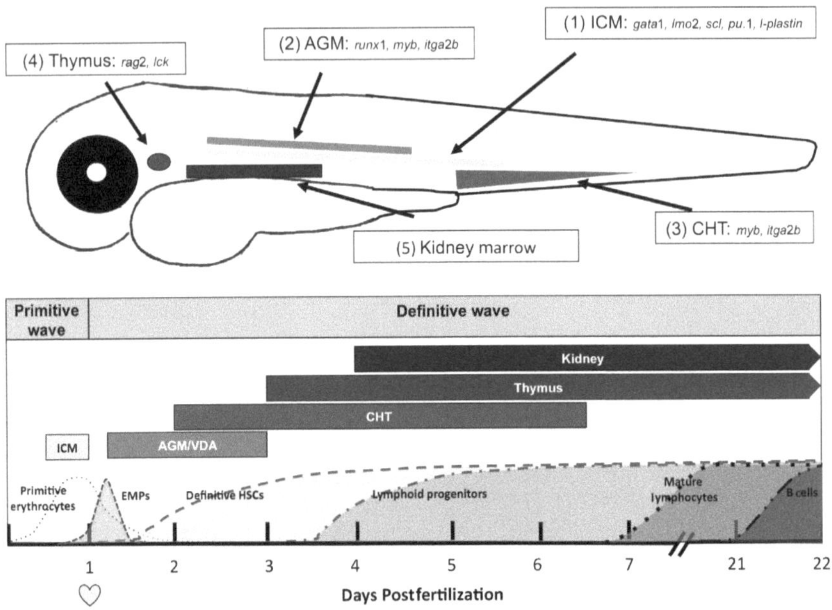

Fig. 1 The developmental stages of hematopoiesis in the zebrafish embryo. Schematic representation of the sites of developmental hematopoiesis (*top*) and the relative windows of production (*bottom*) in zebrafish. The primitive wave, that takes place in the intermediate cell mass (ICM), primarily gives rise to primitive erythrocytes that enter circulation at 24 hpf, coinciding with the initiation of heartbeat. The definitive wave produces HSCs beginning at 28–30 hpf: cells bud from the ventral wall of the dorsal aorta (VDA), in a region known as the aorta-gonad-mesonephros (AGM). HSCs next migrate to the caudal hematopoietic tissue (CHT) to expand and differentiate. By 3 dpf, lymphoid progenitors are found in the thymus; HSCs seed the kidney marrow (KM) at 4 dpf. Mature T-cells emerge by 7 dpf, whereas mature B-cells begin to differentiate by 28 dpf.

Zebrafish with a mutation in the heterodimeric partner of Runx1, *cbfb*, had defective hematopoiesis in which HSCs are specified but unable to emerge from the VDA (Bresciani et al., 2014). Together, these data suggest the existence of an alternative Runx1-independent or "salvage" method for HSC production not seen in mammals.

2.2.2 Essential HSC Regulators: Myb

The transcription factor *myb* is coexpressed with *runx1* in newly committed HSCs in the VDA at 36 hpf. *myb* expression in the VDA is dependent on Runx1 function, reflecting a transcriptional hierarchy. In the mouse, absence of *Myb* leads to embryonic death due to lack of hepatic erythropoiesis

(Mucenski et al., 1991). Myb is also essential for myeloid differentiation (Sakamoto et al., 2006). Use of Tg(*cmyb:GFP)* reporter fish in combination with vascular lines driven by *lmo2* or *flk1* (*kdrl*) has allowed for precise quantification of HSPC specified from hemogenic endothelium (Bertrand et al., 2010; North et al., 2007). Moreover, combinatorial labeling by Flk1 and Myb indicates cells with multipotent functional potential in vitro (Stachura et al., 2011). *myb* mutant zebrafish, *cmyb*^{t25127}, show high evolutionary conservation in definitive HSC production: mutant embryos have no discernible defects in the primitive wave of hematopoiesis, but exhibit defects in all hematopoietic lineages by 20 dpf and die prematurely (Soza-Ried, Hess, Netuschil, Schorpp, & Boehm, 2010).

2.2.3 Essential HSC Regulators: Itg2b (CD41)

As HSCs must possess the ability to migrate and home to alternative niches for development and maintenance, they acquire expression of the cell surface marker *integrin alpha 2b* (*itga2b,* also known as CD41) as they mature (Bertrand et al., 2008). CD41, well known for its function in platelets, was first shown to be required for embryonic HSC function in the mouse (Mikkola, Fujiwara, Schlaeger, Traver, & Orkin, 2003). CD41+ cells isolated from the zebrafish AGM by FACS are capable of sustained engraftment into irradiated adult recipients (Harris et al., 2013), consistent with their functional capabilities in vitro (Stachura et al., 2009). Mature HSCs also express the pan-leukocyte marker Ptprc (previously called CD45), which denotes cells of the hematopoietic lineage, with the exception of mature erythrocytes in all vertebrates.

2.3 Signaling Mechanisms Involved in Definitive Hematopoiesis

Given the diversity of functional effector cell types in the hematopoietic lineage, many signaling pathways are required to actively regulate HSC specification, proliferation, migration, maintenance, and differentiation. Several key signaling pathways, such as Wnt, Hedgehog, Vegf, Notch, BMP, and FGF, have been identified to be critical at precise stages of hematopoietic development (Zhang, Patient, & Liu, 2013). For example, FGF and BMP signaling are required for mesoderm induction and patterning (Pouget et al., 2014). BMP signaling induces mesodermal differentiation into erythrocytes, while FGF activity antagonizes this process, promoting endothelial cell fate, including that of hemogenic endothelium (Pouget et al., 2014).

Notch signaling also plays a critical role in hematovascular regulation. Arterial vs venous fate determination and specification of hemogenic endothelium in the VDA are mediated by the Hedgehog-Vegfa-Notch pathway. Vascular endothelial growth factor A (Vegfa) binds to its receptor, Kdrl (formerly Flk1 or VEGFR2), to induce endothelial cell differentiation and vascular remodeling via PLCγ-ERK activity (Hong, Peterson, Hong, & Peterson, 2006). The Hedgehog pathway acts upstream of Vegf in the somites to regulate vascular commitment, and in turn, Vegf controls local Notch signaling and commitment to hemogenic fate. Therefore, disruption in Hedgehog, Vegf, or Notch signaling causes defects in HSPC formation.

Wnt signaling modulates vascular specification and remodeling in mouse, in part, by β-catenin-mediated control of *Dll4* and Notch activation (Corada et al., 2010). In zebrafish, the noncanonical Wnt ligand, Wnt16, regulates establishment of definitive HSCs via somitic expression of the Notch ligands *DeltaC* (*dlc*) and *DeltaD* (*dld*). Interestingly, their combined action is required for Notch-dependent HSC specification, but not for arterial development (Clements et al., 2011). The signal transduction pathway between Wnt16 and Dlc is mediated by FGF signaling (Lee et al., 2014). Other pathways, including retinoic acid receptor and prostaglandin E2 (PGE2) signaling, can also modify Wnt activity in the AGM to control HSC emergence and proliferative expansion (Goessling et al., 2009). The role of the latter interaction has been confirmed in mammals (Goessling et al., 2009), highlighting the utility of zebrafish to identify and interrogate the interplay of conserved critical HSPC coregulators in vivo. Importantly, as many of these signaling cascades are disrupted in human hematologic disease, a comprehensive understanding of their functional requirements will aid development and application of novel treatments.

3. INTERROGATING HEMATOPOIESIS IN THE ZEBRAFISH MODEL

As reviewed in the preface of this volume (p. xi), the breadth of genetic and genomic tools available in zebrafish is a major advantage of the model (Carroll et al., 2014). Chemical screening has likewise proven to be a well-utilized platform for gene discovery and identification of potential therapeutics. Importantly, the first application of both of these zebrafish assays to the biology of human disease occurred in the hematology field (North et al., 2007; Yeh et al., 2009).

3.1 Genetic Tools for Studying Zebrafish Hematopoiesis

Forward genetic screens have been extremely productive in identifying mutants with defects in specific blood cell types, and have fueled research in hematopoiesis for nearly 20 years (Carroll & North, 2014). In particular, numerous erythrocyte mutants have revolutionized our understanding of genes that are involved in anemia, including the identification of causal human disease mutations. Morpholino oligonucleotide-mediated transient knockdown has also been a valuable screening tool to determine the effects of genes of interest (Carroll & North, 2014); indeed, recent studies have utilized MO knockdown to avoid compensation mechanisms elicited by null mutation (Rossi et al., 2015). Application of gene editing technologies, including TALENS and CRISPR/CAS, along with a variety of inducible systems (i.e., heat-shock, Cre-Lox, Gal4:UAS, nitroreductase), has targeted key regulatory factors, providing a platform for further interrogation in vivo. Together, the ease and diverse means of genetic manipulation have highlighted the utility of the zebrafish model in hematology.

3.2 Chemical Screening for Hematopoietic Regulators

In vivo screening of drug libraries using zebrafish embryos to identify novel pathways involved in vertebrate development is now widely used (Esain, Cortes, & North, 2016; Peterson, Link, Dowling, & Schreiber, 2000). These assays have successfully aimed to isolate novel compound therapies and/or mechanistic insights into pathways regulating hematopoiesis. As zebrafish embryos are semi-permeable to water-soluble chemicals, high-throughput chemical screening can be performed easily in vivo (North et al., 2007). Further, use of precisely timed exposure and/or compounds with known function aids elucidation of novel or interacting pathways that impact hematopoiesis (Esain et al., 2016; North et al., 2007). Hematology screens are not limited to embryonic stages; for example, a recent study treated whole kidney marrow (WKM) to identify compound modifiers of in vivo engraftment following adult competitive transplantation (Li, Lahvic, et al., 2015). Importantly, as discussed above, zebrafish exhibit strong evolutionary conservation with mammals with regard to hematopoietic regulation, giving high potential clinical relevance to the compound modifiers identified.

3.2.1 PGE2 and Related Eicosanoids

The identification of prostaglandin E2 (PGE2) as a hematopoietic modulator through a high-throughput chemical screen in zebrafish was a landmark discovery, as this was the first molecule identified in zebrafish to be successfully translated into clinical trials (North et al., 2007). This study not only opened many other avenues of research using zebrafish to identify therapeutically relevant drugs, but established a paradigm for rapidly applying drug discovery in zebrafish to a clinically important problem. PGE2 is an arachidonic acid derivative, involved in the inflammatory response. We discovered that PGE2 increased HSC number in zebrafish embryos (North et al., 2007). Genetic interaction studies revealed that PGE2 promotes HSC number via cAMP-PKA-mediated regulation of β-catenin stability causing Wnt pathway-associated transcriptional activation (Goessling et al., 2009). Ex vivo exposure to a stable derivative of PGE2 increased the frequency of long-term repopulating HSCs in the bone marrow of irradiated murine recipients (North et al., 2007). Subsequent preclinical in vitro toxicity and xenotransplantation studies (Goessling et al., 2011) led to ongoing clinical trials designed to examine the efficacy of PGE2 for improving the outcome of human HSC transplantation (Cutler et al., 2013; Goessling et al., 2011).

A second drug screen was undertaken in adults to identify drugs that increased HSC production and facilitated bone marrow engraftment, resulting in identification of another eicosanoid, epoxyeicosatrienoic acid (EET) (Li, Lahvic, et al., 2015). Despite the common origin of PGE2 and EETs, the underlying molecular signaling mechanisms and activities for HSC modulation are different, with EETs activating PI3Kγ to promote HSC numbers (Li, Lahvic, et al., 2015). Further, our recent work showed that endocannabinoids, another eicosanoid family member, act upstream of PGE2 during HSC specification (Esain et al., 2015). Pharmacological stimulation of cannabinoid receptor 2 (CNR2), but not CNR1, increased PGE2 synthesis (Esain et al., 2015). Endocannabinoids also have PGE2-independent activity at later phases of hematopoietic development: during colonization of secondary niches, CNR2 stimulation increased migration of HSCs via stimulation of P-selectin activity (Esain et al., 2015). The results of these chemical screens revealed an unexpected but functionally significant role for fatty acids in HSC regulation in both embryos and adults that is conserved across species and has translational value.

3.2.2 Blood Flow and Vascular Function

Chemical screening for embryonic HSC regulators uncovered several classes of compounds with the common ability to regulate blood flow by affecting heart rate, contractility, or vascular diameter of the developing embryo (North et al., 2009). The importance of blood flow for HSC formation was supported by observation of diminished HSC formation in zebrafish and mice lacking heartbeat and blood flow, namely, *silent heart* (*sih*) mutant (mutation in *troponin T2a*) and the *Ncx1*$^{-/-}$ mutants (a Na+/Ca2+ exchanger), respectively (Adamo et al., 2009; North et al., 2009). Compound modifiers of blood flow converged on an ability to impact nitric oxide (NO) production; subsequent studies in zebrafish and mouse embryos showed that direct chemical modulation of NO signaling impacted HSC formation (Adamo et al., 2009; North et al., 2009), including the ability to rescue of flow-associated HSC defects. Further studies have established that endothelial shear stress and/or blood flow leads to extracellular ATP production and stimulation of adenosine 2b receptor (Jing et al., 2015), secretion of BMPs (Kim et al., 2015), and release of PGE2 (Diaz et al., 2015), all mediated by the cAMP-PKA-CREB signaling axis to promote HSC emergence. Together, these results highlight the importance of vascular function in HSC production.

3.2.3 Nuclear Hormone Receptor Signaling

As vasculature architecture and blood flow are essential to HSC production, disruption of vascular specification, including hemogenic endothelial identity, has significant effects on HSCs. Our work indicates that modulation of nuclear hormone receptor signaling has substantial impact on the embryonic hemogenic niche and subsequent HSC development (Carroll et al., 2014; Cortes et al., 2015). Exposure to exogenous 17β-estradiol significantly disrupted *flk1* vessel maturation, *ephrinB2* arterial identity, and specification of *scl*+ hemogenic endothelium, through alterations in Vegf and downstream Notch activity (Carroll et al., 2014). Similarly, elevated levels of vitamin D3 precursor impact hematovascular niche specification as a result of Hedgehog signaling inhibition, leading to reduced HSCs (Cortes et al., 2015). Interestingly, later in development, active Vitamin D3 can cell autonomously stimulate proliferative expansion of HSC through VDR-mediated activation of CXCL8 (Cortes et al., 2016), implying a continuing role for nuclear hormone receptor signaling in HSPC biology.

The positive effect of vitamin D on HSC number is conserved in adults as well as in cell culture of isolated human HSCs (Cortes et al., 2016). Retinoid acid (RA) signaling has also been shown to be essential for HSC development in zebrafish and mice (Chanda, Ditadi, Iscove, & Keller, 2013). Further, RA stimulates increased generation of human hematopoietic progenitor cells from pluripotent stem cells in vitro (Ronn et al., 2015). Altogether, these studies highlight the key impact of nuclear hormone receptor signaling pathways, often indicative of environmental regulation, on HSC formation.

3.3 Interrogating Immunometabolic Impact on HSC Production

During embryogenesis, hematopoiesis must be constantly modified to meet the demands of the growing organism. Environmental sensing and feedback responses must integrate with the master regulatory networks to control HSC production and maintenance, as well as output of differentiated progeny. Recent work in zebrafish and mice has revealed an unexpected but essential role for both metabolic and inflammatory regulation in developmental HSC formation.

3.3.1 Metabolic Control of Hematopoiesis

Metabolic demand is an important regulator of the adult hematopoietic niche. A similar role for energy consumption has been more recently described during embryonic hematopoiesis. Elevated glucose metabolism directly stimulated HSC induction from hemogenic endothelium via stabilization and transcriptional regulation by hypoxia-inducible factor 1α (Hif1α) (Harris et al., 2013); this effect was elicited as a result of a classical "hypoxic response" to local elevation in reactive oxygen species.

Zebrafish *vhl* mutants possess a systemic hypoxic response due to loss of function of the *von Hippel-Lindau* gene, which regulates Hif1α stability; like *VHL* patients, *vhl* zebrafish mutants develop polycythemia (van Rooijen et al., 2009) and they also exhibit increased HSPC number (Harris et al., 2013; van Rooijen et al., 2009). The fundamental role of hypoxic regulation was verified in mammalian studies whereby knockout of *Hif1a* in hemogenic endothelium significantly reduced murine HSC number (Imanirad et al., 2014). Recently, adenosine signaling, important for transduction of ATP to coordinate a myriad of biological pathways, was similarly found to enhance HSC development (Jing et al., 2015).

Energy metabolism is required for nearly all physiological functions including cell proliferation, movement, and growth; as such, coordination of relative energy supply (glucose, ATP) to transcriptional activation by Hif1α, or other factors, is essential to ensure adequate hematopoietic support (blood supply/oxygenation) to adapt to fluctuations in biological demands during tissue development and organismal growth during embryogenesis. Recent reports have indicated that the ability to sense and respond to hematopoietic demand results from regulation via the CNS: serotonergic stress response signaling mediated via the hypothalamic–pituitary–interrenal axis can impact developmental HSC production in responses to hypoxic stimuli via glucocorticoid production (Kwan et al., 2016). As such supply and demand is also pertinent in leukemia models, where rapidly proliferating cells require continuous metabolic supply, these findings may be relevant to hematologic disease.

3.3.2 Inflammatory Regulation of HSC Production

An important role for proinflammatory cytokines has been newly identified in HSC development using both zebrafish and mouse models. Zebrafish embryos have a functional innate immune system by 24 hpf (see also chapter "Modeling Infection Diseases in the Context of a Developing Immune System" by Masud et al.). Four recent independent studies reported a role for immune cells, including primitive neutrophils and macrophages, in HSC formation acting via inflammatory cytokines, such as: tumor necrosis factor α (TNFα), interferons- (IFN-) α, γ, ϕ, and/or interleukins (Espin-Palazon et al., 2014; He et al., 2015; Li et al., 2014; Sawamiphak, Kontarakis, & Stainier, 2014). Loss of select proinflammatory cytokines caused defective HSC emergence and proliferation. Furthermore, genetic knockdown of *pu.1 (spi1)* caused severe loss of HSC number in zebrafish embryos (Li et al., 2014). TNFα, secreted by primitive neutrophils, was found to activate Notch and NF-kB signaling pathways to control vascular niche formation and the HSC program (Espin-Palazon et al., 2014). IFNγ signaling, occurring within the hemogenic endothelium and primitive macrophages, impacted HSC production via modulation of additional downstream inflammatory cascades (Li et al., 2014). Beyond the novel role of the primitive myeloid population, these studies highlight the role of noncell autonomous sterile inflammatory signaling on establishment of HSC number. Future investigations will reveal the impact of inflammatory signaling in hematological disease and potential benefits of proinflammatory cytokines for in vitro HSC expansion.

4. INVESTIGATING ADULT HSC MAINTENANCE AND FUNCTION

Hematopoietic stem cell transplantation (HSCT) is a curative treatment for life-threatening hematopoietic disorders such as bone marrow failure syndromes and leukemias. Nonetheless, donor availability and allogenicity remain major limitations impacting morbidity and mortality (Goessling & North, 2014). A thorough understanding of transplantation biology, including the nascent signals and potential compound modifiers that can enhance HSC number or engraftment efficiency, is of significant clinical interest.

4.1 Irradiation Recovery

Recovery from irradiation injury is a well-established assay to examine the in vivo functional significance of proposed hematopoietic regulators. This method, first developed in mammals in the 1960s, was adapted for use in adult zebrafish, including establishment of maximum tolerated dose for autologous irradiation recovery or transplantation of donor HSPCs (Traver et al., 2004). The first application of this assay in zebrafish confirmed the regulatory impact of Notch signaling on adult HSC numbers, whereby activation accelerated multilineage recovery (Burns, Traver, Mayhall, Shepard, & Zon, 2005). Irradiation recovery was also used to establish the effective dose and in vivo functional mechanism of dmPGE2 on KM HSCs, prior to testing in mammalian models (North et al., 2007), as well as screen for potential radiotherapeutic agents for clinical use (Hwang, Yong, Moretti, & Lu, 2007).

4.2 Transplantation Biology

Zebrafish HSCT assays were developed to study HSC function, including homing, engraftment, and regenerative capacity following genetic or chemical modulation. WKM or embryonic HSC transplants can be used to test cell autonomy: for example, transplantation of WT cells could rescue impairments in multilineage hematopoiesis and lethality that occurs in *gata1* mutant embryos (Traver et al., 2003). Isolated CD41+ HSCs home accurately to the CHT and maintain long-term multilineage potential in embryonic recipients, confirming the functional potential of this developmental niche (Tamplin et al., 2015). CD41+ HSPCs from the embryo have likewise been shown to be competent to engraft adult recipients, mimicking the functional requirements of HSCs in mammals (Harris et al., 2013).

Significant effort had been made to optimize a HSC transplantation assay in adult zebrafish. A dramatic increase in engraftment and donor chimerism was achieved by matching MHC class I genes between donors and recipients (de Jong et al., 2011), minimizing graft-vs-host-disease. Use of the transparent *casper* line facilitated monitoring and evaluation of competitive engraftment in vivo (White et al., 2008). Furthermore, recently developed immunocompromised zebrafish lines, such as the $rag2^{E450fs}$ mutant line, allow robust long-term engraftment of HSCs, similar to that seen in mice (Tang et al., 2014). Finally, competitive transplantation assays have been developed to evaluate the functional impact of novel compound modifiers on donor HSCs compared to labeled matched controls (Li, Lahvic, et al., 2015). The evolution of HSCT assays in zebrafish has continued to aid research expanding our understanding marrow engraftment and identifying novel chemical regulators of HSC function.

4.3 Cell Culture

While it is clearly advantageous to interrogate HSC function in vivo, the development of robust methods for in vitro culture has provided additional means to quantify proliferative and/or differentiation characteristics of defined hematopoietic populations. Zebrafish kidney stromal cell lines have been developed to support WKM culture, resulting in maintenance and expansion of hematopoietic precursor cells in vitro (Stachura et al., 2009). In contrast, zebrafish embryonic stromal trunk cells promote HSC differentiation toward the myeloid, lymphoid, and erythroid fate (Campbell et al., 2015), allowing investigation of lineage potential of cultured HSPCs. Finally, WKM or isolated HSPCs cultured on methylcellulose in the presence of zebrafish-specific cytokines, with or without genetic modulation or addition of compound modifiers, allow direct examination of cell autonomy, proliferative expansion, and erythromyeloid potential (Stachura et al., 2011). The ability to perform in vitro functional characterization of zebrafish HSPCs facilitates translational application of novel signaling pathways or potential therapeutics.

5. MODELING HEMATOLOGICAL DISEASES IN ZEBRAFISH

Blood-related diseases and disorders involve perturbations of a vast array of hematopoietic cell types. In addition to the numerous zebrafish mutants generated using the unbiased screening approaches (Ransom et al., 1996; Weinstein et al., 1996), precise genetic engineering has been utilized to generate models, which express mutations or translocations known to cause hematological disease. These models are invaluable for furthering our understanding of disease onset and progression as well as the discovery of novel mechanisms to control or alleviate hematological defects or deficits.

5.1 Anemia Mutant Models

Zebrafish embryos can survive without red blood cells by passive diffusion of oxygen for ~7 days, allowing study of mutations disrupting erythropoiesis that are otherwise embryonic lethal in mice (Rombough & Drader, 2009). Large-scale forward genetic screens in zebrafish have generated multiple mutants with low red blood cell production or maintenance, reminiscent to human disease states, which can be classified in categories based on their developmental phenotypes: "bloodless," progressive anemia, hypochromic anemia, and photosensitivity (de Jong & Zon, 2005). A subset of representative mutant lines is described below (Table 2).

5.1.1 Bloodless Mutants

Bloodless mutants, such as *cloche*, *moonshine*, *bloodless*, and *vlad tepes*, have no red blood cells, exhibit defects in *gata1* expression, and generally do not develop appropriately (de Jong & Zon, 2005). As these mutations are embryonic lethal, and thus preclude evaluation of their function in adults, we will only briefly summarize the most relevant findings.

Among the first blood mutants identified, *cloche* embryos have disrupted differentiation of both endothelial and hematopoietic structures, supportive of the hemangioblast concept, with almost complete loss of expression for genes essential to hematopoiesis (i.e., *gata1*, *l-plastin*, *lmo2*, *mpx*, *kdrl*) (Liao et al., 1997; Stainier, Weinstein, Detrich, Zon, & Fishman, 1995). The gene mutation underlying *cloche* was recently cloned and identified as a bHLH-PAS transcription factor that drives hematovascular specification (Reischauer et al., 2016). *moonshine* mutants exhibit severe anemia with disruption of both the primitive and definitive wave of hematopoiesis. *moonshine* embryos possess a mutation in *transcriptional intermediary factor 1γ* (*tif1g*) that is highly conserved in humans and is required for erythroid precursors differentiation (Ransom et al., 2004). Finally, *vlad tepes* mutants have a truncation in *gata1* itself, leading to ineffective DNA binding and transactivation of erythroid target genes. The truncation and phenotype of *vlad tepes* mutants closely resembles that seen in mammalian dyserythropoietic anemia and thrombocytopenia (Lyons et al., 2002). Importantly, while these bloodless mutants are not overtly relevant to human disease, as mammalian embryos

Table 2 Zebrafish Lines Modeling Human Hematological Diseases

Classification	Zebrafish Lines	Genetic Mutation	Human Disease
Bloodless mutants	*cloche*	*npas4l*	—
	moonshine	*tif1g*	—
	bloodless		—
	vlad tepes	*gata1*	
Progressive anemia mutants	*chablis and merlot*	*epb41b*	Hereditary elliptocytosis
	riesling	*b-spectrin*	Hereditary spherocytosis
	retsina	*band 3*	Congenital dyserythropoietic anemia type II
Hypochromic mutants	*sauternes*	*alas2*	Congenital sideroblastic anemia
	zinfandel	*globin lcr*	Thalassemia
	chardonnay	*dmnt1*	Microcytic anemia
	chianti	*tfr1a*	
	weissherbst	*ferroportin 1*	Type IV hemochromatosis
Photosensitivity mutants	*dracula*	*ferrochelatase*	Porphyria
Bone marrow failure models	*rps14 and rps29*	*rps14* and *rps29*	Diamond–Blackfan anemia
	rpl11	*rpl11*	Diamond–Blackfan anemia
	dkc1 and *nop10*	*dkc* and *nop10*	Dyskeratosis congenita
	AML-ETO	*runx1*	MDS, AML
	NUP98-HOXA9	*hoxa9*	MPN, AML
	TET2-null	*tet2*	MDS
	crimsonless	*hspa9b*	MDS
	sf3b1	*sf3b1*	MDS
Leukemia models	*Tg(rag2: egfp-myc)*	*myc* (overexpression)	T-ALL
	notch 1 truncation	*notch 1*	T-ALL
	TEL-AML1	*runx1*	pre-B cell ALL

would be unable to survive homozygous loss due to the severity of the erythropoietic defects, they do corroborate strong conservation of the essential functional role of *gata1*.

5.1.2 Progressive Anemia Mutants

Large-scale screens identified a number of progressive anemia mutants where blood formation occurs normally for the first two days of development, followed by subsequent decline in RBC number. These lines provide relevant models of erythroid maturation, expansion, and maintenance. Mutations of cytoskeletal proteins appear to be a common feature of many zebrafish progressive anemia mutants; defects in erythrocyte membrane integrity are considered to be major cause of human anemia.

The zebrafish *merlot* and *chablis* lines possess mutations in the gene coding the 4.1R protein (formerly *band 4.1*), a structural protein expressed in RBCs that anchors the spectrin–actin cytoskeleton to the erythrocyte cell membrane (Shafizadeh et al., 2002). Mutations in this gene result in spiculated erythrocyte membranes and bilobed nuclei in zebrafish, causing severe hemolytic anemia and secondary conditions such as cardiomegaly and splenomegaly (Shafizadeh et al., 2002). Consistent with this phenotype, protein 4.1 deficiencies in humans cause hereditary elliptocytosis, a rare form of hemolytic anemia marked by elliptical-shaped RBCs (de Jong & Zon, 2005) (Fig. 2).

B-spectrin is another membrane protein component important to maintenance of the erythrocyte cytoskeleton; RBCs deficient in *B-spectrin* exhibit cell morphology-induced hemolysis (Liao et al., 2000). The zebrafish *riesling* mutant, which similarly displays hemolysis, has a null mutation in the B-spectrin gene and phenotypically resembles human hereditary spherocytosis (Liao et al., 2000) (Fig. 2). In contrast, *retsina* mutants have defective cytokinesis with an accumulation of erythroblasts, multinuclear erythroblasts, dyserythropoiesis, and anemia, resembling congenital dyserythropoietic anemia type II. This disease is caused by *band 3* mutation, affecting chromosomal segmentation during anaphase (Paw et al., 2003). Together, the identification of these mutants has illustrated conservation of important components controlling normal erythroid differentiation and maintenance, and provides a screening platform to identify drugs that could promote erythropoiesis and control disease burden.

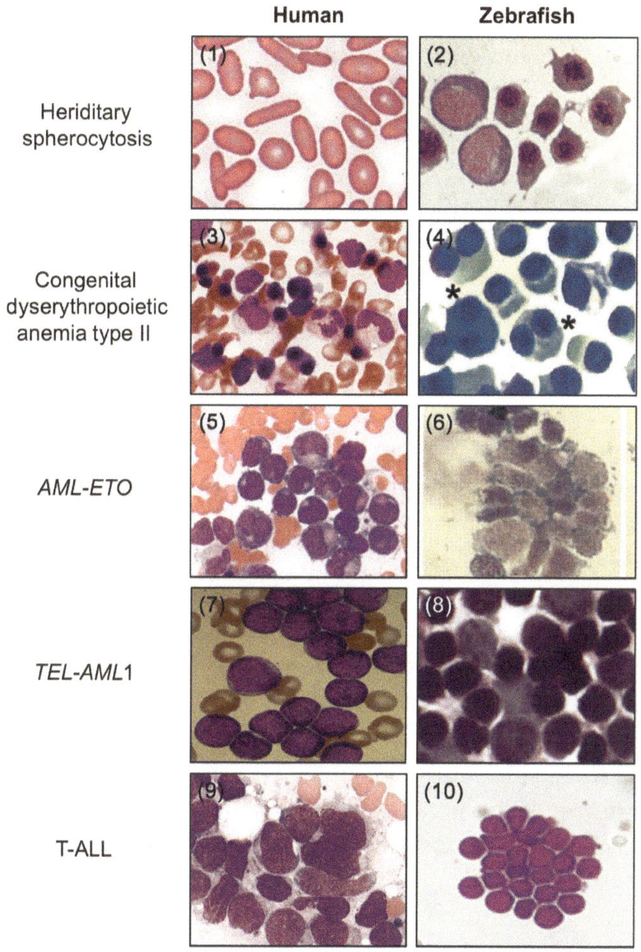

Fig. 2 Histological comparison of hematological disease phenotypes in human and zebrafish. Examples of Wright's and Giemsa-stained blood smears from indicated zebrafish disease models demonstrating recapitulation of some key features observed in the related human disorder or malignancy. (1) Human elliptocytes. This image was originally published in ASH image bank: *Teresa Scordino. Elliptocytes. ASH Image Bank. 2015; Image Number 00060270.* (2) *merlot* mutant. This image was originally published in: *Shafizadeh, E., et al. (2002). Characterization of zebrafish merlot/chablis as non-mammalian vertebrate models for severe congenital anemia due to protein 4.1 deficiency. Development, 129, 4359–4370. (Original published image: Fig. 2A.)* (3) Congenital dyserythropoietic anemia type II. This image was originally published in ASH image bank: *Kristian T. Schafernak. Congenital dyserythropoietic anemia (CDA) type II. ASH Image Bank. 2016; Image Number 00060892.* (4) *retsina* mutant. This image was originally published In: *Paw, B. H., et al. (2003). Cell-specific mitotic defect and dyserythropoiesis associated with erythroid band 3 deficiency. Nature Genetics, 34, 59–64. (Original published image: Fig. 1B.)*

(Continued)

5.1.3 Hypochromic Mutants

Hypochromic mutants can be used for the study of diseases such as congenital sideroblastic anemia, thalassemia, and microcytic anemia (Ransom et al., 1996; Weinstein et al., 1996). These zebrafish mutants are classified by abnormally small erythrocytes with decreased levels of hemoglobin. The *sauternes* mutants have a mutation in *Δ-aminolevulinate synthase* (*ALAS2*), which catalyzes the first step of heme biosynthesis (Brownlie et al., 1998). These erythroid mutants resemble human congenital sideroblastic anemia, where patients develop iron deposition in mitochondria of erythroblasts in the bone marrow. Given the close genetic and physiological resemblance to the human disease phenotype, this mutant line became the first animal model of this particular anemia subtype.

Abnormal hemoglobin production, termed thalassemia, can result from reduction in globin synthesis, iron absorption (i.e., iron deficiency), or vitamin B6 deficiency. The *zinfandel* mutation maps to the major *globin* locus of zebrafish chromosome 3, causing profound anemia, which can serve to model the effects of human thalassemia (Brownlie et al., 2003). Iron metabolism is an essential component of heme biosynthesis. Many human hypochromic anemias result from defective iron absorption, storage, and trafficking. The *chardonnay* mutant has a genetic defect in the iron transporter *DMT1*, which was subsequently found to be a causal mutation in human microcytic anemias (Donovan et al., 2002; Mims et al., 2005). The *chianti* line has a mutation in the *transferrin receptor 1* (*tfr1a*) gene, which is critical for iron acquisition in erythrocytes (Wingert et al., 2004), while the *weissherbst* mutants have a mutation in *ferroportin 1*, an iron transporter (also known as *slc40a1*) that is affected in patients with type IV hemochromatosis (Donovan et al., 2002). Iron-associated anemia can be partially alleviated by

Fig. 2—Cont'd (5) Acute myeloid leukemia with *t(8;21)(q22;22)*. This image was originally published in ASH image bank: *Peter Maslak. AML with t(8;21). ASH Image Bank. 2004; Image Number 00002597.* (6) AML-ETO. This image was originally published in: *Yeh, J. R., et al. (2008). AML1-ETO reprograms hematopoietic cell fate by downregulating scl expression. Development, 135, 401–410. (Original published image: Fig. 3F.)* (7) Pre-B TEL-AML1 ALL. Unpublished. *Kindly provided by Hatem Sabaawy, Rutgers CINJ.* (8) TEL-AML1, Zebrafish. Unpublished. *Kindly provided by Hatem Sabaawy, Rutgers CINJ.* (9) Precursor T-ALL. This image was originally published in ASH image bank: *Peter Maslak. Precursor T-cell acute lymphoblastic leukemia. ASH Image Bank. 2003; Image Number 00002283.* (10) *Tg(rag2:egfp-myc)*. This image was originally published in: *Langenau, D. M., et al. (2003). Myc-induced T cell leukemia in transgenic zebrafish. Science, 299, 887–890. (Original published image: Fig. S1H.)*

supplementation in both zebrafish and human, highlighting conservation of effect across species.

5.1.4 Photosensitivity Mutants

Ferrochelatase catalyzes the formation of heme by transferring iron to protoporphyrin, a heme mediator. The mutation in the human *Ferrochelatase* gene causes protoporphyrin accumulation in the skin, erythrocytes, and liver, resulting in light sensitivity. *dracula* mutants possess light-dependent hemolysis, strong erythrocyte autofluorescence, liver toxicity, and an accumulation of protoporphyrin IX (Childs et al., 2000) caused by a point mutation in *ferrochelatase*, rendering it dysfunctional (Childs et al., 2000). These mutants are now used as a compound-screening platform to alleviate the symptoms of porphyria.

5.2 Bone Marrow Failure Syndromes

Bone marrow failure syndromes are characterized by defects in HSPCs that impair their ability to regenerate and/or differentiate into mature effector cell types. Due to their cellular origin, one or more blood lineages are often affected, with a high propensity for progress to leukemia. A number of human hematopoietic failure disorders have been linked to mutations in ribosomal genes and/or defects in ribosome biogenesis. These include Diamond–Blackfan anemia (DBA) and dyskeratosis congenital, which cause predisposition to cancer in affected patients.

5.2.1 Diamond–Blackfan Anemia

DBA is a rare congenital anemia significantly associated with genetic alterations in ribosomal proteins. The most common mutation in DBA is in *RPS19*, part of the 40S ribosomal subunit (Draptchinskaia et al., 1999). In zebrafish deficient for *rps19* or the related factor *rps14*, L-leucine, a known activator of mRNA translation, has been shown to improve anemia as well as other development defects associated with DBA (Payne et al., 2012); indeed, L-leucine is currently being evaluated as a potential therapy for DBA in clinical trials. *RPL11* mutation also occurs frequently in DBA patients. Morpholino-knockdown and mutant *rpL11* models display hematopoietic defects similar to DBA, along with p53 activation, consistent with their mammalian counterpart (Zhang, Jia, et al., 2013). Treatment with RAP-001, a drug that expands late-stage erythroblasts, can restore hemoglobin levels in DBA zebrafish models (Ear et al., 2015; Zhang, Jia, et al., 2013), representing a potential avenue for further translation. Mutation in *rps29* was also shown to impact hematopoiesis: embryos deficient in *rps29* have erythrocytes defects, increased apoptotic cells, and p53 activation resembling the mechanistic characteristics of DBA, suggesting that this ribosomal protein may also have an important role in disease pathogenesis (Taylor et al., 2012). Overall, these zebrafish studies have provided further mechanistic clarity as well as potential targets for treatment of human bone marrow failure.

5.2.2 *Dyskeratosis Congenita*

Dyskeratosis congenita (DC) is a congenital disease characterized by shortened telomeres and defective stem cell maintenance. Most DC patients experience cytopenia in early adulthood and, as a result, frequently die from bone marrow failure (Zhang, Morimoto, Danilova, Zhang, & Lin, 2012). Mutations in ribonucleoprotein complex genes such as *DKC1* and *NOP10* are known to cause DC (Zhang et al., 2012). Similar to DBA, loss-of-function zebrafish mutants for *dkc1*, *nola1*, and *nop10* have ineffective ribosomal biogenesis and hematopoietic defects (Pereboom, van Weele, Bondt, & MacInnes, 2011). More recently, patients with hematopathology caused by loss-of-function mutations in *poly(A)-specific ribonuclease (PARN)* were identified. *PARN*-deficient cells have shortened telomeres and an aberrant ribosome profile similar to those in DC; genetic knockdown of *PARN* impaired hematopoiesis in zebrafish and human cells, providing evidence for a common genetic effector in blood development and bone marrow failure (Dhanraj et al., 2015).

5.3 Models of Myelodysplastic Syndrome, Myeloproliferative Neoplasms, and Myeloid Malignancies

Myeloid malignancies are heterogeneous disorders characterized by uncontrolled proliferation and/or blockage of differentiation of abnormal myeloid progenitor cells. Myelodysplastic syndrome (MDS) and myeloproliferative neoplasms (MPNs) are often thought to be precursors to myeloid malignancies such as acute myeloid leukemia (AML). Both MDS and MPNs are clonal HSC disorders. MDS is characterized by the simultaneous proliferation and apoptosis of hematopoietic cells, leading to dysplastic changes of developing blood cell progenitors, hence ineffective hematopoiesis. In contrast, MPNs are characterized by proliferation of one or more of the myeloid lineages. As a substantial number of studies indicated MDS or MPN can progress to AML in some patients, zebrafish models with causal gene fusions have been generated and their ongoing contribution to our understanding of human disease is summarized below.

5.3.1 *AML-ETO*

The AML-ETO fusion protein results from the *t(8;21)(q22;q22)* chromosomal translocation often found in patients with AML. Patients expressing the *AML-ETO* fusion gene develop an accumulation of granulocytic precursors in the BM and peripheral blood, indicative of MDS. Injection of human *AML1-ETO* cDNA caused hematopoietic defects in zebrafish embryos, but these transient studies did not allow for cell fate evaluation due to early embryonic lethality, as seen in the mouse models (Kalev-Zylinska et al., 2002). Creation of a heat-shock-inducible transgenic *AML1-ETO* line showed that the fusion protein reprograms erythropoiesis to granulopoiesis, resulting in a robust phenotype that exhibits cytological and transcriptional characteristics similar to human AML (Yeh et al., 2008). *AML1-ETO* embryos were utilized in a chemical screen, leading to the identification of cyclooxygenase (COX) inhibitors and the benzodiazepine Ro5-3335 as potential new treatment options for this devastating disease (Cunningham et al., 2012; Yeh et al., 2009).

5.3.2 *NUP98-HOXA9*

The *NUP98-HOXA9* fusion oncogene is observed in high-risk AML patients. *NUP98-HOXA9* transgenic fish were generated by using the human *t(7;11)(p15;p15)* chromosomal translocation product under the zebrafish *pu.1* myeloid promoter, resulting in a preleukemic state with anemia and myeloid cell expansion in embryos (Forrester et al., 2011). MPN occurs in 23% of adult *NUP98-HOXA9* fish; however, they do not progress to AML (Forrester et al., 2011). *NUP98-HOXA9* transgenic embryos show increased HSC numbers, as well as a lineage bias from erythroid to myeloid fate (Forrester et al., 2011). The NUP98-HOXA9 fusion was recently shown by an unbiased microarray analysis to alter hypermethylation through the DNA methyltransferase, *dnmt1* (Deveau et al., 2015). Use of DNMT inhibitors in transgenic zebrafish embryos normalized genome methylation, reduced HSC numbers, and, in combination with HDAC inhibition, restored the genetic and phenotypic profiles of blood development, providing preclinical evidence for treatment of *NUP98-HOXA9*-induced myeloid disease (Deveau et al., 2015).

5.3.3 *TET2*

The ten-eleven translocation (TET) family of methylcytosine dioxygenases (TET1, TET2, TET3) converts 5-methylcytosine to 5-hydroxymethylcytosine, which initiates the demethylation of genomic CpG islands. TET2 and TET3 are each required for HSC emergence in zebrafish embryos via regulation of Notch signaling (Li, Lan, et al., 2015). Interestingly, *TET2* mutations are frequently observed in MDS (Kosmider et al., 2009). The zebrafish *tet2*-null mutant exhibits normal embryonic and larval hematopoiesis, but develops premyelodysplasia at 11 months of age, and clonal myelodysplasia at 24 months of age (Gjini et al., 2015). These fish have increased progenitors and myelomonocytic cell numbers, and decreased erythrocyte cell numbers with the KM (Gjini et al., 2015) similar to patient profiles.

5.3.4 Other Genetic Models of MDS and MPN

In addition to the targeted generation of known fusion oncogenes zebrafish lines, ENU mutagenesis screens have produced zebrafish mutants exhibiting MDS and MPN phenotypes. One example is the *LDD731* line, which has significantly increased HSCs in the hematopoietic organs, with erythroid/myeloid lineage upregulation during definitive hematopoiesis (Peng et al., 2015). Positional cloning revealed a causal mutation in the *c-cbl* gene, frequently found disrupted in MPN and AML patients (Peng et al., 2015). Zebrafish *crimsonless* mutants display anemia, dysplasia, multilineage cytopenia, and blood cell apoptosis (Craven, French, Ye, de Sauvage, & Rosenthal, 2005). The causal mutation was in *HSPA9B*, a conserved mitochondrial matrix chaperone, that affects mitochondrial function and causes increased oxidative stress and apoptosis specific to hematopoietic cells (Craven et al., 2005). Importantly, this study implicates the loss of *HSPA9B* and mitochondrial dysfunction in the pathogenesis of MDS. More recently, a zebrafish mutant of *SF3B1* (*splicing factor 3b, subunit 1*), one of the most commonly mutated MDS genes, was shown to similarly mimic the macrocytic anemia-like phenotype seen in patients (De La Garza, Cameron, Nik, Payne, & Bowman, 2016). The primitive myeloid cells in *sf3b1* embryos are mature, and definitive hematopoiesis is also defective. As Notch signaling was normal in mutant embryos, *sf3b1* mutation is presumed to act downstream or independent of Notch-mediated control of EHT (De La Garza et al., 2016), suggesting key fate decisions during hematopoietic commitment are controlled at the level of alternative splicing.

5.4 Acute Lymphoblastic Leukemia Zebrafish Models

Acute lymphoblastic leukemia (ALL) is a cancer in which lymphocytic progenitors accumulate and are unable to differentiate into mature T-cells or B-cells; if not treated, ALL is fatal within months of onset. ALL is a clonal disease that evolves through genetic rearrangements and/or mutations within the dominant clone. Based on morphologic, genetic, and immunophenotypic features, lymphoblastic lymphoma (LBL) and ALL are now appreciated to represent the same disease entity, differing in maturation status and extent of bone marrow infiltration. The first zebrafish model of leukemia was a T-cell ALL (T-ALL), in which the *myc* oncogene was overexpressed under the control of the lymphoid-specific *rag2* promoter (Langenau et al., 2003). Many additional models have since been created allowing characterization of clonal evolution, leukemia initiation, and providing platforms for compound screening as detailed below.

5.4.1 Myc-Driven Leukemia

In the Tg(*rag2:egfp-myc*) T-ALL model, induction of leukemia was very efficient, with the majority of injected animals not surviving past sexual maturity, impairing line maintenance (Langenau et al., 2003). Treatment with clinically relevant T-ALL drugs cyclophosphamide and vincristine recapitulated therapeutic responses, thereby validating the relevance of the model to human disease (Mizgirev & Revskoy, 2010). A related conditional model of T-ALL was subsequently established using the *Cre-loxP* system, whereby the heat-shock-inducible *Cre* recombinase (*hsp70:cre*) could trigger the *rag2:loxP-dsred-loxP-GFP-myc* to induce T-ALL (Feng et al., 2007). Conditional mutants first develop T-cell LBL (T-LBL), before progressing to T-ALL (Feng et al., 2007), thus enabling investigation of genetic and mechanistic cues relevant to this process. Overexpression of *bcl-2, s1p1*, and *icam1* in T-LBL were found to accelerate disease onset, but halted frank T-ALL by blocking intravasation and homological dissemination (Feng et al., 2010).

Zebrafish T-ALL models exhibit pre-T-cell origin and clonal expansion, consistent with human disease (Blackburn et al., 2012). Further, the use of transparent immune-compromised $rag2^{E450fs}$ zebrafish as the background strain has allowed dynamic imaging of tumor dissemination and emergence of clonal dominance (Tang et al., 2016). A recent study utilized the Myc-inducible T-ALL model to screen for novel genes contributing to disease onset: heterozygous loss of *dihydrolipoamide S-succinyltransferase* (*dlst*), a tricarboxylic acid (TCA) cycle enzyme, significantly delayed tumor onset without overt effects on zebrafish development (Anderson et al., 2016). DLST functions as a transferase in the α-ketoglutarate dehydrogenase complex, which is critical for energy production and macromolecule synthesis in the TCA cycle. As such, this study highlights the significant metabolic dependence of leukemic cells, which is reminiscent of its prominent regulatory impact during hematopoietic development (Harris et al., 2013).

5.4.2 Notch-Inducible Leukemia

Notch1 regulates many phases of hematopoiesis including hemogenic endothelium commitment, as described above, and the development of early T-cells in the thymus (Chen et al., 2007). *NOTCH1* is also commonly mutated in human T-ALL (Chen et al., 2007). A zebrafish model of human *NOTCH1* truncation develops aggressive and lethal leukemia, which invades tissues throughout the body (Chen et al., 2007). Further, the Notch and Bcl2 antiapoptotic pathways synergize to accelerate leukemia onset (Chen et al., 2007). Interestingly, overexpression of the Notch intracellular domain was sufficient to expand preleukemic clones (Blackburn et al., 2012). Limiting-dilution transplantation further demonstrated that *Notch1* mutation initiated T-ALL via expansion of a premalignant pool of thymocytes, as no effect was observed on the overall leukemic propagating frequency (Blackburn et al., 2012). In contrast, selective activation of Akt pathway and mTORC1 can subsequently increase leukemia propagating cell numbers and synergize with Myc and Notch1 to drive T-ALL (Blackburn et al., 2014). Together these findings point to multiple roles for Notch in the development and progression of T-ALL.

5.4.3 TEL-AML1

The *t(12;21) TEL-AML1* translocation is the most common mutation in childhood cancer. Expression of the TEL-AML1 fusion protein results in B-ALL characterized by accumulation of pre-B-cells due to enhanced self-renewal capacity and impaired differentiation (Sabaawy et al., 2006). Expression of human *TEL-AML1* under a ubiquitous promoter resulted in leukemia development in only a small fraction (~3%) of the transgenic fish in the course of 8–12 months (Sabaawy et al., 2006); however, this low frequency resembles human disease emergence and suggests secondary mutations are required. While long latency and low penetrance have diminished its utility, studies using this line have indicated that the development of leukemia is associated with complete loss of TEL expression and induction of antiapoptotic mechanisms resulting from elevated Bcl/Bax ratios. Altogether, zebrafish models of hematological malignancy provide significant insight into the key pathways that impact cancer development and progression in humans and a promising system to test novel drugs to counteract leukemia.

6. CONCLUDING REMARKS

Zebrafish have become a prominent model in the hematology field for the study of vertebrate blood development and disease. Ex vivo development combined with strong evolutionary conservation of genetic programs and signaling pathways influencing hematopoiesis has proven the power of zebrafish to identify key morphogens and mechanisms controlling HSPC production and function. Importantly, zebrafish are highly amenable to genetic manipulations, facilitating analyses of developmental pathways that are difficult to study in mammalian models due to early embryonic lethality from anemia. Moreover, use of zebrafish as a screening platform to identify compound modifiers which can be directly translated to mammalian biology is a celebrated success derived from investigations on the regulation of the blood system. Over the past 20 years, zebrafish research has yielded an abundance of insights regarding the spatiotemporal regulation and interplay of a large number of pathways important for both primitive and definitive hematopoiesis. The transgenic and mutant zebrafish lines (Table 1) have revolutionized our understanding of the onset and progression of human hematopoietic diseases.

As hematopoiesis is a life-long dynamic process in which HSCs are continuously replenishing all of the blood cells in the entire organism, the signals that guide the emergence and differentiation of erythrocytes and/or HSCs remain an active influence on homeostasis and maintenance. As innovative tools and technologies continue to emerge, the zebrafish model will be uniquely positioned to interrogate precise modulation of multiple regulatory factors in a time- and tissue-specific manner, and, with the further development of antibody markers similar to those utilized in mammalian, will allow unprecedented unbiased in vivo characterization of critical subpopulations associated with specific roles in hematological function or disease. In sum, the knowledge gained from zebrafish has and will continue to benefit our overall understanding of vertebrate hematopoiesis including the identification of causal genetic aberrations and development of rationale therapeutic approaches to treat patients with hematologic disease.

ACKNOWLEDGMENTS

We thank W. Goessling for helpful discussion and editorial support. We thank all authors and corresponding publishers who gave permission for the use of images in Figure 2. In particular, we thank B. Paw for helpful discussion of erythroid disease and image use, and H. Sabaawy for providing the previously unpublished images in Figure 2. W. Kwan was supported by the Canadian Institutes of Health Research; T.E. North is funded by NIH R01DK098241 and the Leukemia and Lymphoma Society.

REFERENCES

Adamo, L., Naveiras, O., Wenzel, P. L., McKinney-Freeman, S., Mack, P. J., Gracia-Sancho, J., et al. (2009). Biomechanical forces promote embryonic haematopoiesis. *Nature, 459*, 1131–1135.

Anderson, N. M., Li, D., Peng, H. L., Laroche, F. J., Mansour, M. R., Gjini, E., et al. (2016). The TCA cycle transferase DLST is important for MYC-mediated leukemogenesis. *Leukemia, 30*, 1365–1374.

Bertrand, J. Y., Chi, N. C., Santoso, B., Teng, S., Stainier, D. Y., & Traver, D. (2010). Haematopoietic stem cells derive directly from aortic endothelium during development. *Nature, 464*, 108–111.

Bertrand, J. Y., Kim, A. D., Teng, S., & Traver, D. (2008). CD41+ cmyb+ precursors colonize the zebrafish pronephros by a novel migration route to initiate adult hematopoiesis. *Development, 135*, 1853–1862.

Blackburn, J. S., Liu, S., Raiser, D. M., Martinez, S. A., Feng, H., Meeker, N. D., et al. (2012). Notch signaling expands a pre-malignant pool of T-cell acute lymphoblastic leukemia clones without affecting leukemia-propagating cell frequency. *Leukemia, 26*, 2069–2078.

Blackburn, J. S., Liu, S., Wilder, J. L., Dobrinski, K. P., Lobbardi, R., Moore, F. E., et al. (2014). Clonal evolution enhances leukemia-propagating cell frequency in T cell acute lymphoblastic leukemia through Akt/mTORC1 pathway activation. *Cancer Cell, 25*, 366–378.

Bresciani, E., Carrington, B., Wincovitch, S., Jones, M., Gore, A. V., Weinstein, B. M., et al. (2014). CBFbeta and RUNX1 are required at 2 different steps during the development of hematopoietic stem cells in zebrafish. *Blood, 124*, 70–78.

Brownlie, A., Donovan, A., Pratt, S. J., Paw, B. H., Oates, A. C., Brugnara, C., et al. (1998). Positional cloning of the zebrafish sauternes gene: A model for congenital sideroblastic anaemia. *Nature Genetics, 20*, 244–250.

Brownlie, A., Hersey, C., Oates, A. C., Paw, B. H., Falick, A. M., Witkowska, H. E., et al. (2003). Characterization of embryonic globin genes of the zebrafish. *Developmental Biology, 255*, 48–61.

Burns, C. E., DeBlasio, T., Zhou, Y., Zhang, J., Zon, L., & Nimer, S. D. (2002). Isolation and characterization of runxa and runxb, zebrafish members of the runt family of transcriptional regulators. *Experimental Hematology, 30*, 1381–1389.

Burns, C. E., Traver, D., Mayhall, E., Shepard, J. L., & Zon, L. I. (2005). Hematopoietic stem cell fate is established by the Notch-Runx pathway. *Genes & Development, 19*, 2331–2342.

Campbell, C., Su, T., Lau, R. P., Shah, A., Laurie, P. C., Avalos, B., et al. (2015). Zebrafish embryonic stromal trunk (ZEST) cells support hematopoietic stem and progenitor cell (HSPC) proliferation, survival, and differentiation. *Experimental Hematology, 43*, 1047–1061.

Carroll, K. J., Esain, V., Garnaas, M. K., Cortes, M., Dovey, M. C., Nissim, S., et al. (2014). Estrogen defines the dorsal-ventral limit of VEGF regulation to specify the location of the hemogenic endothelial niche. *Developmental Cell, 29*, 437–453.

Carroll, K. J., & North, T. E. (2014). Oceans of opportunity: Exploring vertebrate hematopoiesis in zebrafish. *Experimental Hematology, 42*, 684–696.

Chanda, B., Ditadi, A., Iscove, N. N., & Keller, G. (2013). Retinoic acid signaling is essential for embryonic hematopoietic stem cell development. *Cell, 155*, 215–227.

Chen, J., Jette, C., Kanki, J. P., Aster, J. C., Look, A. T., & Griffin, J. D. (2007). NOTCH1-induced T-cell leukemia in transgenic zebrafish. *Leukemia, 21*, 462–471.

Chen, A. T., & Zon, L. I. (2009). Zebrafish blood stem cells. *Journal of Cellular Biochemistry, 108*, 35–42.

Childs, S., Weinstein, B. M., Mohideen, M. A., Donohue, S., Bonkovsky, H., & Fishman, M. C. (2000). Zebrafish dracula encodes ferrochelatase and its mutation provides a model for erythropoietic protoporphyria. *Current Biology: CB, 10*, 1001–1004.

Clements, W. K., Kim, A. D., Ong, K. G., Moore, J. C., Lawson, N. D., & Traver, D. (2011). A somitic Wnt16/Notch pathway specifies haematopoietic stem cells. *Nature, 474*, 220–224.

Colle-Vandevelde, A. (1963). Blood anlage in teleostei. *Nature, 198*, 1223.

Corada, M., Nyqvist, D., Orsenigo, F., Caprini, A., Giampietro, C., Taketo, M. M., et al. (2010). The Wnt/beta-catenin pathway modulates vascular remodeling and specification by upregulating Dll4/Notch signaling. *Developmental Cell, 18*, 938–949.

Cortes, M., Chen, M. J., Stachura, D. L., Liu, S. Y., Kwan, W., Wright, F., et al. (2016). Developmental vitamin D availability impacts hematopoietic stem cell production. *Cell Reports, 17*, 458–468.

Cortes, M., Liu, S. Y., Kwan, W., Alexa, K., Goessling, W., & North, T. E. (2015). Accumulation of the vitamin D precursor cholecalciferol antagonizes hedgehog signaling to impair hemogenic endothelium formation. *Stem Cell Reports, 5*, 471–479.

Craven, S. E., French, D., Ye, W., de Sauvage, F., & Rosenthal, A. (2005). Loss of Hspa9b in zebrafish recapitulates the ineffective hematopoiesis of the myelodysplastic syndrome. *Blood, 105*, 3528–3534.

Cunningham, L., Finckbeiner, S., Hyde, R. K., Southall, N., Marugan, J., Yedavalli, V. R., et al. (2012). Identification of benzodiazepine Ro5-3335 as an inhibitor of CBF leukemia through quantitative high throughput screen against RUNX1-CBFbeta interaction. *Proceedings of the National Academy of Sciences of the United States of America, 109*, 14592–14597.

Cutler, C., Multani, P., Robbins, D., Kim, H. T., Le, T., Hoggatt, J., et al. (2013). Prostaglandin-modulated umbilical cord blood hematopoietic stem cell transplantation. *Blood, 122*, 3074–3081.

de Jong, J. L., Burns, C. E., Chen, A. T., Pugach, E., Mayhall, E. A., Smith, A. C., et al. (2011). Characterization of immune-matched hematopoietic transplantation in zebrafish. *Blood, 117*, 4234–4242.

de Jong, J. L., & Zon, L. I. (2005). Use of the zebrafish system to study primitive and definitive hematopoiesis. *Annual Review of Genetics, 39*, 481–501.

De La Garza, A., Cameron, R. C., Nik, S., Payne, S. G., & Bowman, T. V. (2016). The spliceosomal component Sf3b1 is essential for hematopoietic differentiation in zebrafish. *Experimental Hematology, 44*(9), 826–837.

Deveau, A. P., Forrester, A. M., Coombs, A. J., Wagner, G. S., Grabher, C., Chute, I. C., et al. (2015). Epigenetic therapy restores normal hematopoiesis in a zebrafish model of NUP98-HOXA9-induced myeloid disease. *Leukemia, 29*, 2086–2097.

Dhanraj, S., Gunja, S. M., Deveau, A. P., Nissbeck, M., Boonyawat, B., Coombs, A. J., et al. (2015). Bone marrow failure and developmental delay caused by mutations in poly(A)-specific ribonuclease (PARN). *Journal of Medical Genetics, 52*, 738–748.

Diaz, M. F., Li, N., Lee, H. J., Adamo, L., Evans, S. M., Willey, H. E., et al. (2015). Biomechanical forces promote blood development through prostaglandin E2 and the cAMP-PKA signaling axis. *The Journal of Experimental Medicine, 212*, 665–680.

Donovan, A., Brownlie, A., Dorschner, M. O., Zhou, Y., Pratt, S. J., Paw, B. H., et al. (2002). The zebrafish mutant gene chardonnay (cdy) encodes divalent metal transporter 1 (DMT1). *Blood, 100*, 4655–4659.

Draptchinskaia, N., Gustavsson, P., Andersson, B., Pettersson, M., Willig, T. N., Dianzani, I., et al. (1999). The gene encoding ribosomal protein S19 is mutated in Diamond-Blackfan anaemia. *Nature Genetics, 21*, 169–175.

Ear, J., Huang, H., Wilson, T., Tehrani, Z., Lindgren, A., Sung, V., et al. (2015). RAP-011 improves erythropoiesis in zebrafish model of Diamond-Blackfan anemia through antagonizing lefty1. *Blood*, *126*, 880–890.

Esain, V., Cortes, M., & North, T. E. (2016). Enumerating hematopoietic stem and progenitor cells in zebrafish embryos. *Methods in Molecular Biology*, *1451*, 191–206.

Esain, V., Kwan, W., Carroll, K. J., Cortes, M., Liu, S. Y., Frechette, G. M., et al. (2015). Cannabinoid receptor-2 regulates embryonic hematopoietic stem cell development via prostaglandin E2 and P-selectin activity. *Stem Cells*, *33*, 2596–2612.

Espin-Palazon, R., Stachura, D. L., Campbell, C. A., Garcia-Moreno, D., Del Cid, N., Kim, A. D., et al. (2014). Proinflammatory signaling regulates hematopoietic stem cell emergence. *Cell*, *159*, 1070–1085.

Feng, H., Langenau, D. M., Madge, J. A., Quinkertz, A., Gutierrez, A., Neuberg, D. S., et al. (2007). Heat-shock induction of T-cell lymphoma/leukaemia in conditional Cre/lox-regulated transgenic zebrafish. *British Journal of Haematology*, *138*, 169–175.

Feng, H., Stachura, D. L., White, R. M., Gutierrez, A., Zhang, L., Sanda, T., et al. (2010). T-lymphoblastic lymphoma cells express high levels of BCL2, S1P1, and ICAM1, leading to a blockade of tumor cell intravasation. *Cancer Cell*, *18*, 353–366.

Forrester, A. M., Grabher, C., McBride, E. R., Boyd, E. R., Vigerstad, M. H., Edgar, A., et al. (2011). NUP98-HOXA9-transgenic zebrafish develop a myeloproliferative neoplasm and provide new insight into mechanisms of myeloid leukaemogenesis. *British Journal of Haematology*, *155*, 167–181.

Galloway, J. L., Wingert, R. A., Thisse, C., Thisse, B., & Zon, L. I. (2005). Loss of gata1 but not gata2 converts erythropoiesis to myelopoiesis in zebrafish embryos. *Developmental Cell*, *8*, 109–116.

Gjini, E., Mansour, M. R., Sander, J. D., Moritz, N., Nguyen, A. T., Kesarsing, M., et al. (2015). A zebrafish model of myelodysplastic syndrome produced through tet2 genomic editing. *Molecular and Cellular Biology*, *35*, 789–804.

Goessling, W., Allen, R. S., Guan, X., Jin, P., Uchida, N., Dovey, M., et al. (2011). Prostaglandin E2 enhances human cord blood stem cell xenotransplants and shows long-term safety in preclinical nonhuman primate transplant models. *Cell Stem Cell*, *8*, 445–458.

Goessling, W., & North, T. E. (2014). Repairing quite swimmingly: Advances in regenerative medicine using zebrafish. *Disease Models & Mechanisms*, *7*, 769–776.

Goessling, W., North, T. E., Loewer, S., Lord, A. M., Lee, S., Stoick-Cooper, C. L., et al. (2009). Genetic interaction of PGE2 and Wnt signaling regulates developmental specification of stem cells and regeneration. *Cell*, *136*, 1136–1147.

Harris, J. M., Esain, V., Frechette, G. M., Harris, L. J., Cox, A. G., Cortes, M., et al. (2013). Glucose metabolism impacts the spatiotemporal onset and magnitude of HSC induction in vivo. *Blood*, *121*, 2483–2493.

He, Q., Zhang, C., Wang, L., Zhang, P., Ma, D., Lv, J., et al. (2015). Inflammatory signaling regulates hematopoietic stem and progenitor cell emergence in vertebrates. *Blood*, *125*, 1098–1106.

Herbomel, P., Thisse, B., & Thisse, C. (2001). Zebrafish early macrophages colonize cephalic mesenchyme and developing brain, retina, and epidermis through a M-CSF receptor-dependent invasive process. *Developmental Biology*, *238*, 274–288.

Hong, C. C., Peterson, Q. P., Hong, J. Y., & Peterson, R. T. (2006). Artery/vein specification is governed by opposing phosphatidylinositol-3 kinase and MAP kinase/ERK signaling. *Current Biology: CB*, *16*, 1366–1372.

Hwang, M., Yong, C., Moretti, L., & Lu, B. (2007). Zebrafish as a model system to screen radiation modifiers. *Current Genomics*, *8*, 360–369.

Imanirad, P., Solaimani Kartalaei, P., Crisan, M., Vink, C., Yamada-Inagawa, T., de Pater, E., et al. (2014). HIF1alpha is a regulator of hematopoietic progenitor and stem cell development in hypoxic sites of the mouse embryo. *Stem Cell Research*, *12*, 24–35.

Jing, L., Tamplin, O. J., Chen, M. J., Deng, Q., Patterson, S., Kim, P. G., et al. (2015). Adenosine signaling promotes hematopoietic stem and progenitor cell emergence. *The Journal of Experimental Medicine*, *212*, 649–663.

Kalev-Zylinska, M. L., Horsfield, J. A., Flores, M. V., Postlethwait, J. H., Vitas, M. R., Baas, A. M., et al. (2002). Runx1 is required for zebrafish blood and vessel development and expression of a human RUNX1-CBF2T1 transgene advances a model for studies of leukemogenesis. *Development*, *129*, 2015–2030.

Kierdorf, K., Erny, D., Goldmann, T., Sander, V., Schulz, C., Perdiguero, E. G., et al. (2013). Microglia emerge from erythromyeloid precursors via Pu.1- and Irf8-dependent pathways. *Nature Neuroscience*, *16*, 273–280.

Kim, P. G., Nakano, H., Das, P. P., Chen, M. J., Rowe, R. G., Chou, S. S., et al. (2015). Flow-induced protein kinase A-CREB pathway acts via BMP signaling to promote HSC emergence. *The Journal of Experimental Medicine*, *212*, 633–648.

Kosmider, O., Gelsi-Boyer, V., Cheok, M., Grabar, S., Della-Valle, V., Picard, F., et al. (2009). TET2 mutation is an independent favorable prognostic factor in myelodysplastic syndromes (MDSs). *Blood*, *114*, 3285–3291.

Kwan, W., Cortes, M., Frost, I., Esain, V., Theodore, L. N., Liu, S. Y., et al. (2016). The central nervous system regulates embryonic HSPC production via stress-responsive glucocorticoid receptor signaling. *Cell Stem Cell*, *19*, 370–382.

Langenau, D. M., Traver, D., Ferrando, A. A., Kutok, J. L., Aster, J. C., Kanki, J. P., et al. (2003). Myc-induced T cell leukemia in transgenic zebrafish. *Science*, *299*, 887–890.

Lee, Y., Manegold, J. E., Kim, A. D., Pouget, C., Stachura, D. L., Clements, W. K., et al. (2014). FGF signalling specifies haematopoietic stem cells through its regulation of somitic notch signalling. *Nature Communications*, *5*, 5583.

Li, Y., Esain, V., Teng, L., Xu, J., Kwan, W., Frost, I. M., et al. (2014). Inflammatory signaling regulates embryonic hematopoietic stem and progenitor cell production. *Genes & Development*, *28*, 2597–2612.

Li, P., Lahvic, J. L., Binder, V., Pugach, E. K., Riley, E. B., Tamplin, O. J., et al. (2015). Epoxyeicosatrienoic acids enhance embryonic haematopoiesis and adult marrow engraftment. *Nature*, *523*, 468–471.

Li, C., Lan, Y., Schwartz-Orbach, L., Korol, E., Tahiliani, M., Evans, T., et al. (2015). Overlapping requirements for Tet2 and Tet3 in normal development and hematopoietic stem cell emergence. *Cell Reports*, *12*, 1133–1143.

Liao, W., Bisgrove, B. W., Sawyer, H., Hug, B., Bell, B., Peters, K., et al. (1997). The zebrafish gene cloche acts upstream of a flk-1 homologue to regulate endothelial cell differentiation. *Development*, *124*, 381–389.

Liao, E. C., Paw, B. H., Peters, L. L., Zapata, A., Pratt, S. J., Do, C. P., et al. (2000). Hereditary spherocytosis in zebrafish riesling illustrates evolution of erythroid beta-spectrin structure, and function in red cell morphogenesis and membrane stability. *Development*, *127*, 5123–5132.

Lyons, S. E., Lawson, N. D., Lei, L., Bennett, P. E., Weinstein, B. M., & Liu, P. P. (2002). A nonsense mutation in zebrafish gata1 causes the bloodless phenotype in vlad tepes. *Proceedings of the National Academy of Sciences of the United States of America*, *99*, 5454–5459.

Medvinsky, A., & Dzierzak, E. (1996). Definitive hematopoiesis is autonomously initiated by the AGM region. *Cell*, *86*, 897–906.

Mikkola, H. K., Fujiwara, Y., Schlaeger, T. M., Traver, D., & Orkin, S. H. (2003). Expression of CD41 marks the initiation of definitive hematopoiesis in the mouse embryo. *Blood*, *101*, 508–516.

Mims, M. P., Guan, Y., Pospisilova, D., Priwitzerova, M., Indrak, K., Ponka, P., et al. (2005). Identification of a human mutation of DMT1 in a patient with microcytic anemia and iron overload. *Blood*, *105*, 1337–1342.

Mizgirev, I. V., & Revskoy, S. (2010). A new zebrafish model for experimental leukemia therapy. *Cancer Biology & Therapy*, *9*, 895–902.
Mucenski, M. L., McLain, K., Kier, A. B., Swerdlow, S. H., Schreiner, C. M., Miller, T. A., et al. (1991). A functional c-myb gene is required for normal murine fetal hepatic hematopoiesis. *Cell*, *65*, 677–689.
North, T. E., de Bruijn, M. F., Stacy, T., Talebian, L., Lind, E., Robin, C., et al. (2002). Runx1 expression marks long-term repopulating hematopoietic stem cells in the midgestation mouse embryo. *Immunity*, *16*, 661–672.
North, T. E., Goessling, W., Peeters, M., Li, P., Ceol, C., Lord, A. M., et al. (2009). Hematopoietic stem cell development is dependent on blood flow. *Cell*, *137*, 736–748.
North, T. E., Goessling, W., Walkley, C. R., Lengerke, C., Kopani, K. R., Lord, A. M., et al. (2007). Prostaglandin E2 regulates vertebrate haematopoietic stem cell homeostasis. *Nature*, *447*, 1007–1011.
North, T., Gu, T. L., Stacy, T., Wang, Q., Howard, L., Binder, M., et al. (1999). Cbfa2 is required for the formation of intra-aortic hematopoietic clusters. *Development*, *126*, 2563–2575.
Paw, B. H., Davidson, A. J., Zhou, Y., Li, R., Pratt, S. J., Lee, C., et al. (2003). Cell-specific mitotic defect and dyserythropoiesis associated with erythroid band 3 deficiency. *Nature Genetics*, *34*, 59–64.
Payne, E. M., Virgilio, M., Narla, A., Sun, H., Levine, M., Paw, B. H., et al. (2012). L-Leucine improves the anemia and developmental defects associated with Diamond-Blackfan anemia and del(5q) MDS by activating the mTOR pathway. *Blood*, *120*, 2214–2224.
Peng, X., Dong, M., Ma, L., Jia, X. E., Mao, J., Jin, C., et al. (2015). A point mutation of zebrafish c-cbl gene in the ring finger domain produces a phenotype mimicking human myeloproliferative disease. *Leukemia*, *29*, 2355–2365.
Pereboom, T. C., van Weele, L. J., Bondt, A., & MacInnes, A. W. (2011). A zebrafish model of dyskeratosis congenita reveals hematopoietic stem cell formation failure resulting from ribosomal protein-mediated p53 stabilization. *Blood*, *118*, 5458–5465.
Peterson, R. T., Link, B. A., Dowling, J. E., & Schreiber, S. L. (2000). Small molecule developmental screens reveal the logic and timing of vertebrate development. *Proceedings of the National Academy of Sciences of the United States of America*, *97*, 12965–12969.
Pouget, C., Peterkin, T., Simoes, F. C., Lee, Y., Traver, D., & Patient, R. (2014). FGF signalling restricts haematopoietic stem cell specification via modulation of the BMP pathway. *Nature Communications*, *5*, 5588.
Ransom, D. G., Bahary, N., Niss, K., Traver, D., Burns, C., Trede, N. S., et al. (2004). The zebrafish moonshine gene encodes transcriptional intermediary factor 1gamma, an essential regulator of hematopoiesis. *PLoS Biology*, *2*, E237.
Ransom, D. G., Haffter, P., Odenthal, J., Brownlie, A., Vogelsang, E., Kelsh, R. N., et al. (1996). Characterization of zebrafish mutants with defects in embryonic hematopoiesis. *Development*, *123*, 311–319.
Reischauer, S., Stone, O. A., Villasenor, A., Chi, N., Jin, S. W., Martin, M., et al. (2016). Cloche is a bHLH-PAS transcription factor that drives haemato-vascular specification. *Nature*, *535*, 294–298.
Rombough, P., & Drader, H. (2009). Hemoglobin enhances oxygen uptake in larval zebrafish (Danio rerio) but only under conditions of extreme hypoxia. *The Journal of Experimental Biology*, *212*, 778–784.
Ronn, R. E., Guibentif, C., Moraghebi, R., Chaves, P., Saxena, S., Garcia, B., et al. (2015). Retinoic acid regulates hematopoietic development from human pluripotent stem cells. *Stem Cell Reports*, *4*, 269–281.

Rossi, A., Kontarakis, Z., Gerri, C., Nolte, H., Holper, S., Kruger, M., et al. (2015). Genetic compensation induced by deleterious mutations but not gene knockdowns. *Nature, 524*, 230–233.

Sabaawy, H. E., Azuma, M., Embree, L. J., Tsai, H. J., Starost, M. F., & Hickstein, D. D. (2006). TEL-AML1 transgenic zebrafish model of precursor B cell acute lymphoblastic leukemia. *Proceedings of the National Academy of Sciences of the United States of America, 103*, 15166–15171.

Sakamoto, H., Dai, G., Tsujino, K., Hashimoto, K., Huang, X., Fujimoto, T., et al. (2006). Proper levels of c-Myb are discretely defined at distinct steps of hematopoietic cell development. *Blood, 108*, 896–903.

Sawamiphak, S., Kontarakis, Z., & Stainier, D. Y. (2014). Interferon gamma signaling positively regulates hematopoietic stem cell emergence. *Developmental Cell, 31*, 640–653.

Shafizadeh, E., Paw, B. H., Foott, H., Liao, E. C., Barut, B. A., Cope, J. J., et al. (2002). Characterization of zebrafish merlot/chablis as non-mammalian vertebrate models for severe congenital anemia due to protein 4.1 deficiency. *Development, 129*, 4359–4370.

Sood, R., English, M. A., Belele, C. L., Jin, H., Bishop, K., Haskins, R., et al. (2010). Development of multilineage adult hematopoiesis in the zebrafish with a runx1 truncation mutation. *Blood, 115*, 2806–2809.

Soza-Ried, C., Hess, I., Netuschil, N., Schorpp, M., & Boehm, T. (2010). Essential role of c-myb in definitive hematopoiesis is evolutionarily conserved. *Proceedings of the National Academy of Sciences of the United States of America, 107*, 17304–17308.

Stachura, D. L., Reyes, J. R., Bartunek, P., Paw, B. H., Zon, L. I., & Traver, D. (2009). Zebrafish kidney stromal cell lines support multilineage hematopoiesis. *Blood, 114*, 279–289.

Stachura, D. L., Svoboda, O., Lau, R. P., Balla, K. M., Zon, L. I., Bartunek, P., et al. (2011). Clonal analysis of hematopoietic progenitor cells in the zebrafish. *Blood, 118*, 1274–1282.

Stainier, D. Y., Weinstein, B. M., Detrich, H. W., 3rd, Zon, L. I., & Fishman, M. C. (1995). Cloche, an early acting zebrafish gene, is required by both the endothelial and hematopoietic lineages. *Development, 121*, 3141–3150.

Tamplin, O. J., Durand, E. M., Carr, L. A., Childs, S. J., Hagedorn, E. J., Li, P., et al. (2015). Hematopoietic stem cell arrival triggers dynamic remodeling of the perivascular niche. *Cell, 160*, 241–252.

Tang, Q., Abdelfattah, N. S., Blackburn, J. S., Moore, J. C., Martinez, S. A., Moore, F. E., et al. (2014). Optimized cell transplantation using adult rag2 mutant zebrafish. *Nature Methods, 11*, 821–824.

Tang, Q., Moore, J. C., Ignatius, M. S., Tenente, I. M., Hayes, M. N., Garcia, E. G., et al. (2016). Imaging tumour cell heterogeneity following cell transplantation into optically clear immune-deficient zebrafish. *Nature Communications*, 7, 10358.

Taylor, A. M., Humphries, J. M., White, R. M., Murphey, R. D., Burns, C. E., & Zon, L. I. (2012). Hematopoietic defects in rps29 mutant zebrafish depend upon p53 activation. *Experimental Hematology, 40*, 228–237.e5.

Traver, D., Paw, B. H., Poss, K. D., Penberthy, W. T., Lin, S., & Zon, L. I. (2003). Transplantation and in vivo imaging of multilineage engraftment in zebrafish bloodless mutants. *Nature Immunology, 4*, 1238–1246.

Traver, D., Winzeler, A., Stern, H. M., Mayhall, E. A., Langenau, D. M., Kutok, J. L., et al. (2004). Effects of lethal irradiation in zebrafish and rescue by hematopoietic cell transplantation. *Blood, 104*, 1298–1305.

van Rooijen, E., Voest, E. E., Logister, I., Korving, J., Schwerte, T., Schulte-Merker, S., et al. (2009). Zebrafish mutants in the von Hippel-Lindau tumor suppressor display a hypoxic response and recapitulate key aspects of Chuvash polycythemia. *Blood, 113*, 6449–6460.

Weinstein, B. M., Schier, A. F., Abdelilah, S., Malicki, J., Solnica-Krezel, L., Stemple, D. L., et al. (1996). Hematopoietic mutations in the zebrafish. *Development, 123*, 303–309.
White, R. M., Sessa, A., Burke, C., Bowman, T., LeBlanc, J., Ceol, C., et al. (2008). Transparent adult zebrafish as a tool for in vivo transplantation analysis. *Cell Stem Cell, 2*, 183–189.
Wingert, R. A., Brownlie, A., Galloway, J. L., Dooley, K., Fraenkel, P., Axe, J. L., et al. (2004). The chianti zebrafish mutant provides a model for erythroid-specific disruption of transferrin receptor 1. *Development, 131*, 6225–6235.
Yeh, J. R., Munson, K. M., Chao, Y. L., Peterson, Q. P., Macrae, C. A., & Peterson, R. T. (2008). AML1-ETO reprograms hematopoietic cell fate by downregulating scl expression. *Development, 135*, 401–410.
Yeh, J. R., Munson, K. M., Elagib, K. E., Goldfarb, A. N., Sweetser, D. A., & Peterson, R. T. (2009). Discovering chemical modifiers of oncogene-regulated hematopoietic differentiation. *Nature Chemical Biology, 5*, 236–243.
Zhang, Z., Jia, H., Zhang, Q., Wan, Y., Zhou, Y., Jia, Q., et al. (2013). Assessment of hematopoietic failure due to Rpl11 deficiency in a zebrafish model of Diamond-Blackfan anemia by deep sequencing. *BMC Genomics, 14*, 896.
Zhang, Y., Morimoto, K., Danilova, N., Zhang, B., & Lin, S. (2012). Zebrafish models for dyskeratosis congenita reveal critical roles of p53 activation contributing to hematopoietic defects through RNA processing. *PLoS One*, 7, e30188.
Zhang, C., Patient, R., & Liu, F. (2013). Hematopoietic stem cell development and regulatory signaling in zebrafish. *Biochimica et Biophysica Acta, 1830*, 2370–2374.

CHAPTER FIVE

Making It New Again: Insight Into Liver Development, Regeneration, and Disease From Zebrafish Research

Shuang Wang*, Sophie R. Miller†, Elke A. Ober†, Kirsten C. Sadler*,‡,1

*Icahn School of Medicine at Mount Sinai, New York, NY, United States
†Danish Stem Cell Center (DanStem), University of Copenhagen, Copenhagen N, Denmark
‡New York University Abu Dhabi, Abu Dhabi, United Arab Emirates
[1]Corresponding author: e-mail address: kirsten.edepli@nyu.edu

Contents

Abstract

The adult liver of most vertebrates is predominantly comprised of hepatocytes. However, these cells must work in concert with biliary, stellate, vascular, and immune cells to accomplish the vast array of hepatic functions required for physiological homeostasis. Our understanding of liver development was accelerated as zebrafish emerged as an ideal vertebrate system to study embryogenesis. Through work in zebrafish and other models, it is now clear that the cells in the liver develop in a coordinated fashion during

Current Topics in Developmental Biology, Volume 124
ISSN 0070-2153
http://dx.doi.org/10.1016/bs.ctdb.2016.11.012

embryogenesis through a complex yet incompletely understood set of molecular guidelines. Zebrafish research has uncovered many key players that govern the acquisition of hepatic potential, cell fate, and plasticity. Although rare, some hepatobiliary diseases—especially biliary atresia—are caused by developmental defects; we discuss how research using zebrafish to study liver development has informed our understanding of and approaches to liver disease. The liver can be injured in response to an array of stressors including viral, mechanical/surgical, toxin-induced, immune-mediated, or inborn defects in metabolism. The liver has thus evolved the capacity to efficiently repair and regenerate. We discuss the emerging field of using zebrafish to study liver regeneration and highlight recent advances where zebrafish genetics and imaging approaches have provided novel insights into how cell plasticity contributes to liver regeneration.

1. INTRODUCTION

The liver carries out a diverse array of functions that are essential to physiological homeostasis. Among these are the ability to synthesize serum proteins including clotting factors, lipoproteins, and proteins such as albumin that generate colloidal pressure of serum, urea detoxification, energy storage, metabolism of xenobiotics, including most drugs, toxins, and alcohol, the ability to respond to insulin to maintain energy homeostasis and the production of bile which aids in digestion. All of these functions are conserved from mammals to fish and work in zebrafish on liver development and disease over the last decade has complimented nearly a century of work in mammalian systems. These topics have recently been reviewed by experts in the field (Goessling & Sadler, 2015; Goessling & Stainier, 2016; Wilkins & Pack, 2013).

The breadth of knowledge distilled from research across the vertebrate lineage has provided a rich understanding of how the liver forms, functions, regenerates, and what processes can cause diseases in this organ. In recent years, there is an increasing appreciation of how pathways important for hepatogenesis also contribute to liver regeneration and disease. Here, we describe work to uncover the basis of hepatobiliary development and regeneration in zebrafish and discuss how this work has provided insight into several diseases.

Loss of liver function is fatal and hepatic failure is a leading cause of death, accounting for 1.4% of all deaths in the United States annually (Murphy, Xu, & Kochanek, 2012). Liver-related mortality is even higher in regions of the world where viral hepatitis infection is poorly controlled or health care options are limited (Lozano et al., 2012). Mammals have a notoriously poor ability to regenerate just about every organ, except for the liver, which can mobilize either differentiated hepatocytes or progenitor cells to repeatedly repopulate the liver following injury. Despite this, the level of liver injury caused by toxin overload, metabolic imbalance, infection or other insults can overwhelm its regenerative capacity, leading to liver failure. Thus far, the only curative treatment for liver failure is liver transplantation, which is severely limited by the availability of liver donors (Martin, DiMartini, Feng, Brown, & Fallon, 2013). Devising alternatives to liver transplant or expanding the potential donor pool is a major goal of hepatic regenerative medicine. Promoting regeneration by reinitiating liver development, generating progenitor-like cells that can repopulate the damaged liver or boosting the growth of a small piece of transplanted liver are all areas that hold promise for achieving this goal. Progress will be expedited by research that provides a better understanding of the genes and pathways important for liver development and regeneration. In addition to the liver, the heart, fin, and spinal cord can also regenerate in zebrafish (reviewed by Gemberling, Bailey, Hyde, & Poss, 2013), making research in this model useful for providing insight into the general mechanisms that prevent effective regeneration in mammals. Here, we focus on the discoveries in the field of liver development and regeneration using zebrafish that have provided novel insights into how liver disease occurs and strategies to defend against it.

2. LIVER DEVELOPMENT: ZEBRAFISH AS A PARADIGM

As part of the digestive system, the liver is derived from endoderm that also gives rise to the other digestive organs (pancreas, gut, gall bladder). Hepatocytes derive from bipotential hepatobiliary progenitor cells and recent studies in zebrafish and other systems indicate that cell plasticity is important both during development and in repopulating the adult liver after injury. Liver development has been studied extensively in zebrafish and this topic has recently been thoroughly reviewed (Goessling & Stainier, 2016; Wilkins & Pack, 2013).

Compared to other organs discussed in chapters in this volume, the liver has a relatively simple structure. The hepatocyte comprises up to 60% of total cells of the liver—even higher in zebrafish—and nearly all of the critical hepatic functions are carried out by these cells. Indeed, every parameter of hepatic function examined has shown that the zebrafish liver performs the same functions ascribed to the mammalian liver, making this an excellent system to study hepatic physiology. There are many other structural and functional similarities between fish and mammals: the biliary network composed of biliary epithelial cells (BECs) arranged in a ductal system transports bile from hepatocytes to the gall bladder, a unique vasculature with a fenestrated endothelium facilitates transport of molecules from the blood to the hepatocytes, a set of resident immune cells, and the hepatic stellate cells (HSCs), which respond to injury by forming a fibrotic scar. All of these cell types, with the exception of Kupffer cells, the resident hepatic macrophage, have been identified in zebrafish. Table 1 outlines the transgenic lines commonly used to mark each of the hepatic cell types.

Work using zebrafish has uncovered many of the genes and pathways that are essential for hepatocyte and biliary development, and this has been extensively reviewed (Chu & Sadler, 2009; Goessling & Stainier, 2016). We provide an overview of the fundamental aspects of liver development that have been discovered using zebrafish with a focus on work showing that defects in biliary development can give rise to disease.

Table 1 Transgenic Zebrafish Provide Markers of Hepatic Cell Types

Cell Type	Transgenic Marker	Gene Marker	References
Hepatocytes	*Tg(fabp10a:dsRed)* *Tg(fabp10a:GFP)*	*Fatty acid-binding protein 10a*	Korzh, Emelyanov, and Korzh (2001), Mudumana, Wan, Singh, Korzh, and Gong (2004), Nguyen et al. (2012), Ober, Verkade, Field, and Stainier (2006), and Her, Yeh, and Wu (2003)
Biliary epithelial cells	*Tg(tp1bglob:eGFP)* *Tg(krt18:eGFP)*	*notch1* *keratin 18*	Parsons et al. (2009) Wilkins, Gong, and Pack (2014)
Stellate cells	*Tg(hand2:eGFP)*	*hand2*	Yin, Evason, Maher, and Stainier (2012)
Endothelial cells	*Tg(kdlr1:mCherry)*	*Kinase insert domain receptor like (also known as flk1)*	Sakaguchi, Sadler, Crosnier, and Stainier (2008)

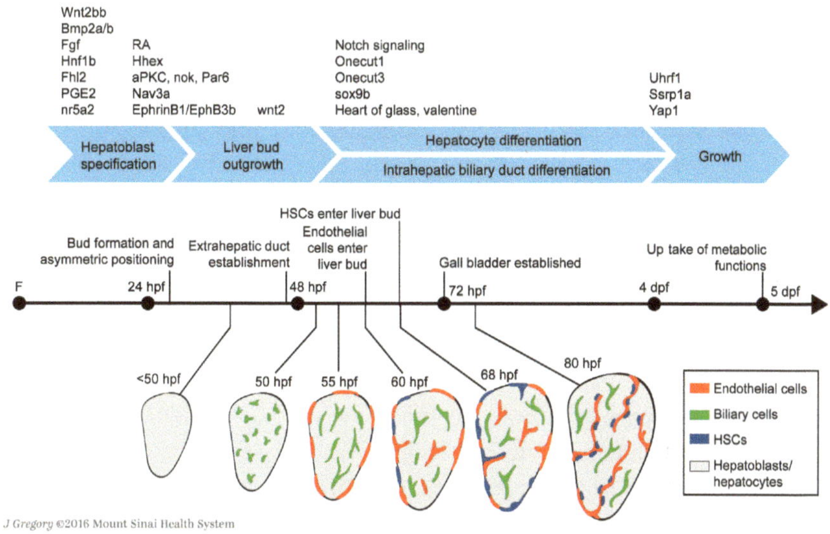

Fig. 1 Zebrafish develop a functional liver within 4 days. Detailed timeline and schematics of the key cellular process driving the differentiation of liver architecture are indicated below the phase-line. Molecular regulators mentioned in the text are listed above the developmental phase they are associated with; for a detailed review of the molecular mechanisms controlling liver development, see Goessling and Stainier (2016) and Wilkins and Pack (2013). *dpf*, days postfertilization; *hpf*, hours postfertilization. Printed with permission of ©Mount Sinai Health System.

2.1 The Origin of Hepatic Cell Types

Liver development in all species proceeds through three general phases: progenitor specification, cell type and tissue differentiation, and outgrowth (Fig. 1). While this process proceeds much faster in zebrafish, where it is complete in a few days compared to a few weeks in mammals, the genes and pathways that control each step are largely conserved (Chu & Sadler, 2009; Goessling & Stainier, 2016).

Hepatic specification in zebrafish begins early in development after the formation of the foregut endoderm around 24 h postfertilization (hpf). Hepatoblasts, the liver progenitors, are initially located at the embryonic midline and subsequently aggregate into a liver bud left to the midline (Field, Ober, Roeser, & Stainier, 2003). During specification, hepatoblasts, the liver progenitor cells, are marked by the expression of two key transcription factors: *hhex* and *prox1* (Ober et al., 2006; Wallace, Yusuff, Sonntag, Chin, & Pack, 2001). Hepatoblasts are bipotential and give rise to hepatocytes and to cholangiocytes, the BECs that form the intrahepatic bile ducts. Differentiation of hepatoblasts into mature hepatocytes and BECs begins between 40 and 60 hpf. Progressive differentiation is marked by the initiation of genes that participate in key hepatocyte functions, such as *ceruloplasmin* (Korzh et al., 2001), *fabp10a* (Venkatachalam, Thisse, Thisse, & Wright, 2009), and *gc* (Chu & Sadler, 2009; Goessling & Stainier, 2016; Noel, Reis, Arain, & Ober, 2010). Of note, the *fabp10a* promoter (also called *l-fabp*; Her et al., 2003) is an invaluable resource used by the zebrafish community to drive high expression of transgenes specifically in hepatocytes (see Table 1).

The biliary system in zebrafish is analogous to that of mammals, with a few unique features which are largely based on the difference in hepatic architecture (Yao et al., 2012) and the presence of a preductal epithelial biliary (PDEC) cell in zebrafish. PDECs connect the bile canaliculi of hepatocytes to the smaller branches of the biliary network, analogous to the Canal of Hering in mammals. PDECs then form larger bile ducts to transport the bile produced by the hepatocytes out of the liver (Hardman, Volz, Kullman, & Hinton, 2007; Yao et al., 2012). PDECs are too small to see histologically, but ductal structures distinguishable by 70 hpf by staining (Lorent et al., 2004) and through use of transgenics (Table 1). A reporter line that turns on GFP in cells with active Notch signaling (*Tg(Tp1bglob:eGFP)*; Parsons et al., 2009) highlights the first differentiating biliary cells as early as 45 hpf (Lorent, Moore, Siekmann, Lawson, & Pack, 2010) and *Tg(keratin18: GFP)* visualizes the forming of intra- and extrahepatic ducts from 3 days postfertilization (dpf) onwards (Wilkins et al., 2014). The forming biliary network can transport bile efficiently to the gallbladder (Farber et al., 2001) as early as 4 dpf.

Concomitant with the formation of the biliary network, vascular endothelial cells form a complementary network of their own (Fig. 1). The vascular network is not only essential for the outgrowth of this developing organ, it also provides instructive architectural cues to the developing hepatocytes and biliary cells. Hepatic endothelial cells originate from the adjacent posterior cardinal vein (Hen et al., 2015) and migrate to first surround by 54 hpf and subsequently invade the liver resulting in the formation of hepatic sinusoids and vessels by 72 hpf. Endothelial *Tg(kdrl:eGFP)* expression suggests that the endothelial cells migrate along the basal side of hepatocytes to form a branched vascular network by 80 hpf, coinciding with and promoting hepatocyte polarization (Sakaguchi et al., 2008). This intimate interaction between hepatocytes and endothelial cells allows for maximal exchange area between the blood and hepatocytes, an aspect of hepatic architecture that is important for the hepatocytes serving as the first pass at metabolism of compounds that arrive at the liver immediately after being absorbed in the gut. In mammals, hepatoblasts delaminate from the foregut epithelium and migrate as cords into the mesenchyme that is subsequently invaded by hematopoietic progenitor cells to form the liver bud (Medlock & Haar, 1983). Analysis of zebrafish mutants with vascular specification defects has uncovered a requirement for endothelial precursors during early liver development, which is conserved in mouse (Field et al., 2003; Korzh et al., 2008; Matsumoto, Yoshitomi, Rossant, & Zaret, 2001; Sakaguchi et al., 2008).

Prospective HSCs (see Fig. 1), expressing *Tg(hand2:eGFP),* invade the liver from the periphery and move along the developing vascular network starting around 68 hpf (Yin et al., 2012). Although HSCs associate with the vascular network, analysis of *cloche* mutants lacking sinusoidal endothelial cells (Stainier, Weinstein, Detrich, Zon, & Fishman, 1995) revealed that they are dispensable for HSC invasion into the liver and its differentiation (Yin et al., 2012). Whether HSCs are of neural crest, endothelial, or myofibroblast/lateral plate mesoderm (LPM) origin has been debated for a long time. Recent lineage-tracing experiments in mouse revealed that HSCs are generally mesoderm derived (Asahina et al., 2009). This was confirmed and further refined by work in zebrafish showing that HSCs are derived from the LPM and not endothelial cells, as *Tg(hand2:eGFP)*-expressing HSCs are present in *cloche* mutants (Yin et al., 2012). Thus, the interaction between several cell types is required for proper liver formation

The differentiation of the hepatic architecture is largely completed by 5 dpf (Lorent et al., 2004; Pack et al., 1996) while organ growth continues until adulthood. At 5 dpf, the liver takes up metabolic functions (Farber et al., 2001; Howarth, Yin, Yeh, & Sadler, 2013) and liver development is essentially complete. A detailed timeline of key cellular processes driving differentiation of the liver architecture is shown in Fig. 1.

2.2 Key Decisions: Liver or Pancreas?

The liver and pancreas share common progenitors in the anterior foregut endoderm. As discussed in chapter "Zebrafish Pancreas Development and Regeneration: Fishing for Diabetes Therapies" by Prince et al., the pancreas and liver both derive from foregut endoderm and organ fates are determined by both intrinsic and extrinsic factors that promote hepatic lineage and suppress the pancreatic fate. Single-cell lineage-tracing experiments in zebrafish using caged fluorescent dyes in six- to eight-somite stage embryos showed that the most lateral foregut cells give rise to liver and ventral/exocrine pancreas, while more medial cells will form the intestine and dorsal pancreas (Chung, Shin, & Stainier, 2008). Likewise, bipotential hepatopancreatic progenitors were identified in the mouse by embryonic tissue explant experiments and single-cell lineage tracing (Deutsch, Jung, Zheng, Lora, & Zaret, 2001; Miki et al., 2012; Rossi, Dunn, Hogan, & Zaret, 2001; Tremblay & Zaret, 2005).

Vertebrate liver specification represents a two-step process, in which anteroposterior patterning events in the early embryo establish in a broad endodermal region the competence to adopt a hepatic fate and the subsequent release of inductive signals by a tightly controlled signaling cascade. Hepatic competence has primarily been studied in mice, where pioneer transcription factors that bind chromatin and make it accessible for lineage-specific transcription factors have been extensively studied (Zaret & Mango, 2016). Two such pioneering transcription factors that have been studied in detail in zebrafish are Gata4 and Gata6. Depletion of either of these factors results in significantly smaller liver buds, while the double knockdown of both factors leads to the complete absence of hepatic *transferrin* expression, a marker of differentiated hepatocytes (Holtzinger & Evans, 2005). Bmp signaling promotes hepatic competence in vertebrates, in part by regulating Gata expression (Shin et al., 2007). In zebrafish, *bmp2b* is expressed in the LPM adjacent to the foregut endoderm preceding hepatic specification, and hepatoblast *hhex* and *prox1* expression is reduced when Bmp signaling is blocked during this period (Chung et al., 2008; Shin et al., 2007). Intriguingly, overexpression of *bmp2b* increases hepatoblasts at the expense of Pdx1-expressing ventral pancreas and intestinal progenitors (Chung et al., 2008), providing strong support for the fate switching functions of Bmp signaling. Subsequent transcriptome analysis uncovered the transcription factor *four and a half LIM domains 1b* (*fhl1b*) as a target of Bmp2b signaling and a critical mediator of its role in promoting hepatic fate vs the pancreas fate choice (Xu, Cui, Del Campo, & Shin, 2016). The description of how *pdx1* promotes pancreatic fate is described in chapter "Zebrafish in toxicology and environmental health" by Bambino and Chu.

2.3 Delivering a Fully Formed Organ

Both intrinsic and extrinsic signals are critical for formation of all organs, as discussed in nearly all chapters in this volume. In the liver, a breakthrough discovery identified liver-inductive signals following the establishment of hepatic competence in the early embryo through a forward genetic screen where the *prometheus/wnt2bb* mutant was discovered to have severely impaired hepatoblast formation (Ober et al., 2006). Other work showed that enhanced Wnt signaling in *apc* mutants or via Wnt2bb/Wnt8 overexpression induces ectopic liver progenitors at the expense of ventral pancreas and intestinal progenitors, placing Wnt signaling at the center of the hepatic induction phase of liver development (Goessling et al., 2008; Poulain & Ober, 2011; Shin, Lee, Poss, & Stainier, 2011). Wnt signaling, which requires Hnf1b and EpCam activity during hepatoblasts specification (Lancman et al., 2013; Lu, Ma, Yang, Shi, & Luo, 2013), is suppressed prior to hepatic induction (Goessling et al., 2008). It is then activated through Wnt2bb during hepatic induction, and further activated through additional Wnt ligands to promote proliferation of hepatoblasts following hepatic specification (Poulain & Ober, 2011). In support of this paradigm, *wnt2bb* is expressed in the LPM after *bmp2b* (Chung et al., 2008; Ober et al., 2006), supporting the model where Bmp signaling has a permissive function and Wnt signaling has an inductive function during hepatic specification (Goessling & Stainier, 2016).

Cytokine prostaglandin E_2 (PGE_2) signaling likely also has switch-like activity upstream of Bmp2b (Nissim et al., 2014). PGE_2 receptors *ep2a* and *ep4a* are expressed bilaterally in the foregut domain during early somitogenesis stages and *bmp2b* expression is reduced when these receptors are depleted. Intriguingly, treatment of mouse embryonic stem cells with high or low amounts of PGE_2 leads to fate switching between liver and pancreas differentiation (Nissim et al., 2014). This hepatogenic potential of PGE_2 signaling similarly promotes liver regeneration following partial hepatectomy in zebrafish and mouse (Goessling et al., 2009). Additional work from the Goessling group demonstrated that *nuclear receptor subfamily 5, group A, member 2* (*nr5a2*) is also important for determining liver vs pancreas fate. By using a combination of morpholinos, mutants, and drugs to target Nr5a2 in developing zebrafish, they identified a requirement during exocrine, but not endocrine, pancreas development, and for liver growth (Nissim et al., 2016). This suggests that Nr5a2 has different roles at different states of development, first, in priming the hepatopancreatic progenitors, and then in their outgrowth.

These findings are not only important for providing a full understanding of liver development, but are also of practical utility, as they can be utilized to design effective protocols to promote hepatic fate in culture. Similar to work described in chapter "Zebrafish in toxicology and environmental health" by Bambino and Chu where factors that drive pancreas development have been used to generate pancreatic beta cells in vitro, the progress in understanding the molecular switches regulating this early fate decision in the embryo has influenced protocols to differentiate mammalian embryonic or induced pluripotent stem cells into hepatocytes. In these protocols, many of the above described key signaling events occurring in embryonic liver development have been recapitulated (Gouon-Evans et al., 2006; Nissim et al., 2014; Si-Tayeb et al., 2010). Thus, work on the development of multiple organs can be combined and translated to clinical utility.

2.4 Organ Asymmetry: Positioning the Liver in the Body

A hallmark of the liver and other internal organs is their asymmetric arrangement with respect to the anteroposterior body axis allowing their efficient packing within the body cavity. Hepatoblasts are specified at the midline of the embryo and are placed on the left side in zebrafish and on the right side in mammals. We will focus on work in zebrafish that has established a detailed model of vertebrate liver positioning.

Seminal experiments in zebrafish embryos first identified the asymmetric migration of the bilateral neighboring LPM as a critical step for the leftward positioning of hepatoblasts and gut looping (Horne-Badovinac, Rebagliati, & Stainier, 2003). During gut looping, left LPM moves dorsal of the endoderm and the right LPM moves ventrolateral. In *heart and soul* mutants (Stainier et al., 1996), where protein kinase C, iota (*prkci*), is inactivated, the epithelial structure of the LPM and its asymmetric movement are disrupted, resulting in a midline liver and gut (Horne-Badovinac et al., 2003). Likewise, mutants with impaired extracellular matrix remodeling, specifically, laminin degradation, display defective LPM migration, and midline livers (Hochgreb-Hägele, Yin, Koo, Bronner, & Stainier, 2013; Yin, Kikuchi, Hochgreb, Poss, & Stainier, 2010). Together, these findings suggest that as the right LPM migrates left, it displaces the midline endoderm to the left by a motive force.

We found by *live*-imaging of sparsely labeled hepatoblasts that they migrate actively to form the liver bud on the left and that their directionality is provided by the repulsive activity in the right LPM (Cayuso et al., 2016). These cell–cell interactions involve long-cell protrusions and are coordinated by EphrinB1 and its receptor EphB3b, located in hepatoblasts and the LPM, respectively. The precise contributions of active hepatoblast migration and the LPMs pushing force to liver positioning remain to be determined. Another study used chemical screening identifying retinoic acid and specifically its receptor Rargb in control of organ laterality as early as specification stages (Garnaas et al., 2012), which has also been shown to play a role in laterality in mammals (Yasui, Morishima, Nakazawa, & Aikawa, 1998).

These phenotypes are highly reminiscent of human disorders such as heterotaxy (OMIM: 601086), in which the majority of internal organs are misaligned. The consequences can be disastrous: patients have cardiac defects (as discussed in chapter "Modeling syndromic congenital heart defects in zebrafish" by Grant et al.), cystic kidneys (discussed in chapter "Using zebrafish to study kidney development and disease" by Jerman and Sun), and some have biliary defects. Work using zebrafish has provided insight into the basic developmental biology of the liver and other endodermal organs has influenced stem cell protocols and has laid the foundation for identifying molecular mechanisms underlying left–right organ morphogenesis and associated disorders.

2.5 Liver Outgrowth

A number of aspects contribute to liver growth and size such as signaling pathways promoting proliferation, DNA replication, and the regulation of cell survival at the level of progenitors as well as differentiated cell types. A working hypothesis in the field is that there are some shared factors that control hepatic outgrowth in embryos and liver regeneration in adults and that some of these factors may be coopted by hepatocytes as they are transformed into malignant cells during hepatocarcinogenesis.

Following specification of the progenitor cell pool, the appropriate number of cells to form the full organ is dictated by the concerted action of Wnt, prostaglandin PGE_2, and nitric oxide signaling promoting proliferation (Cox et al., 2014; Goessling et al., 2008; Nissim et al., 2014; Poulain & Ober, 2011). In parallel, Id2a maintains hepatoblast number and survival (Khaliq, Choi, So, & Shin, 2015), whereas cell survival independent of proliferation is controlled through Anxa4 and Sorting nexin7 (Zhang et al., 2014). Loss-of-function mutants revealed that biliary cell survival requires Snapc4 (Schaub et al., 2012), while a set of genes required for mitochondrial processes and homeostasis, including *tomm22*, *subv3L1*, *trx2*, and *mtch2,* are vital for hepatocyte survival (Curado et al., 2010; Landgraf, Strobach, Kiess, & Korner, 2016; Schaub et al., 2012; Zhang et al., 2014). Whether these same genes are important in maintaining cell survival in adults is not yet known.

Understanding the mechanisms by which organ-specific programs regulate proliferation remains a central unanswered question in developmental biology. Given the complexity of cell proliferation, a significant number of factors have been identified mediating this cellular process at different levels. The transcription factors Cebpa and Klf6/Copeb regulate cell proliferation, with or without cell death (Yuan et al., 2015; Zhao et al., 2010). The *def* gene controls G1 to S-phase transition and small livers observed in mutants are due to cell cycle arrest (Chen et al., 2005). The loss or impaired function of several chromatin remodeling factors including Hdac1, Hdac3, and Ssrp1a results in small livers by 5 dpf generally due to reduced cell proliferation and DNA replication defects (Anderson et al., 2009; Farooq et al., 2008; Koltowska et al., 2013; Noel et al., 2008). Our work has also found that cell cycle arrest due to widespread epigenetic changes is important for hepatic outgrowth. Mutation of *the ubiquitin like with phd and ring finger domains 1* gene (*uhrf1*) does not affect liver specification (Sadler, Krahn, Gaur, & Ukomadu, 2007), but during outgrowth, arrests hepatocytes in S-phase of the cell cycle (Jacob et al., 2015). Uhrf1 functions as a cofactor for DNA methyltransferase 1 (*dnmt1*) to copy the pattern of DNA methylation during cell division (Bostick et al., 2007; Sharif et al., 2007). We showed that *uhrf1* mutant hepatocytes continually replicate their DNA, yet fail to divide and ultimately undergo apoptosis. Mutation in *dnmt1* shows the same phenotype (Jacob et al., 2015), indicating that reprogramming of the DNA methylome is the likely underlying mechanism of this unusual cell cycle phenotype.

Compelling evidence from zebrafish emphasizes that key genes involved in development also cause liver cancer. The involvement of the Hippo pathway in regulation of hepatic size and in liver cancer has been intensely studied in mice (Yimlamai, Fowl, & Camargo, 2015) and a recent study in zebrafish identified that Yap signaling promotes liver growth by supplying the cellular building blocks enabling cell proliferation, leading to both hepatomegaly and liver tumors in adults (Cox et al., 2016). Thus, in both the developing liver and liver cancer, the Hippo pathway is essential. Similarly, our work has shown that while loss of *uhrf1* blocks liver development (Sadler et al., 2007), overexpression of human UHRF1 in zebrafish hepatocytes causes liver cancer (Mudbhary et al., 2014). It will be interesting to determine if these progrowth and epigenetic pathways cooperate to promote liver cancer.

We speculate that many of the factors described here are also important for the continuous growth of the liver during larval maturation and adulthood. Several of these factors have been shown to also control liver regeneration in adult zebrafish and mammals, highlighting the parallels between hepatic outgrowth in embryos and regeneration in adults.

2.6 Biliary Development

When the liver anlage is established, bipotential hepatoblasts differentiate into hepatocytes and BECs. Notch signaling plays an important and recurrent role in the formation of the tubular biliary network (Lorent et al., 2010, 2004). Knockdown of Notch receptors or Jagged ligands at early stages of biliary cell development inhibits the formation of BECs leading to bile duct paucity (Lorent et al., 2004), similar to the dominant Jagged1 phenotypes in human Alagille Syndrome (OMIM: 118450), discussed later. Experiments to inhibit Notch during discrete developmental times showed that Notch regulates the remodeling of BECs into a functional network, including the formation of hepatocyte bile canaliculi (Lorent et al., 2010). Specifically, the assembly of specified BECs into a functional network requires Sox9b, which acts both as a target and mediator of Notch signaling (Delous et al., 2012; Manfroid et al., 2012; Xu et al., 2016). Biliary network malformations observed in *sox9b* mutant larvae do not recover and adult mutants exhibit signs of cholestasis with bile accumulated in liver and pancreas (Delous et al., 2012).

In addition the transcription factors Onecut1/Hnf6, Onecut3, and Vhnf1 (Matthews, Lorent, & Pack, 2008; Matthews, Lorent, Russo, & Pack, 2004) are all integral to development of the zebrafish hepatobiliary network, as in mammals (Lemaigre, 2009). The establishment of cell polarity and junctions is critical for hepatobiliary development in mammals (see review by Lemaigre, 2009) and work with zebrafish mutants confirms this conserved requirement for tight junctions in biliary development. Mutation of the tight junction protein *claudin15-like b* in zebrafish causes abnormal biliary ducts, canalicular malformations, and impaired hepatocyte polarity (Cheung et al., 2012).

Subsequent maintenance of the biliary network is regulated by Snapc4, a subunit of a small nuclear RNA-activating complex (Schaub et al., 2012). As hepatocytes seem unaffected in *snapc4* mutants, indicating that the function of Snapc4 is specific to biliary cells. In several of these studies, *live*-imaging of GFP-expressing biliary cells in cultured livers or in embryos provided invaluable insights into cell behaviors driving this dynamic morphogenetic process (Lorent et al., 2004; Ningappa et al., 2015). There are several tools to study the zebrafish biliary system (Table 1); most of them utilize confocal imaging of the liver followed by 3D rendering to obtain a comprehensive view of the biliary network.

In summary, despite the relative simplicity of the liver, the coordinated development of the different hepatic cell types is complex. By integrating studies in zebrafish with other models, including rodents and reptiles, we are achieving a deeper understanding of how the liver is patterned, how fate choices are made, and how liver growth is controlled. In a later section, we discuss how similar paradigms are being studied in the context of liver regeneration.

2.7 Biliary Disease: Alagille Syndrome and Beyond

Biliary atresia (BA) is a rare pediatric disease typically diagnosed in infancy that is caused by progressive obliteration of the bile ducts. This is the best-characterized relationship between a developmental defect and hepatobiliary disease. Loss of bile ducts results in accumulation of bile in the liver, which is highly toxic to hepatocytes, and BA is fatal without transplant. The prevailing theory in the field is that a failure in bile duct development is the fundamental cause of BA (Asai, Miethke, & Bezerra, 2015). Alagille syndrome is an inherited cause of BA that is also associated with cardiac and skeletal malformations caused by mutations in Jagged-1. Extensive studies in humans, mice, and zebrafish have demonstrated how Jagged/Notch signaling pathway functions in biliary specification and differentiation, and a subset of BA patients has mutations in one of the genes in this pathway (Flynn et al., 2004; Kodama, Hijikata, Kageyama, Shimotohno, & Chiba, 2004).

Other human genes altered in biliary disease have also been identified through work in zebrafish. Sequencing of the zebrafish genome revealed that about 70% of human genes have at least one obvious zebrafish counterpart, and this number increases to 82% when focusing on human disease-related genes (Howe et al., 2013). Together, with the ease for studying gene function and the high morphological and molecular similarity of tissue and organ development in zebrafish to that in humans makes zebrafish an ideal system for testing candidate disease factors identified by genome-wide association studies (GWAS). This has been particularly successful when studying genes regulating biliary development.

GWAS have identified several candidate genes for BA, including Adducin 3 (*ADD3*) and X-prolyl aminopeptidase 1 (XPNPEP1). A recent study in zebrafish found that only depletion of *add3* caused biliary defects similar to BA, prioritizing ADD3 as a BA susceptibility gene over XPNPEP1

(Tang et al., 2016). Another study of Caucasian BA cases identified a susceptibility locus on human chromosome 14q21.3, including the ARF6 gene (Ningappa et al., 2015). The two zebrafish homologs, *arf6a* and *arf6b* are expressed in the liver and knock down experiments showed they are required for early biliary development. In another study, knockdown of *vsp33b* and *vipar,* genes commonly mutated in arthrogryposis, renal dysfunction, and cholestasis syndrome (ARC) patients in zebrafish reproduced the biliary defects observed in ARC patients and established *vsp33b* and Vipar as drivers of this disease (Cullinane et al., 2010; Matthews et al., 2005). Finally, analysis of over 60 BA patients identified glypican 1 (GPC1) as a BA susceptibility gene, and BA patients have reduced GPC1 expression. The finding that bile duct development was defective in zebrafish *gpc1* morphants (Cui, Leyva-Vega, et al., 2013) provides support to this as a candidate gene for BA. However, many BA patients have no detectable mutations in any of the genes described earlier, leading to a second theory that a viral infection prevents biliary development.

Support for the viral basis of BA came from studies in zebrafish which found that activation of the interferon response in the liver was sufficient to cause both a reduction of number and shortening of biliary branches (Cui, EauClaire, & Matthews, 2013; Matthews et al., 2011). The interferon response was induced by injecting zebrafish larvae with a drug (5-azacytidine) that both blocks DNA methylation and causes DNA damage, and examination of samples from BA patients suggests that some have reduced DNA methylation in the biliary cells that remain in these diseased livers (Cui, EauClaire, et al., 2013; Matthews et al., 2011). Together, these data support a compelling hypothesis that either an intrinsic defect in the ability of biliary cells to faithfully copy their DNA methylome during cell division or that an external agent, such as an environmental toxin (see chapter "Zebrafish in toxicology and environmental health" by Bambino and Chu) can mimic a viral infection and potentially promote immune response-mediated destruction of biliary cells.

One of the most exciting studies in recent years in the field of biliary disease was carried out using zebrafish to identify a naturally occurring plant compound that can destroy the extrahepatic biliary duct (Lorent et al., 2015). Reports in the livestock literature indicated that in regions of Australia, lambs born to mothers who ate a specific plant would develop fatal BA. A team of basic and translational investigators set out to identify the toxic compound produced by this plant. Using zebrafish to screen extract from these plants, the damaging compound was identified as 4-methoxy-*seco*-betavulgarin, which they termed Biliatresone

(Lorent et al., 2015). Transcriptional profiling of livers with Biliatresone-induced liver injury revealed an increased expression of genes associated with redox stress (Zhao et al., 2016). A fluorescent sensor showed that extrahepatic biliary duct cells are particularly sensitive to redox stress in zebrafish, supported by a parallel study in mouse providing synergistic results (Waisbourd-Zinman et al., 2016; Zhao et al., 2016). Together, both studies outline a possible mechanism for human pathologies and highlight the power of cross-species investigations. Intriguingly, the mechanism of Biliatresone action may hold the key to identifying how other toxins or infections, which are more common in humans, can cause biliary disease. In summary, BA represents the most common liver disease caused by developmental defects in the liver. Zebrafish have provided not only an optimal system to test the functional relevance of candidate BA genes, but also as a system to discover new pathways that are important for biliary development and for BA.

3. LIVER REGENERATION

The liver is the body's primary defense against endogenous and environmental stressors, and as such, it harbors an amazing intrinsic ability to regenerate upon injury. Mammals are fantastic failures when it comes to regeneration, with the liver among the few exceptions. In contrast, zebrafish easily regrow damaged or resected organs including the heart, pancreas, spinal cord, and fins (Gemberling et al., 2013). The features distinguishing organisms with robust regenerative capacity from mammals, who show a poor regenerative response, is intensely investigated (Stoick-Cooper, Moon, & Weidinger, 2007). A major goal in regenerative medicine is to manipulate mammalian cells so that they can be more like fish and repair or replace damaged organs.

3.1 Mechanisms of Liver Regeneration

The almost unlimited proliferative capacity of hepatocytes is central to the liver's regenerative potential. Hepatocytes are quiescent under normal physiological conditions but can be stimulated to grow and regain liver function when the liver is damaged and new liver mass is needed to maintain homeostasis. This property is demonstrated through serial hepatocyte transplantation experiments using rodent models of liver injury (Overturf, al-Dhalimy, Ou, Finegold, & Grompe, 1997; Rhim, Sandgren, Degen, Palmiter, & Brinster, 1994) whereby mature hepatocytes repopulate the damaged liver. This stands in stark contrast to tissues like skin or gut, where lineage-specific stem cells proliferate and differentiate to replace the loss of mature cells. However, in response to severe liver injury that renders hepatocytes incapable of efficient cell division, it is thought that bipotential hepatobiliary progenitor-like cells are activated to replace the damaged liver. This response has recently been elegantly investigated in zebrafish (Choi, Ninov, Stainier, & Shin, 2014; He, Lu, Zou, & Luo, 2014). Some evidence shows that parenchymal cells can be induced to transdifferentiate to regenerate functional hepatocytes (Michalopoulos & Khan, 2015), yet it is unclear whether this mechanism of regeneration is conserved across species.

Decades of studies on liver regeneration in mammals have revealed a complex network of mitogens and downstream signaling pathways as essential for sustaining liver function and stimulating proliferation when hepatocytes are compromised or their mass is reduced. This has been extensively reviewed (Michalopoulos, 2014; Taub, 2004) and Fig. 2 summarizes the basic mechanisms of liver regeneration in mice and zebrafish. We discuss the common mechanisms that guide liver regeneration in vertebrates where zebrafish research has provided unique insights.

Compared to rodent models, zebrafish embryos and larvae provide several advantages such as ease of manipulation, state-of-the-art imaging, the ability to carry out chemical screens and generate mutants and transgenics (Table 1). These characteristics of zebrafish have been applied to the study of liver regeneration, and provided a unique lens with which to view this intensively studied and complex process. Enhancing the ability of the liver to regenerate, in turn, could improve outcomes for patients awaiting transplantation, for those receiving a small-for-size transplant, and could enhance the regrowth of the remaining liver tissue in living donors.

3.2 Stimulating Regeneration: Cut or Kill

Since the hepatocyte proliferative capacity can be exhausted in end-stage liver failure, an alternative to patient-derived hepatocytes is sought. Two potential different mechanisms utilized by the liver to regenerate lost mass have been identified through work in zebrafish and mammals, outlined in Fig. 2. First, hepatocytes serve as the source of more hepatocytes, and can be stimulated to proliferate. The fact that fully differentiated hepatocytes can in effect serve as their own "stem cells" under most physiological conditions suggests that replacing failed livers with mature hepatocytes could augment liver function.

A number of experimental approaches have been developed in mammals and zebrafish to study liver regeneration. These can be classified into two

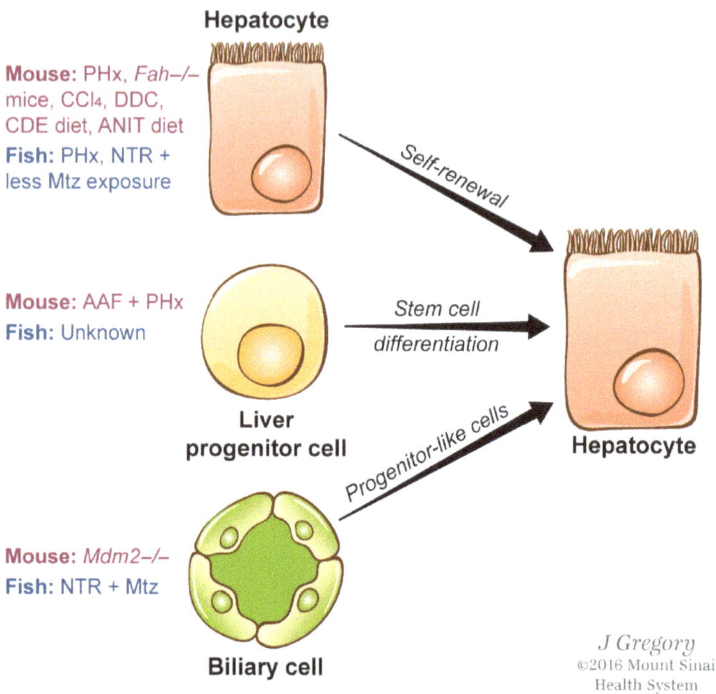

Fig. 2 The liver regenerates itself through three possible mechanisms. Under physiological conditions or mild liver injury, mature hepatocytes proliferate to replace lost hepatocytes. Under severe stress that inhibits or exhausts the proliferative capacity of hepatocytes, progenitor-like cells, which may be derived from the biliary lineage, proliferate, and differentiate into mature hepatocytes. *ANIT*, alpha-naphthyl-isothiocyanate; *CCl_4*, carbon tetrachloride; *CDE*, choline-deficient, ethionine-supplemented; *DDC*, 3,5-diethoxycarbonyl-1,4-dihydrocollidine; *Fah−/−*, FAH-deficient mice; *Mdm2−/−*, MDM2-deficient mice; *NTR+Mtz*, nitroreductase transgenic zebrafish treated with metronidazole; *PH+AAF*, partial hepatectomy followed by 2-acetylaminofluorene treatment; *PHx*, partial hepatectomy. Printed with permission of ©Mount Sinai Health System.

categories: (1) surgical or (2) injury-induced, and a few studies combining both strategies. In surgical models, parts of the liver are surgically removed to trigger hypertrophy and proliferation of hepatocytes in the remnant lobes (Higgins & Anderson, 1931). In injury-induced models of liver regeneration, liver injury is introduced chemically, genetically, or by a combination of both. In chemically induced injury models, animals are exposed to hepatotoxins that trigger cell death in the liver followed by proliferation of remaining hepatocytes to replace dead cells. Similar principles are adopted in genetically induced liver injury models; where instead of chemicals, engineered genetic cytotoxins in hepatocytes produce continuous or trigger-induced cell death that lead to heightened hepatocyte proliferation. While injury-induced models more closely resemble the etiology of liver disease patients, surgical models provide a simplified model of regeneration that arguably led to the most mechanistic insights into a highly complex biological process.

3.2.1 Liver Resection in Mammals

One major advantage of the surgical model is the nearly century old observation that following partial hepatectomy, the remaining hepatocytes synchronously emerge from a state of quiescence and reenter the cell cycle (Higgins & Anderson, 1931). Within minutes of removing 70% of the liver mass in mammals, the enormous change in blood pressure due to loss of tissue mass initiates a series of signaling pathways, including induction in urokinase activity, Wnt/β-catenin, and Notch signaling (Michalopoulos, 2010). Within hours, cytokines and growth factors are activated to stimulate hepatocyte exit from quiescence (G_0) to enter the cell cycle.

Of all the cytokines that have been investigated, Interleukin-6 signaling alone controls the expression of ~36% of over 100 "immediate-early genes," a unique gene signature that characterize this early regenerative phase (Li et al., 2001). In a simplified system of cultured primary hepatocytes, a number of growth factors can promote hepatocyte proliferation including Hepatocyte growth factor, ligands of Epithelial growth factor receptors such as epithelial growth factor, transforming growth factor alpha, amphiregulin, hepatocyte growth factor, and heparin-binding EGF-like growth factor and fibroblast growth factors. These have been extensively studied. At around 24–30 h posthepatectomy, hepatocytes begin to prepare for S-phase by expressing DNA replication-related genes under the control of the retinoblastoma protein and various cyclins, which is followed by maximum DNA synthesis at around 48 h posthepatectomy. In this model, each hepatocyte proliferates roughly 1.6 times on average to make up for lost liver mass. A recent study of the response to removing different amounts of mouse liver discovered two distinct regenerative strategies adopted by the liver: following 30% hepatectomy, hepatocytes mainly expand in size to make up for the lost liver mass, whereas following 70% hepatectomy, hepatocytes both hypertroph and proliferate to recover lost mass (Miyaoka et al., 2012).

The physiological, metabolic, and hemodynamic changes that occur following partial hepatectomy are rapid and dynamic. This is a complex process that despite intense investigation by teams of outstanding scientists over several decades, a master regulator of regeneration has yet to be identified. This suggests that there is no "magic bullet" for promoting regeneration. Instead, attention is now focused on defining a strategy to improve regeneration that relies on coordinated manipulation of multiple pathways.

3.2.2 Liver Resection in Zebrafish

The impressive capacity to undergo epimorphic regeneration in several tissues such as fin, heart, and liver in zebrafish stands in stark contrast to mammals, where few organs can regenerate. In contrast, mammals regenerate the liver by compensatory regeneration, whereby hepatocytes proliferate in the remaining lobes of the liver. Compensation by moderate enlargement of the remaining lobes has been also documented in zebrafish, so that in this system, regeneration occurs by epimorphic regrowth of the resected lobe and compensatory enlargement of the remaining lobe to restore liver mass by 7 days following surgery (Goessling et al., 2008; Kan, Junghans, & Izpisua Belmonte, 2009; Sadler et al., 2007).

Consistent with the critical role of Wnt signaling in both development and cancer, this pathway is essential for hepatocyte proliferation during liver regeneration in both mammals and zebrafish. Mice deficient in β-catenin have decreased markers of cell proliferation after partial hepatectomy (Tan, Behari, Cieply, Michalopoulos, & Monga, 2006). In line with these findings, zebrafish mutants deficient in Wnt signaling have reduced liver progenitor populations (Poulain & Ober, 2011), whereas increased Wnt signaling as a result of *apc* mutations in zebrafish promoted both hepatic specification and liver regeneration in adult zebrafish following partial hepatectomy (Goessling et al., 2008). A chemical screen using zebrafish embryos identified a nonselective Cox inhibitor, indomethacin, which inhibited Wnt signaling and blocked liver specification in embryos and regeneration in adults through the suppression of PGE_2 production (Goessling et al., 2009; North et al., 2007). This suggests a novel link between inflammation, PGE_2 accumulation, and cellular proliferation. This work also highlights the reiterative use of pathways in liver development and regeneration.

The high level of conservation of the regenerative process exhibited by mouse and zebrafish livers following partial hepatectomy allowed investigators to apply zebrafish genetics to test the in vivo contribution of several pathways that have thus far only been shown to regulate hepatocyte proliferation in cell culture systems. Striking examples are transgenic zebrafish where transient and inducible overexpression of dominant negative variants of Fibroblast growth factor receptor (Fgfr) or Bone morphogenetic protein receptor (Bmpr) that can almost completely block signaling through these receptors. With this strategy, Kan et al. were able to overcome both the challenges of ligand/receptor multiplicity and establish the importance of Fgfr and Bmpr signaling in liver regeneration. Blocking either Fgfr or Bmpr signaling dramatically impaired liver regeneration in zebrafish following partial hepatectomy (Kan et al., 2009). These pathways are also essential for liver specification during development and promote proliferation in cultured hepatocytes (Taub, 2004) highlighting these as common mechanisms that regulate liver growth under physiological conditions.

3.2.3 Killing Hepatocytes in Mammals Using Toxins and Genetics

Hepatic failure occurs when the replicative capacity of hepatocytes is insufficient to replace massive damage. If the liver suffers widespread damage and hepatocyte proliferation is limited or if hepatocyte loss is so massive that the remaining cells cannot adequately and quickly repopulate the liver, an alternative mechanism to regenerate lost mass is activated: bipotential liver progenitor cells that can differentiate into both cells of the biliary and the hepatic lineage (i.e., oval cells in mammals) are activated to repopulate the liver (Michalopoulos & Khan, 2015). In this scenario, liver regeneration is dictated not only by hepatocyte proliferation but also on the degree of continuous cell death, by the extent of fibrosis, which may physically restrain liver growth, and by the ability to mobilize progenitor cells. We review the extensive use of mouse models of fulminate hepatic failure here and then discuss how work in zebrafish has extended the understanding of progenitor-like cells and regeneration (see Fig. 2).

Carbon tetrachloride is the most frequently used chemical to induce liver injury in mammals, as a single dose induces massive hepatocyte necrosis within 24 h (Wong, Chan, & Lee, 1998), which is efficiently repaired by replacement of the damaged hepatocytes. Repeated doses cause chronic hepatic injury with the associated inflammation, fibrosis, and activation of the immune response. Thus, chemically induced liver injury creates an environment that more closely resemble patients with chronic liver diseases, but use of this approach is complicated by the myriad of factors inherent in the response. In addition, systemically applied chemicals such as carbon tetrachloride may have off-target effects on nonliver tissues that indirectly influence the regenerative response in the liver.

Genetically induced liver injury models ameliorate both off-target effects and the nonuniform distribution associated with chemically induced liver injury. In both mice and humans with fumarylacetoacetate hydrolase (FAH) deficiency (OMIM: 613871), the lack of this critical enzyme for tyrosine metabolism in hepatocytes leads to accumulation of toxic metabolic intermediates and massive liver injury. If the hepatocytes are not replaced, mice die within 12 h after birth (Grompe et al., 1993) unless the production of the toxic metabolite is blocked with a drug. However, when the drug is removed, all hepatocytes die and the mice succumb to liver failure (Grompe et al., 1995) unless *Fah* expressing hepatocytes competent to engraft and proliferate are reintroduced to the liver. This model serves as a template to introduce alternative sources of hepatocytes to evaluate the ability of grafting and repopulating the liver and rescuing the mouse. This system allows direct comparison of the in vivo regenerative potential of the different cell types in the liver and establishes the available sources of new hepatocytes to regenerate the liver under different stressors.

In the FAH-deficient mouse model, lineage tracing of donor cells revealed that almost all the hepatocytes that repopulated the *Fah*−/− mouse liver originated from hepatocytes whereas only <1% originated from Sox9+ liver stem cells (Tarlow, Finegold, & Grompe, 2014). This is supported by lineage-tracing experiments in a range of chemically induced liver damage (Schaub, Malato, Gormond, & Willenbring, 2014; Yanger et al., 2014). In vivo lineage-tracing experiments, in another case of severe liver damage in Mdm2-deficient mice showed that liver was regenerated by expansion of biliary-derived progenitor cells (Lu et al., 2015). Thus, progenitors are a likely source of new hepatocytes in cases where regeneration cannot be accomplished by hepatocytes alone.

A unique attribute of hepatocytes suggested to promote their proliferative capacity is their ploidy. An elegant study using Fah-deficient mice demonstrated that under stressed conditions, the different ploidy states of hepatocytes dubbed the "ploidy conveyor," provide an additional source of genetic diversity that can be selected for to promote organism survival (Duncan et al., 2010, 2012). It is not clear whether zebrafish utilize this same mechanism, as hepatocyte ploidy has not been thoroughly investigated in this model system. It will be interesting to see whether similar mechanisms apply to the handful of other cell types in mammals that demonstrate polyploidy (Schoenfelder & Fox, 2015).

3.2.4 *Progenitor-Mediated Regeneration in Zebrafish: Seeing Is Believing*

Studies in zebrafish and mice demonstrated that biliary cells themselves may transdifferentiate into hepatocytes, which would occur through a dedifferentiated progenitor-like state. Two of these studies combined the beauty of zebrafish genetics with state-of-the-art imaging techniques to perform in vivo lineage-tracing experiments that unequivocally established biliary cells as the source of hepatocytes at least in one model of extensive hepatocyte loss.

Fig. 3 shows a zebrafish model analogous to the *Fah−/−* mouse, whereby massive hepatocyte death was induced in transgenic zebrafish by adding the nontoxic prodrug metronidazole to transgenic zebrafish that selectively express nitroreductase in hepatocytes (*Tg(fabp10:mChNTR)*) and converts metronidazole into a cytotoxic metabolite, followed by liver regeneration within days (Choi et al., 2014; Curado et al., 2007; He et al., 2014). By crossing *Tg(fabp10:mChNTR)* and other similar lines with transgenic zebrafish that express lineage-specific fluorescent markers

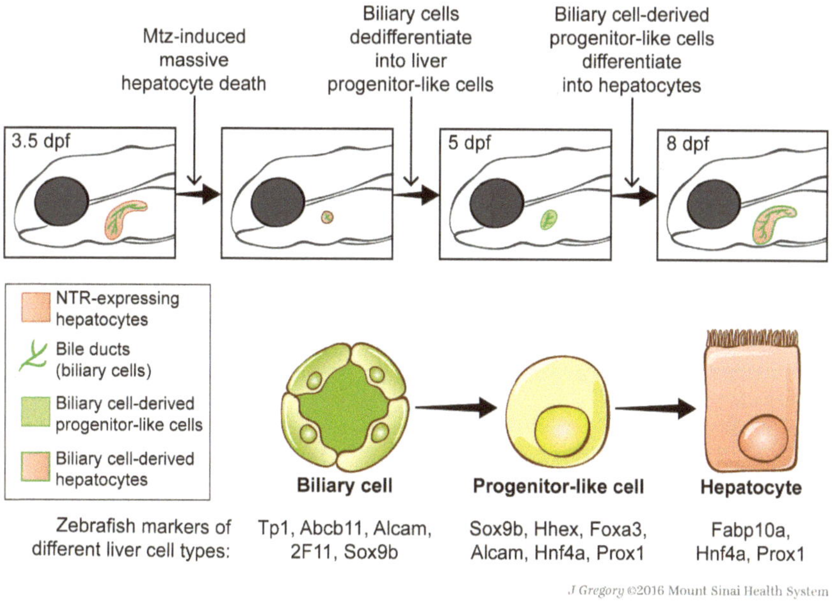

Fig. 3 Biliary cells as a potential source of hepatocytes in zebrafish with massive hepatocyte loss. Molecular markers used in these studies are listed in the *bottom* (Choi et al., 2014; He et al., 2014). *dpf,* days postfertilization; *Mtz,* metronidazole; *NTR,* nitroreductase. Printed with permission of ©Mount Sinai Health System.

(Table 1), elegant imaging studies during the entire time period of regeneration performed by two groups showed that transdifferentiation of biliary cells into hepatocytes occurred (Choi et al., 2014; He et al., 2014). Another study showed that this transdifferentiation occurred through transient Wnt and Notch signaling in cells which appeared to be an intermediate hepatobiliary cell (Huang et al., 2014).

Thus, it appears that under normal conditions or mild stress, hepatocytes serve as their own progenitors, but under extreme circumstances when the proliferation of hepatocytes is impaired or insufficient, a progenitor-like cell population likely derived from the biliary lineage can expand to fill that void (Fig. 2). By taking advantage of the unique attributes of zebrafish that allow imaging in real time during regeneration, zebrafish has been used to its best advantage to promote discoveries in liver regeneration.

3.3 Epigenetics Mechanisms Regulate Hepatocyte Proliferation

Most of the work to identify key genes involved in liver regeneration in mice and zebrafish has relied on manipulation of single genes. However, genome-wide gene expression studies established that the coordinated expression of thousands of genes is required for the liver to regenerate even in the comparatively simple partial hepatectomy model (Li et al., 2001; Su, Guidotti, Pezacki, Chisari, & Schultz, 2002). Current data suggest that no single signaling molecule or pathway alone is sufficient to drive this massive change in the gene expression landscape, suggesting that either complex cross-talk between different signaling pathways or yet uncharacterized higher-order processes can influence multiple pathways simultaneously. Epigenetics represent such higher-order processes that can alter the expression of many genes simultaneously, as is found during liver regeneration, yet its role in liver regeneration has only begun to be investigated.

The involvement of epigenetic mechanisms in the regeneration of the zebrafish fin, arguably the most widely studied model for tissue regeneration in zebrafish have uncovered a role for DNA methylation in this process. For example, de novo DNA methyltransferases are induced during zebrafish fin regeneration (Takayama, Shimoda, Takanaga, Hozumi, & Kikuchi, 2014) and DNA hypomethylation occurs early during fin regeneration, which was attributed to active DNA demethylation (Hirose, Shimoda, & Kikuchi, 2013). Whether repatterning of the DNA methylome is conserved in other regenerating tissues is not known; however, a recent exciting study demonstrated that epigenetic regulation of *leptin b* is important for the function of this gene in heart and fin regeneration (Kang et al., 2016). The "tissue regeneration enhancer element (TREE)" that was identified in this study was found to be upstream of other genes with known roles in zebrafish fin (*Fgf* effectors) and heart (*nrg1*) regeneration. The fact that TREEs are activated during the regeneration of different tissues and are conserved between zebrafish and mouse suggests that they may also be involved in liver regeneration, which represents an exciting future avenue of investigations.

Despite the ability to facilitate massive changes in gene expression profiles such as those found during liver regeneration, very few studies have investigated the involvement of DNA and histone modifications in this process. In fact, the most comprehensive studies that investigated the involvement of epigenetic processes on liver regeneration came from studies using zebrafish. Through a forward genetic screen in zebrafish embryos, we found that *uhrf1* mutants fail in hepatic outgrowth during development and heterozygote mutation for *uhrf1* represses liver regeneration following partial hepatectomy in adults. Transcriptome analysis of *uhrf1* mutants revealed a dramatic across-the-board induction of cell cycle genes that corresponded with cell cycle defects such as aberrant DNA replication, cell cycle arrest in S-phase, and cell death, leading to the observed liver growth defect (Jacob et al., 2015). Moreover, overexpression of Uhrf1 in zebrafish hepatocytes led to cellular senescence followed by uncontrolled cellular proliferation and rapid onset of liver cancer (Mudbhary et al., 2014), placing the epigenetic regulator Uhrf1 and DNA methylation at the center of the regulatory network in control of hepatocyte proliferation. How global changes in DNA methylation preferentially affect the expression of genes controlling cell proliferation during hepatic outgrowth in the developing zebrafish embryo, and whether the same epigenetic mechanisms apply to liver regeneration in adult mammals has yet to be determined.

Further evidence in support of epigenetic factors driving liver regeneration has emerged from work with zebrafish with metronidazole-induced liver injury in *Tg(fabp10a:CFP-NTR)* zebrafish (Ko et al., 2016). In this study, a screen of 44 compounds with known targets identified two distinct bromodomain and extraterminal domain (BET) inhibitors that almost completely blocked liver regeneration BET inhibitors, which prevent the binding of transcription factors and chromatin modifiers to acetylated histone tails (Shi & Vakoc, 2014), prevented the conversion of biliary cells to hepatoblasts and the proliferation of hepatoblasts (Ko et al., 2016). These results suggest that transcription factor binding to acetylated histone tails plays important roles in liver regeneration and are reminiscent of the requirement of TREEs during regeneration of the zebrafish fin. The finding that BET inhibitors similarly inhibited oval cell proliferation in mice suggests a conserved role for BET proteins and histone acetylation in the control of liver progenitor cell expansion. These studies highlight how the cutting edge in liver regeneration research is accelerated through work in zebrafish and other models.

4. STRENGTHS AND CHALLENGES OF USING ZEBRAFISH TO STUDY LIVER REGENERATION

The zebrafish represents one of the simplest vertebrate organisms that can be used to model the human liver. In zebrafish embryos, which conveniently develop ex vivo, mature hepatocytes can be observed as early as 72 hpf and the liver is fully functional by 5 dpf (Goessling & Sadler, 2015). This short timeframe for liver development makes zebrafish embryos a very cost-effective model to study liver processes such as liver regeneration. Additionally, fast development combined with the fact that each successful mating produces hundreds of embryos together make the zebrafish model highly adaptable to large-scale unbiased chemical/genetic screening that often yield novel and unexpected insights into the underlying biology, as demonstrated by the surprising discovery of epigenetic regulators in liver regeneration (Ko et al., 2016).

As the liver represents our body's first line of defense against environmental stressors, the scalability of the zebrafish system also provides a unique opportunity to screen single and/or complex mixtures of environmental toxicants for their impact on liver regeneration. Another major advantage of using zebrafish embryos to study liver regeneration lies within the imaging prowess of this model. In addition to wild-type zebrafish having optically transparent embryos, transgenic zebrafish lines have been established that fluorescently label every major cell population in the liver (Table 1), providing a comprehensive toolkit to evaluate the contribution of each cell type to liver regeneration in live animals. This is unmatched in any other model organism.

Like all experimental models, there are limitations to using zebrafish. As most major advantages of the zebrafish model are associated with the embryonic and larval stage, the majority of zebrafish studies investigating liver regeneration are carried out in young animals. Even though the liver by 5 dpf carries out all functions found in the adult mammalian liver, it is possible that the liver regeneration program in young zebrafish has unique properties. For example, recent lineage-tracing experiments in adult mice have failed to observe transdifferentiation of biliary cells into hepatocytes that is readily observed in the embryonic zebrafish liver following severe liver injury (Grompe, 2014). One potential explanation for the observed discrepancies is that embryonic biliary cells are less differentiated and thus harbor more stem cell-like, transdifferentiation potential than adult biliary cells, an important difference that needs to be taken into account for future investigations using the zebrafish embryonic system.

The small body size of zebrafish embryos makes them cost-effective and ideal for large-scale chemical/genetic screens; yet, the same attribute makes surgical manipulation highly unfeasible in this model and renders it difficult to directly compare to the mouse partial hepatectomy model for liver regeneration. The recent development of *Tg(fabp10a:mChNTR)* zebrafish may provide an opportunity to bridge this gap. Using partial ablation of hepatocytes which appears to activate a liver regeneration program that parallels the response to total ablation in larvae has suggested that similar programs are activated at both stages (Choi et al., 2014). This indicates that properties governing liver regeneration in larvae may be utilized throughout the zebrafish lifespan.

5. CONCLUDING REMARKS: HOW STUDYING LIVER DEVELOPMENT AND REGENERATION PROVIDES AN UNDERSTANDING OF LIVER DISEASE

Zebrafish have historically been used to study embryonic development, and, as a result of this focus for the first few decades when zebrafish were used as a model organism, a majority of studies on the zebrafish liver focused on development. Here, we have reviewed liver development and regeneration in zebrafish and compared the process of regeneration in zebrafish to that in mammals. From these basic studies, several insights have emerged that have informed our understanding of disease. There are extensive recent reviews on the use of zebrafish to study liver disease (Goessling & Sadler, 2015; Grompe, 2014; Wilkins & Pack, 2013). Here, we focus on the intersection between development and regeneration and focus on biliary disease as an illustrative example of how the basic processes of development have promoted our understanding of liver disease.

Work on biliary disease, in particular BA, serves as illustrative example of how use of zebrafish combined with our growing knowledge of developmental processes can be utilized to make exciting new discoveries in pathophysiology of liver and biliary disease. We have discussed how zebrafish research on hepatobiliary development has provided a framework to understand how the liver recovers from massive injury. Yet the most common causes of liver disease do not appear to have a developmental origin. Can studies in zebrafish address these, too? The answer is a most decidedly yes. Many labs are actively investigating the basis of diabetes and its impact on the liver, the process of hepatocellular transformation which generates liver cancer and the effects of hepatocyte injury on the wound healing response that when overstimulated, leads to liver fibrosis and cirrhosis. In addition, the growing community of toxicology researchers using zebrafish as discussed in chapter "Zebrafish in toxicology and environmental health" by Bambino and Chu is leading the charge to expand use of this system to identify causes and cures for toxin-induced liver injury.

The field is rapidly making gains into the fundamental in vivo biology of liver disease, as evidenced by the outstanding growth of the Zebrafish Disease Models Society (http://zdmsociety.org), in particular, in the area of liver and other gastroenterological diseases. We look forward to the impact that translational zebrafish studies will make in human disease.

ACKNOWLEDGMENTS

This work is supported by grants from the NIH (5R01DK080789 and 5R01AA018886 to K.C.S.) and the Novo Nordisk Foundation (NNF15CC0018344 to E.A.O.). We are grateful to Drs. Jaime Chu and Valerie Gouon-Evans for their critical reading of the manuscript and comments and especially to Jill Gregory for illustrations.

REFERENCES

Anderson, R. M., Bosch, J. A., Goll, M. G., Hesselson, D., Dong, P. D., Shin, D., et al. (2009). Loss of Dnmt1 catalytic activity reveals multiple roles for DNA methylation during pancreas development and regeneration. *Developmental Biology*, *334*, 213–223.

Asahina, K., Tsai, S. Y., Li, P., Ishii, M., Maxson, R. E., Sucov, H. M., et al. (2009). Mesenchymal origin of hepatic stellate cells, submesothelial cells, and perivascular mesenchymal cells during mouse liver development. *Hepatology*, *49*, 998–1011.

Asai, A., Miethke, A., & Bezerra, J. A. (2015). Pathogenesis of biliary atresia: Defining biology to understand clinical phenotypes. *Nature Reviews. Gastroenterology & Hepatology*, *12*, 342–352.

Bostick, M., Kim, J. K., Esteve, P. O., Clark, A., Pradhan, S., & Jacobsen, S. E. (2007). UHRF1 plays a role in maintaining DNA methylation in mammalian cells. *Science*, *317*, 1760–1764.

Cayuso, J., Dzementsei, A., Fischer, J. C., Karemore, G., Caviglia, S., Bartholdson, J., et al. (2016). EphrinB1/EphB3b coordinate bidirectional epithelial-mesenchymal interactions controlling liver morphogenesis and laterality. *Developmental Cell*, *39*, 316–328.

Chen, J., Ruan, H., Ng, S. M., Gao, C., Soo, H. M., Wu, W., et al. (2005). Loss of function of def selectively up-regulates Delta113p53 expression to arrest expansion growth of digestive organs in zebrafish. *Genes & Development*, *19*, 2900–2911.

Cheung, I. D., Bagnat, M., Ma, T. P., Datta, A., Evason, K., Moore, J. C., et al. (2012). Regulation of intrahepatic biliary duct morphogenesis by Claudin 15-like b. *Developmental Biology, 361*, 68–78.

Choi, T. Y., Ninov, N., Stainier, D. Y., & Shin, D. (2014). Extensive conversion of hepatic biliary epithelial cells to hepatocytes after near total loss of hepatocytes in zebrafish. *Gastroenterology, 146*, 776–788.

Chu, J., & Sadler, K. C. (2009). New school in liver development: Lessons from zebrafish. *Hepatology, 50*, 1656–1663.

Chung, W.-S., Shin, C. H., & Stainier, D. Y. R. (2008). Bmp2 signaling regulates the hepatic versus pancreatic fate decision. *Developmental Cell, 15*, 738–748.

Cox, A. G., Hwang, K. L., Brown, K. K., Evason, K. J., Beltz, S., Tsomides, A., et al. (2016). Yap reprograms glutamine metabolism to increase nucleotide biosynthesis and enable liver growth. *Nature Cell Biology, 18*, 886–896.

Cox, A. G., Saunders, D. C., Kelsey, P. B., Jr., Conway, A. A., Tesmenitsky, Y., Marchini, J. F., et al. (2014). S-nitrosothiol signaling regulates liver development and improves outcome following toxic liver injury. *Cell Reports, 6*, 56–69.

Cui, S., EauClaire, S. F., & Matthews, R. P. (2013). Interferon-gamma directly mediates developmental biliary defects. *Zebrafish, 10*, 177–183.

Cui, S., Leyva-Vega, M., Tsai, E. A., EauClaire, S. F., Glessner, J. T., Hakonarson, H., et al. (2013). Evidence from human and zebrafish that GPC1 is a biliary atresia susceptibility gene. *Gastroenterology, 144*, 1107–1115.e3.

Cullinane, A. R., Straatman-Iwanowska, A., Zaucker, A., Wakabayashi, Y., Bruce, C. K., Luo, G., et al. (2010). Mutations in VIPAR cause an arthrogryposis, renal dysfunction and cholestasis syndrome phenotype with defects in epithelial polarization. *Nature Genetics, 42*, 303–312.

Curado, S., Anderson, R. M., Jungblut, B., Mumm, J., Schroeter, E., & Stainier, D. Y. (2007). Conditional targeted cell ablation in zebrafish: A new tool for regeneration studies. *Developmental Dynamics, 236*, 1025–1035.

Curado, S., Ober, E. A., Walsh, S., Cortes-Hernandez, P., Verkade, H., Koehler, C. M., et al. (2010). The mitochondrial import gene tomm22 is specifically required for hepatocyte survival and provides a liver regeneration model. *Disease Models & Mechanisms, 3*, 486–495.

Delous, M., Yin, C., Shin, D., Ninov, N., Debrito Carten, J., Pan, L., et al. (2012). Sox9b is a key regulator of pancreaticobiliary ductal system development. *PLoS Genetics, 8*, e1002754.

Deutsch, G., Jung, J., Zheng, M., Lora, J., & Zaret, K. S. (2001). A bipotential precursor population for pancreas and liver within the embryonic endoderm. *Development, 128*, 871.

Duncan, A. W., Hanlon Newell, A. E., Bi, W., Finegold, M. J., Olson, S. B., Beaudet, A. L., et al. (2012). Aneuploidy as a mechanism for stress-induced liver adaptation. *Journal of Clinical Investigation, 122*, 3307–3315.

Duncan, A. W., Taylor, M. H., Hickey, R. D., Hanlon Newell, A. E., Lenzi, M. L., Olson, S. B., et al. (2010). The ploidy conveyor of mature hepatocytes as a source of genetic variation. *Nature, 467*, 707–710.

Farber, S. A., Pack, M., Ho, S.-Y., Johnson, I. D., Wagner, D. S., Dosch, R., et al. (2001). Genetic analysis of digestive physiology using fluorescent phospholipid reporters. *Science, 292*, 1385.

Farooq, M., Sulochana, K. N., Pan, X., To, J., Sheng, D., Gong, Z., et al. (2008). Histone deacetylase 3 (hdac3) is specifically required for liver development in zebrafish. *Developmental Biology, 317*, 336–353.

Field, H. A., Ober, E. A., Roeser, T., & Stainier, D. Y. (2003). Formation of the digestive system in zebrafish I. Liver morphogenesis. *Developmental Biology, 253*, 279–290.

Flynn, D. M., Nijjar, S., Hubscher, S. G., de Goyet, J. d. V., Kelly, D. A., Strain, A. J., et al. (2004). The role of Notch receptor expression in bile duct development and disease. *The Journal of Pathology*, *204*, 55–64.

Garnaas, M. K., Cutting, C. C., Meyers, A., Kelsey, P. B., Jr., Harris, J. M., North, T. E., et al. (2012). Rargb regulates organ laterality in a zebrafish model of right atrial isomerism. *Developmental Biology*, *372*, 178–189.

Gemberling, M., Bailey, T. J., Hyde, D. R., & Poss, K. D. (2013). The zebrafish as a model for complex tissue regeneration. *Trends in Genetics*, *29*, 611–620.

Goessling, W., North, T. E., Loewer, S., Lord, A. M., Lee, S., Stoick-Cooper, C. L., et al. (2009). Genetic interaction of PGE2 and Wnt signaling regulates developmental specification of stem cells and regeneration. *Cell*, *136*, 1136–1147.

Goessling, W., North, T. E., Lord, A. M., Ceol, C., Lee, S., Weidinger, G., et al. (2008). APC mutant zebrafish uncover a changing temporal requirement for wnt signaling in liver development. *Developmental Biology*, *320*, 161–174.

Goessling, W., & Sadler, K. C. (2015). Zebrafish: An important tool for liver disease research. *Gastroenterology*, *149*, 1361–1377.

Goessling, W., & Stainier, D. Y. (2016). Endoderm specification and liver development. *Methods in Cell Biology*, *134*, 463–483.

Gouon-Evans, V., Boussemart, L., Gadue, P., Nierhoff, D., Koehler, C. I., Kubo, A., et al. (2006). BMP-4 is required for hepatic specification of mouse embryonic stem cell-derived definitive endoderm. *Nature Biotechnology*, *24*, 1402–1411.

Grompe, M. (2014). Liver stem cells, where art thou? *Cell Stem Cell*, *15*, 257–258.

Grompe, M., al-Dhalimy, M., Finegold, M., Ou, C. N., Burlingame, T., Kennaway, N. G., et al. (1993). Loss of fumarylacetoacetate hydrolase is responsible for the neonatal hepatic dysfunction phenotype of lethal albino mice. *Genes & Development*, *7*, 2298–2307.

Grompe, M., Lindstedt, S., Al-Dhalimy, M., Kennaway, N. G., Papaconstantinou, J., Torres-Ramos, C. A., et al. (1995). Pharmacological correction of neonatal lethal hepatic dysfunction in a murine model of hereditary tyrosinaemia type I. *Nature Genetics*, *10*, 453–460.

Hardman, R. C., Volz, D. C., Kullman, S. W., & Hinton, D. E. (2007). An in vivo look at vertebrate liver architecture: Three-dimensional reconstructions from medaka (Oryzias latipes). *Anatomical Record (Hoboken, N.J.: 2007)*, *290*, 770–782.

He, J., Lu, H., Zou, Q., & Luo, L. (2014). Regeneration of liver after extreme hepatocyte loss occurs mainly via biliary transdifferentiation in zebrafish. *Gastroenterology*, *146*, 789–800. e788.

Hen, G., Nicenboim, J., Mayseless, O., Asaf, L., Shin, M., Busolin, G., et al. (2015). Venous-derived angioblasts generate organ-specific vessels during zebrafish embryonic development. *Development*, *142*, 4266–4278.

Her, G. M., Yeh, Y. H., & Wu, J. L. (2003). 435-bp liver regulatory sequence in the liver fatty acid binding protein (L-FABP) gene is sufficient to modulate liver regional expression in transgenic zebrafish. *Developmental Dynamics*, *227*, 347–356.

Higgins, G. M., & Anderson, R. M. (1931). Experimental pathology of the liver. I. Restoration of the liver of the white rat following partial surgical removal. *Archives of Pathology*, *12*, 186–202.

Hirose, K., Shimoda, N., & Kikuchi, Y. (2013). Transient reduction of 5-methylcytosine and 5-hydroxymethylcytosine is associated with active DNA demethylation during regeneration of zebrafish fin. *Epigenetics*, *8*, 899–906.

Hochgreb-Hägele, T., Yin, C., Koo, D. E. S., Bronner, M. E., & Stainier, D. Y. R. (2013). Laminin β1a controls distinct steps during the establishment of digestive organ laterality. *Development*, *140*, 2734.

Holtzinger, A., & Evans, T. (2005). Gata4 regulates the formation of multiple organs. *Development*, *132*, 4005.

Horne-Badovinac, S., Rebagliati, M., & Stainier, D. Y. R. (2003). A cellular framework for gut-looping morphogenesis in zebrafish. *Science*, *302*, 662.

Howarth, D. L., Yin, C., Yeh, K., & Sadler, K. C. (2013). Defining hepatic dysfunction parameters in two models of fatty liver disease in zebrafish larvae. *Zebrafish*, *10*, 199–210.

Howe, K., Clark, M. D., Torroja, C. F., Torrance, J., Berthelot, C., Muffato, M., et al. (2013). The zebrafish reference genome sequence and its relationship to the human genome. *Nature*, *496*, 498–503.

Huang, M., Chang, A., Choi, M., Zhou, D., Anania, F. A., & Shin, C. H. (2014). Antagonistic interaction between Wnt and Notch activity modulates the regenerative capacity of a zebrafish fibrotic liver model. *Hepatology*, *60*, 1753–1766.

Jacob, V., Chernyavskaya, Y., Chen, X., Tan, P. S., Kent, B., Hoshida, Y., et al. (2015). DNA hypomethylation induces a DNA replication-associated cell cycle arrest to block hepatic outgrowth in uhrf1 mutant zebrafish embryos. *Development*, *142*, 510–521.

Kan, N. G., Junghans, D., & Izpisua Belmonte, J. C. (2009). Compensatory growth mechanisms regulated by BMP and FGF signaling mediate liver regeneration in zebrafish after partial hepatectomy. *The FASEB Journal*, *23*, 3516–3525.

Kang, J., Hu, J., Karra, R., Dickson, A. L., Tornini, V. A., Nachtrab, G., et al. (2016). Modulation of tissue repair by regeneration enhancer elements. *Nature*, *532*, 201–206.

Khaliq, M., Choi, T. Y., So, J., & Shin, D. (2015). Id2a is required for hepatic outgrowth during liver development in zebrafish. *Mechanisms of Development*, *138*(Pt. 3), 399–414.

Ko, S., Choi, T. Y., Russell, J. O., So, J., Monga, S. P., & Shin, D. (2016). Bromodomain and extraterminal (BET) proteins regulate biliary-driven liver regeneration. *Journal of Hepatology*, *64*, 316–325.

Kodama, Y., Hijikata, M., Kageyama, R., Shimotohno, K., & Chiba, T. (2004). The role of notch signaling in the development of intrahepatic bile ducts. *Gastroenterology*, *127*, 1775–1786.

Koltowska, K., Apitz, H., Stamataki, D., Hirst, E. M., Verkade, H., Salecker, I., et al. (2013). Ssrp1a controls organogenesis by promoting cell cycle progression and RNA synthesis. *Development*, *140*, 1912–1918.

Korzh, S., Emelyanov, A., & Korzh, V. (2001). Developmental analysis of ceruloplasmin gene and liver formation in zebrafish. *Mechanisms of Development*, *103*, 137–139.

Korzh, S., Pan, X., Garcia-Lecea, M., Winata, C., Pan, X., Wohland, T., et al. (2008). Requirement of vasculogenesis and blood circulation in late stages of liver growth in zebrafish. *BMC Developmental Biology*, *8*, 84.

Lancman, J. J., Zvenigorodsky, N., Gates, K. P., Zhang, D., Solomon, K., Humphrey, R. K., et al. (2013). Specification of hepatopancreas progenitors in zebrafish by hnf1ba and wnt2bb. *Development*, *140*, 2669.

Landgraf, K., Strobach, A., Kiess, W., & Korner, A. (2016). Loss of mtch2 function impairs early development of liver, intestine and visceral adipocytes in zebrafish larvae. *FEBS Letters*, *590*, 2852–2861.

Lemaigre, F. P. (2009). Mechanisms of liver development: Concepts for understanding liver disorders and design of novel therapies. *Gastroenterology*, *137*, 62–79.

Li, W., Liang, X., Leu, J. I., Kovalovich, K., Ciliberto, G., & Taub, R. (2001). Global changes in interleukin-6–dependent gene expression patterns in mouse livers after partial hepatectomy. *Hepatology*, *33*, 1377–1386.

Lorent, K., Gong, W., Koo, K. A., Waisbourd-Zinman, O., Karjoo, S., Zhao, X., et al. (2015). Identification of a plant isoflavonoid that causes biliary atresia. *Science Translational Medicine*, 7, 286ra267.

Lorent, K., Moore, J. C., Siekmann, A. F., Lawson, N., & Pack, M. (2010). Reiterative use of the notch signal during zebrafish intrahepatic biliary development. *Developmental Dynamics*, *239*, 855–864.

Lorent, K., Yeo, S.-Y., Oda, T., Chandrasekharappa, S., Chitnis, A., Matthews, R. P., et al. (2004). Inhibition of Jagged-mediated Notch signaling disrupts zebrafish biliary development and generates multi-organ defects compatible with an Alagille syndrome phenocopy. *Development*, *131*, 5753.

Lozano, R., Naghavi, M., Foreman, K., Lim, S., Shibuya, K., Aboyans, V., et al. (2012). Global and regional mortality from 235 causes of death for 20 age groups in 1990 and 2010: A systematic analysis for the Global Burden of Disease Study 2010. *Lancet*, *380*, 2095–2128.

Lu, W. Y., Bird, T. G., Boulter, L., Tsuchiya, A., Cole, A. M., Hay, T., et al. (2015). Hepatic progenitor cells of biliary origin with liver repopulation capacity. *Nature Cell Biology*, *17*, 971–983.

Lu, H., Ma, J., Yang, Y., Shi, W., & Luo, L. (2013). EpCAM is an endoderm-specific Wnt derepressor that licenses hepatic development. *Developmental Cell*, *24*, 543–553.

Manfroid, I., Ghaye, A., Naye, F., Detry, N., Palm, S., Pan, L., et al. (2012). Zebrafish sox9b is crucial for hepatopancreatic duct development and pancreatic endocrine cell regeneration. *Developmental Biology*, *366*, 268–278.

Martin, P., DiMartini, A., Feng, S., Brown, R., & Fallon, M. (2013). Evaluation for liver transplantation in adults: 2013 practice guideline by the AASLD and the American Society of Transplantation. *Hepatology (Baltimore, Md.)*, *59*, 1144–1165.

Matsumoto, K., Yoshitomi, H., Rossant, J., & Zaret, K. S. (2001). Liver organogenesis promoted by endothelial cells prior to vascular function. *Science*, *294*, 559.

Matthews, R. P., Eauclaire, S. F., Mugnier, M., Lorent, K., Cui, S., Ross, M. M., et al. (2011). DNA hypomethylation causes bile duct defects in zebrafish and is a distinguishing feature of infantile biliary atresia. *Hepatology*, *53*, 905–914.

Matthews, R. P., Lorent, K., & Pack, M. (2008). Transcription factor onecut3 regulates intrahepatic biliary development in zebrafish. *Developmental Dynamics*, *237*, 124–131.

Matthews, R. P., Lorent, K., Russo, P., & Pack, M. (2004). The zebrafish onecut gene hnf-6 functions in an evolutionarily conserved genetic pathway that regulates vertebrate biliary development. *Developmental Biology*, *274*, 245–259.

Matthews, R. P., Plumb-Rudewiez, N., Lorent, K., Gissen, P., Johnson, C. A., Lemaigre, F., et al. (2005). Zebrafish vps33b, an ortholog of the gene responsible for human arthrogryposis-renal dysfunction-cholestasis syndrome, regulates biliary development downstream of the onecut transcription factor hnf6. *Development*, *132*, 5295–5306.

Medlock, E. S., & Haar, J. L. (1983). The liver hemopoietic environment: I. Developing hepatocytes and their role in fetal hemopoiesis. *The Anatomical Record*, *207*, 31–41.

Michalopoulos, G. K. (2010). Liver regeneration after partial hepatectomy: Critical analysis of mechanistic dilemmas. *The American Journal of Pathology*, *176*, 2–13.

Michalopoulos, G. K. (2014). Advances in liver regeneration. *Expert Review of Gastroenterology & Hepatology*, *8*, 897–907.

Michalopoulos, G. K., & Khan, Z. (2015). Liver stem cells: Experimental findings and implications for human liver disease. *Gastroenterology*, *149*, 876–882.

Miki, R., Yoshida, T., Murata, K., Oki, S., Kume, K., & Kume, S. (2012). Fate maps of ventral and dorsal pancreatic progenitor cells in early somite stage mouse embryos. *Mechanisms of Development*, *128*, 597–609.

Miyaoka, Y., Ebato, K., Kato, H., Arakawa, S., Shimizu, S., & Miyajima, A. (2012). Hypertrophy and unconventional cell division of hepatocytes underlie liver regeneration. *Current Biology*, *22*, 1166–1175.

Mudbhary, R., Hoshida, Y., Chernyavskaya, Y., Jacob, V., Villanueva, A., Fiel, M. I., et al. (2014). UHRF1 overexpression drives DNA hypomethylation and hepatocellular carcinoma. *Cancer Cell*, *25*, 196–209.

Mudumana, S. P., Wan, H., Singh, M., Korzh, V., & Gong, Z. (2004). Expression analyses of zebrafish transferrin, ifabp, and elastaseB mRNAs as differentiation markers for the three major endodermal organs: Liver, intestine, and exocrine pancreas. *Developmental Dynamics, 230*, 165–173.

Murphy, S. L., Xu, J., & Kochanek, K. D. (2012). Deaths: Final data for 2012. *Nationalvital statistics reports:* Vol. 63. Hyattsville, MD: National Center for Health Statistics.

Nguyen, A. T., Emelyanov, A., Koh, C. H., Spitsbergen, J. M., Parinov, S., & Gong, Z. (2012). An inducible kras(V12) transgenic zebrafish model for liver tumorigenesis and chemical drug screening. *Disease Models & Mechanisms, 5*, 63–72.

Ningappa, M., So, J., Glessner, J., Ashokkumar, C., Ranganathan, S., Min, J., et al. (2015). The role of ARF6 in biliary atresia. *PLoS One, 10*, e0138381.

Nissim, S., Sherwood, R. I., Wucherpfennig, J., Saunders, D., Harris, J. M., Esain, V., et al. (2014). Prostaglandin E2 regulates liver versus pancreas cell-fate decisions and endodermal outgrowth. *Developmental Cell, 28*, 423–437.

Nissim, S., Weeks, O., Talbot, J. C., Hedgepeth, J. W., Wucherpfennig, J., Schatzman-Bone, S., et al. (2016). Iterative use of nuclear receptor Nr5a2 regulates multiple stages of liver and pancreas development. *Developmental Biology, 418*, 108–123.

Noel, E. S., Casal-Sueiro, A., Busch-Nentwich, E., Verkade, H., Dong, P. D., Stemple, D. L., et al. (2008). Organ-specific requirements for Hdac1 in liver and pancreas formation. *Developmental Biology, 322*, 237–250.

Noel, E. S., Reis, M. D., Arain, Z., & Ober, E. A. (2010). Analysis of the albumin/alpha-fetoprotein/afamin/group specific component gene family in the context of zebrafish liver differentiation. *Gene Expression Patterns, 10*, 237–243.

North, T. E., Goessling, W., Walkley, C. R., Lengerke, C., Kopani, K. R., Lord, A. M., et al. (2007). Prostaglandin E2 regulates vertebrate haematopoietic stem cell homeostasis. *Nature, 447*, 1007–1011.

Ober, E. A., Verkade, H., Field, H. A., & Stainier, D. Y. R. (2006). Mesodermal Wnt2b signalling positively regulates liver specification. *Nature, 442*, 688–691.

Overturf, K., al-Dhalimy, M., Ou, C. N., Finegold, M., & Grompe, M. (1997). Serial transplantation reveals the stem-cell-like regenerative potential of adult mouse hepatocytes. *The American Journal of Pathology, 151*, 1273–1280.

Pack, M., Solnica-Krezel, L., Malicki, J., Neuhauss, S. C., Schier, A. F., Stemple, D. L., et al. (1996). Mutations affecting development of zebrafish digestive organs. *Development, 123*, 321.

Parsons, M. J., Pisharath, H., Yusuff, S., Moore, J. C., Siekmann, A. F., Lawson, N., et al. (2009). Notch-responsive cells initiate the secondary transition in larval zebrafish pancreas. *Mechanisms of Development, 126*, 898–912.

Poulain, M., & Ober, E. A. (2011). Interplay between Wnt2 and Wnt2bb controls multiple steps of early foregut-derived organ development. *Development, 138*, 3557.

Rhim, J., Sandgren, E., Degen, J., Palmiter, R., & Brinster, R. (1994). Replacement of diseased mouse liver by hepatic cell transplantation. *Science, 263*, 1149–1152.

Rossi, J. M., Dunn, N. R., Hogan, B. L. M., & Zaret, K. S. (2001). Distinct mesodermal signals, including BMPs from the septum transversum mesenchyme, are required in combination for hepatogenesis from the endoderm. *Genes & Development, 15*, 1998–2009.

Sadler, K. C., Krahn, K. N., Gaur, N. A., & Ukomadu, C. (2007). Liver growth in the embryo and during liver regeneration in zebrafish requires the cell cycle regulator, uhrf1. *Proceedings of the National Academy of Sciences of the United States of America, 104*, 1570–1575.

Sakaguchi, T. F., Sadler, K. C., Crosnier, C., & Stainier, D. Y. R. (2008). Endothelial signals modulate hepatocyte apicobasal polarization in zebrafish. *Current Biology, 18*, 1565–1571.

Schaub, J. R., Malato, Y., Gormond, C., & Willenbring, H. (2014). Evidence against a stem cell origin of new hepatocytes in a common mouse model of chronic liver injury. *Cell Reports*, *8*, 933–939.

Schaub, M., Nussbaum, J., Verkade, H., Ober, E. A., Stainier, D. Y., & Sakaguchi, T. F. (2012). Mutation of zebrafish Snapc4 is associated with loss of the intrahepatic biliary network. *Developmental Biology*, *363*, 128–137.

Schoenfelder, K. P., & Fox, D. T. (2015). The expanding implications of polyploidy. *Journal of Cell Biology*, *209*, 485–491.

Sharif, J., Muto, M., Takebayashi, S., Suetake, I., Iwamatsu, A., Endo, T. A., et al. (2007). The SRA protein Np95 mediates epigenetic inheritance by recruiting Dnmt1 to methylated DNA. *Nature*, *450*, 908–912.

Shi, J., & Vakoc, C. R. (2014). The mechanisms behind the therapeutic activity of BET bromodomain inhibition. *Molecular Cell*, *54*, 728–736.

Shin, D., Lee, Y., Poss, K. D., & Stainier, D. Y. R. (2011). Restriction of hepatic competence by Fgf signaling. *Development*, *138*, 1339.

Shin, D., Shin, C. H., Tucker, J., Ober, E. A., Rentzsch, F., Poss, K. D., et al. (2007). Bmp and Fgf signaling are essential for liver specification in zebrafish. *Development*, *134*, 2041.

Si-Tayeb, K., Noto, F. K., Nagaoka, M., Li, J., Battle, M. A., Duris, C., et al. (2010). Highly efficient generation of human hepatocyte-like cells from induced pluripotent stem cells. *Hepatology*, *51*, 297–305.

Stainier, D. Y., Fouquet, B., Chen, J. N., Warren, K. S., Weinstein, B. M., Meiler, S. E., et al. (1996). Mutations affecting the formation and function of the cardiovascular system in the zebrafish embryo. *Development*, *123*, 285–292.

Stainier, D. Y., Weinstein, B. M., Detrich, H. W., Zon, L. I., & Fishman, M. C. (1995). Cloche, an early acting zebrafish gene, is required by both the endothelial and hematopoietic lineages. *Development*, *121*, 3141.

Stoick-Cooper, C. L., Moon, R. T., & Weidinger, G. (2007). Advances in signaling in vertebrate regeneration as a prelude to regenerative medicine. *Genes & Development*, *21*, 1292–1315.

Su, A. I., Guidotti, L. G., Pezacki, J. P., Chisari, F. V., & Schultz, P. G. (2002). Gene expression during the priming phase of liver regeneration after partial hepatectomy in mice. *Proceedings of the National Academy of Sciences of the United States of America*, *99*, 11181–11186.

Takayama, K., Shimoda, N., Takanaga, S., Hozumi, S., & Kikuchi, Y. (2014). Expression patterns of dnmt3aa, dnmt3ab, and dnmt4 during development and fin regeneration in zebrafish. *Gene Expression Patterns*, *14*, 105–110.

Tan, X., Behari, J., Cieply, B., Michalopoulos, G. K., & Monga, S. P. (2006). Conditional deletion of beta-catenin reveals its role in liver growth and regeneration. *Gastroenterology*, *131*, 1561–1572.

Tang, V., Cofer, Z. C., Cui, S., Sapp, V., Loomes, K. M., & Matthews, R. P. (2016). Loss of a candidate biliary atresia susceptibility gene, add3a, causes biliary developmental defects in zebrafish. *Journal of Pediatric Gastroenterology and Nutrition*, *63*, 524–530.

Tarlow, B. D., Finegold, M. J., & Grompe, M. (2014). Clonal tracing of Sox9+ liver progenitors in mouse oval cell injury. *Hepatology*, *60*, 278–289.

Taub, R. (2004). Liver regeneration: From myth to mechanism. *Nature Reviews. Molecular Cell Biology*, *5*, 836–847.

Tremblay, K. D., & Zaret, K. S. (2005). Distinct populations of endoderm cells converge to generate the embryonic liver bud and ventral foregut tissues. *Developmental Biology*, *280*, 87–99.

Venkatachalam, A. B., Thisse, C., Thisse, B., & Wright, J. M. (2009). Differential tissue-specific distribution of transcripts for the duplicated fatty acid-binding protein

10 (fabp10) genes in embryos, larvae and adult zebrafish (Danio rerio). *The FEBS Journal, 276*, 6787–6797.

Waisbourd-Zinman, O., Koh, H., Tsai, S., Lavrut, P. M., Dang, C., Zhao, X., et al. (2016). The toxin biliatresone causes mouse extrahepatic cholangiocyte damage and fibrosis through decreased glutathione and SOX17. *Hepatology, 64*, 880–893.

Wallace, K. N., Yusuff, S., Sonntag, J. M., Chin, A. J., & Pack, M. (2001). Zebrafish hhex regulates liver development and digestive organ chirality. *Genesis, 30*, 141–143.

Wilkins, B. J., Gong, W., & Pack, M. (2014). A novel keratin18 promoter that drives reporter gene expression in the intrahepatic and extrahepatic biliary system allows isolation of cell-type specific transcripts from zebrafish liver. *Gene Expression Patterns, 14*, 62–68.

Wilkins, B. J., & Pack, M. (2013). Zebrafish models of human liver development and disease. *Comprehensive Physiology, 3*, 1213–1230.

Wong, F. W., Chan, W. Y., & Lee, S. S. (1998). Resistance to carbon tetrachloride-induced hepatotoxicity in mice which lack CYP2E1 expression. *Toxicology and Applied Pharmacology, 153*, 109–118.

Xu, J., Cui, J., Del Campo, A., & Shin, C. H. (2016). Four and a half LIM domains 1b (Fhl1b) is essential for regulating the liver versus pancreas fate decision and for β-cell regeneration. *PLoS Genetics, 12*, e1005831.

Yanger, K., Knigin, D., Zong, Y., Maggs, L., Gu, G., Akiyama, H., et al. (2014). Adult hepatocytes are generated by self-duplication rather than stem cell differentiation. *Cell Stem Cell, 15*, 340–349.

Yao, Y., Lin, J., Yang, P., Chen, Q., Chu, X., Gao, C., et al. (2012). Fine structure, enzyme histochemistry, and immunohistochemistry of liver in zebrafish. *Anatomical Record (Hoboken, N.J.: 2007), 295*, 567–576.

Yasui, H., Morishima, M., Nakazawa, M., & Aikawa, E. (1998). Anomalous looping, atrioventricular cushion dysplasia, and unilateral ventricular hypoplasia in the mouse embryos with right isomerism induced by retinoic acid. *Anatomical Record, 250*, 210–219.

Yimlamai, D., Fowl, B. H., & Camargo, F. D. (2015). Emerging evidence on the role of the Hippo/YAP pathway in liver physiology and cancer. *Journal of Hepatology, 63*, 1491–1501.

Yin, C., Evason, K. J., Maher, J. J., & Stainier, D. Y. (2012). The basic helix-loop-helix transcription factor, heart and neural crest derivatives expressed transcript 2, marks hepatic stellate cells in zebrafish: Analysis of stellate cell entry into the developing liver. *Hepatology, 56*, 1958–1970.

Yin, C., Kikuchi, K., Hochgreb, T., Poss, K. D., & Stainier, D. Y. R. (2010). Hand2 regulates extracellular matrix remodeling essential for gut-looping morphogenesis in zebrafish. *Developmental Cell, 18*, 973–984.

Yuan, H., Wen, B., Liu, X., Gao, C., Yang, R., Wang, L., et al. (2015). CCAAT/enhancer-binding protein alpha is required for hepatic outgrowth via the p53 pathway in zebrafish. *Scientific Reports, 5*, 15838.

Zaret, K. S., & Mango, S. E. (2016). Pioneer transcription factors, chromatin dynamics, and cell fate control. *Current Opinion in Genetics & Development, 37*, 76–81.

Zhang, D., Golubkov, V. S., Han, W., Correa, R. G., Zhou, Y., Lee, S., et al. (2014). Identification of Annexin A4 as a hepatopancreas factor involved in liver cell survival. *Developmental Biology, 395*, 96–110.

Zhao, X., Lorent, K., Wilkins, B. J., Marchione, D. M., Gillespie, K., Waisbourd-Zinman, O., et al. (2016). Glutathione antioxidant pathway activity and reserve determine toxicity and specificity of the biliary toxin biliatresone in zebrafish. *Hepatology, 64*, 894–907.

Zhao, X., Monson, C., Gao, C., Gouon-Evans, V., Matsumoto, N., Sadler, K. C., et al. (2010). Klf6/copeb is required for hepatic outgrowth in zebrafish and for hepatocyte specification in mouse ES cells. *Developmental Biology, 344*, 79–93.

CHAPTER SIX

"Muscling" Throughout Life: Integrating Studies of Muscle Development, Homeostasis, and Disease in Zebrafish

Michelle F. Goody*, Erin V. Carter†, Elisabeth A. Kilroy†, Lisa Maves‡,§, Clarissa A. Henry*,†,1

*School of Biology and Ecology, University of Maine, Orono, ME, United States
†Graduate School of Biomedical Sciences and Engineering, University of Maine, Orono, ME, United States
‡Center for Developmental Biology and Regenerative Medicine, Seattle Children's Research Institute, Seattle, WA, United States
§Division of Cardiology, University of Washington, Seattle, WA, United States
[1]Corresponding author: e-mail address: Clarissa_Henry@umit.maine.edu

Contents

Current Topics in Developmental Biology, Volume 124
ISSN 0070-2153
http://dx.doi.org/10.1016/bs.ctdb.2016.11.002

Abstract

The proper development and function of skeletal muscle is vital for health throughout the lifespan. Skeletal muscle function enables posture, breathing, and locomotion; and also impacts systemic processes—such as metabolism, thermoregulation, and immunity. Diseases of skeletal muscle (myopathies, muscular dystrophies) and even some neurological, age-related, and metabolic diseases compromise muscle function and negatively affect health span and quality of life. There have been numerous, recent examples of studies on skeletal muscle development with exciting, therapeutic implications for muscle diseases. The zebrafish (*Danio rerio*) is a vertebrate model organism well accepted for developmental biology and biomedical research and thus an ideal system in which to elucidate the translational implications of mechanisms regulating skeletal muscle development and homeostasis. Muscle fiber types (slow- vs fast-twitch) are spatially segregated in zebrafish allowing for the opportunity to identify distinct mechanisms regulating fiber type specification during development as well as observe fiber type-specific effects in zebrafish models of muscle diseases. Accessible genetics coupled with transparent zebrafish embryos has enabled in vivo cell biology experiments allowing for the visualization and understanding of never-before-seen cellular processes occurring in muscle development, regeneration, and disease. In addition, high-throughput drug screening provides a platform for efficient drug discovery. The purpose of this chapter is to review the studies in zebrafish that significantly contributed to our understanding of cellular and molecular mechanisms regulating skeletal muscle development, homeostasis, or disease in vertebrates, with a particular emphasis on the basic developmental biology studies with promising therapeutic implications.

1. INTRODUCTION

This review focuses on areas where understanding mechanisms underlying skeletal muscle development has led to insights into muscle diseases (Table 2). Muscle diseases, such as muscular dystrophies and myopathies, are stereotypically characterized by muscle wasting, weakness, and impaired locomotion. One of the major challenges regarding therapy development is the phenotypic variability within and between dystrophies and myopathies. For example, within a subset of muscular dystrophies called dystroglycanopathies, phenotypes span the clinical spectrum from the most severe (congenital muscular dystrophy (CMD)–dystroglycanopathy with eye and brain abnormalities) to the mildest (adult onset limb-girdle muscular dystrophy–dystroglycanopathy). We predict that use of zebrafish models for these diseases will not only provide new insight into the connection between muscle development and disease, but also should identify the cellular and molecular etiology of phenotypic variability, which then could help in therapy design.

2. FROM MESODERM TO MUSCULATURE: OVERVIEW OF SKELETAL MUSCLE DEVELOPMENT

The first step in vertebrate muscle development is segmentation of the paraxial mesoderm into somites. Somites give rise to multiple tissues: muscle (fast-twitch, slow-twitch, and muscle progenitor cells that contribute to muscle growth), elements of the skeleton and tendons, and hematopoietic stem cells (Murayama et al., 2015; Nguyen et al., 2014; Qiu et al., 2016). After somites form, the anterior border cells (ABCs) of each somite give rise to the zebrafish equivalent of the amniote dermomyotome (Hollway et al., 2007; Stellabotte, Dobbs-McAuliffe, Fernández, Feng, & Devoto, 2007) (Fig. 1, red cells). The amniote dermomyotome contains multipotent progenitor cells that proliferate and express Pax3 and Pax7 (Buckingham et al., 2003). In zebrafish, ABCs migrate laterally after somitogenesis and form an extremely thin layer of cells, the external cell layer (ECL). The ECL expresses Pax3 and Pax7 and contributes to muscle growth (Hollway et al., 2007; Stellabotte et al., 2007); thus, the zebrafish ECL is considered the functional equivalent of the amniote dermomyotome.

Next, muscle cells elongate and attach to myotendinous junctions (MTJs) (Fig. 1). After elongation, the terminal ends of muscle fibers attach to somite boundaries, which become myotome boundaries and give rise to MTJs (Long, Adcock, & Root, 2002) (Fig. 1, black chevrons). Then, fast-twitch muscle cells fuse into multinucleate myofibers and myofibrillogenesis generates sarcomeres, the contractile units of muscle.

Slow-twitch muscle progenitors are initially located medially but, shortly after somite formation, they migrate laterally and become the most superficial layer of muscle (Devoto, Melançon, Eisen, & Westerfield, 1996) (Fig. 1B, blue cells). Slow-twitch fiber migration is necessary and sufficient to trigger elongation of the fast-twitch fibers (Henry & Amacher, 2004) (Fig. 1, dark green cells).

2.1 The Tortoise or the Hare? Fiber Type Specification During Development

Muscle fibers are frequently classified as slow-twitch/type I or fast-twitch/type II, but intermediate fiber types provide for an almost continuous spectrum of physiological possibilities (Schiaffino & Reggiani, 2011). Fiber types are defined by the particular myosin heavy chain isoforms that they express. Many other factors also contribute to a fiber's phenotype—such

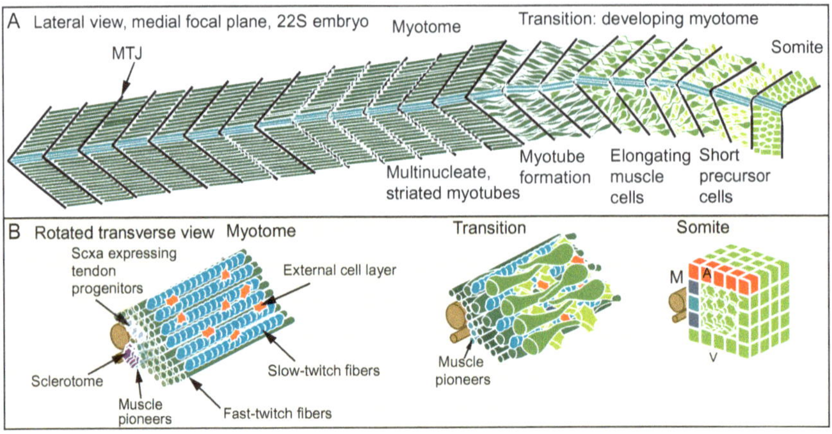

Fig. 1 Zebrafish skeletal muscle structure and development. (A) Medial focal plane, side oriented, anterior left, dorsal top, 22 somite stage zebrafish embryo showing stages of skeletal muscle development along the anterior–posterior axis. The anterior most segments contain long muscle cells and are called myotomes (*blue* cells, slow-twitch muscle pioneers; *dark green* cells, fast-twitch fibers (multinucleate); *black chevrons*, myotendinous junctions (MTJs)). Muscle segments toward the posterior of the embryo contain cells in different phases of fast-twitch muscle morphogenesis (called the transition zone). From left to right within the transition zone, *dark-medium green* cells are elongated but irregularly shaped myotubes, *medium green* cells are elongating via protrusion extension and filling, and *light green* cells are short and extending protrusions in all directions. All muscle cells within the transition zone are mononucleate. At this developmental stage, muscle segments in the very posterior of the embryo are not yet undergoing fast muscle morphogenesis and are called somites. (B) Rotated, transverse, 3D views of one myotome, one muscle segment in the transition zone, and one somite (from left to right, respectively). *Brown cylinders* represent the neural tube on top and the notochord below. Myotomes contain slow-twitch fibers that have completed their migration (*blue*) and fast-twitch fibers that have completed their morphogenesis (*dark green*). Also shown are tendon progenitors (*gray* cells), sclerotome (*purple* cells), and satellite cells in the external cell layer (*red*). Muscle segments in the transition zone contain slow-twitch fibers part way through their migration (*blue*), fast-twitch fibers medial to the migrating slow fibers that have undergone morphogenesis (*dark green*), presumptive fast cells undergoing morphogenesis (*lighter shades of green*), and satellite cells of the external cell layer (*red*). Somites consist of epithelial cells surrounding an inner mesenchymal group of cells (both shown in *green*), medially located slow muscle cells that have not yet migrated (*blue*), and anterior border cells (*red*) that will become satellite cells of the external cell layer. *M*, medial; *A*, anterior; *V*, ventral.

as contraction rate, response to neural stimulation, mode of metabolism, and ATP usage. Understanding development and regeneration of different fiber types is important because some genetic/acquired muscle diseases have more deleterious effects on particular muscle fiber types (Ciciliot, Rossi, Dyar,

Blaauw, & Schiaffino, 2013; Schiaffino & Reggiani, 2011). Thus, it is possible that some muscle and metabolic diseases could be treated by shifting fiber type proportions or by supporting the development of a specific fiber type (Ljubicic, Burt, & Jasmin, 2014). Toward this end, we need to understand the developmental processes that generate muscle fiber types in embryos.

The molecular mechanisms of zebrafish muscle fiber type specification have been reviewed recently (Jackson & Ingham, 2013; Talbot & Maves, 2016). Four genetically, morphologically, and/or physiologically distinct fiber types have been identified in zebrafish embryos: superficial slow-twitch fibers, medial slow-twitch muscle pioneers, medial fast fibers, and fast-twitch fibers (Wolff, Roy, & Ingham, 2003) (Table 1). High levels of hedgehog signaling early in zebrafish myogenesis are needed to specify slow fiber types as well as medial fast fibers (Currie & Ingham, 1996; Wolff et al., 2003); and hedgehog is important for later fast muscle differentiation (Feng, Adiarte, & Devoto, 2006; Hammond et al., 2007). Hedgehog signaling also promotes slow muscle fiber identity in mouse and chick embryonic limb muscles (Cann, Lee, & Stockdale, 1999; Hu, McGlinn, Harfe, Kardon, & Tabin, 2012; Li et al., 2004). Recent studies in zebrafish have identified roles for new factors in slow- and/or fast-twitch fiber

Table 1 Markers for Cell Types in Zebrafish Skeletal Muscle

Cell Type	Transgenic Marker	Gene Marker	Antibody
Satellite cells	*Tg(pax7a:eGFP)*, *Tg(pax7a:GFP)*, *Tg(pax7b:gal4;UAS:GFP)*, *Tg(pax3a:GFP)*	*pax7a*	Pax3, Pax7
Superficial slow-twitch fibers	*Tg(smyhc1:GFP)*, *Tg(prdm1a:GFP)*, *Tg(tnnc1b:eGFP)*	*tnnt1*, *mybpc1*, *smyhc1*, *prdm1a*, *tnnc1b*	F59, S58, Smyhc, Prox1
Muscle pioneers	*Tg(eng2a:eGFP)*	*tnnt1*, *mybpc1*, *smyhc1*, *eng1a*, *eng1b*, *eng2a*, *eng2b*	Engrailed, F59, S58, Smyhc, Prox1
Medial fast-twitch fibers	*Tg(eng2a:eGFP)*, *Tg(mylpfa:H2B-GFP)*, *Tg(mylpfa:GFP)*	*tpma*, *tnnt3a/b*, *tnnc2*, *myhz2*, *mylpfa*, *eng1a*, *eng1b*, *eng2a*, *eng2b*	Engrailed, F310
Fast-twitch fibers	*Tg(mylpfa:GFP)*, *Tg(mylpfa:H2B-GFP)*	*tpma*, *tnnt3a/b*, *tnnc2*, *myhz2*, *mylpfa*	F310

Genes common to slow- and fast-twitch fibers: *acta1*, *myhz1*, *myod*, *myf5*, *myog*.
Antibodies or stains common to slow- and fast-twitch fibers: MF20, phalloidin.

specification, including transcriptional regulators (Devakanmalai, Zumrut, & Ozbudak, 2013; Yao, Farr, Tapscott, & Maves, 2013), components of signaling pathways (Tu, Tsao, Lee, & Yang, 2014), and microRNAs (Ketley et al., 2013; Lin et al., 2013; O'Brien, Hernandez-Lagunas, Artinger, & Ford, 2014). Whether these new factors play conserved roles in mammalian fiber type specification remains to be determined.

2.2 Early Muscle Fiber Morphogenesis

2.2.1 Slow-Twitch Muscle Elongation

Presumptive slow-twitch muscle cells are initially stacked like short bricks: they are cuboidal cells arranged in a sheet of approximately 4×5 cells. Next, this sheet of epithelial cells transforms into a stack, about 20 deep, of elongated slow-twitch muscle cells. Live imaging elucidated the cellular behaviors that underlie this transformation (Daggett, Domingo, Currie, & Amacher, 2007; Yin & Solnica-Krezel, 2007). Slow-twitch cells first elongate in the dorsal–ventral direction (perpendicular to their final anterior–posterior alignment). Next, slow-twitch cells rotate such that they are aligned anterior-posteriorly (Yin & Solnica-Krezel, 2007). The actin regulatory protein Cap1 is specifically expressed in these cells and is required for elongation (Daggett et al., 2007). In the future, it will be interesting to determine the molecular mechanisms that mediate slow-twitch fiber elongation.

2.2.2 Slow-Twitch Muscle Migration

After they have elongated, slow-twitch fibers migrate laterally through the presumptive fast-twitch muscle domain. There are two notable aspects of this cellular behavior. One is that slow-twitch fibers are both necessary and sufficient to trigger fast-twitch muscle cell elongation (Henry & Amacher, 2004). It will be important to identify the molecular cues involved, because that knowledge could be applied to promote the growth and/or regeneration of fast-twitch fibers. Another interesting aspect of slow-twitch fiber migration is that these cells migrate after they have differentiated. These aspects have made slow-twitch muscle migration an enticing phenomenon.

One mechanism that regulates slow-twitch muscle migration is differential cell–cell adhesion (Cortes et al., 2003). N-cadherin is initially expressed in both presumptive slow and presumptive fast muscle cells (Cortes et al., 2003). M-cadherin is initially expressed only in slow-twitch muscle fibers but is upregulated in the fast-twitch muscle fibers medial to slow-twitch

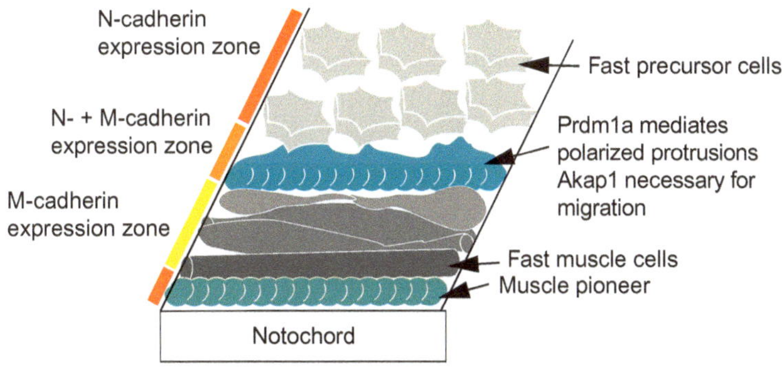

Fig. 2 Mechanisms regulating slow-twitch muscle migration in zebrafish. A medial focal plane of the dorsal half of one muscle segment in the transition zone of a side oriented, anterior left, dorsal top zebrafish embryo is cartooned. Cell–cell adhesion mediated by N- and M-cadherin plays a role in efficient and coordinated slow muscle migration. Dynamic zones of N- and/or M-cadherin expression instruct migrating slow-twitch muscle cells through the presumptive fast-twitch muscle domain toward the lateral surface of the embryo. Adjacent to the notochord, a zone of N-cadherin expression (*red*) occurs where the nonmigratory slow-twitch muscle pioneer fibers (*light blue*) reside. Fast-twitch muscle cells undergo morphogenesis (*dark gray*) medial to the migrating front of slow-twitch fibers in a zone of M-cadherin expression (*yellow*). A zone of M- and N-cadherin expression (*orange*) marks the migrating front of slow-twitch fibers (*blue*). The transcription factor Prdm1a is required cell autonomously to polarize filopodia to the leading edge of migrating slow-twitch fibers and the scaffolding protein Akap12 is also required for slow-twitch muscle migration in zebrafish. Presumptive fast-twitch muscle cells (*light gray*) lateral to the migrating front of slow-twitch fibers (*blue*) have not yet undergone morphogenesis and are in a zone of N-cadherin expression (*red*).

muscle migration (Cortes et al., 2003). This results in a "wave" of cadherin expression as slow-twitch muscle fibers migrate (Fig. 2). Ectopic expression of N-cadherin traps individual slow muscle cells in the fast muscle domain (Cortes et al., 2003). Taken together, these data suggest that differential cell adhesion is one mechanism that contributes to proper slow-twitch muscle cell migration.

Evidence from experiments with zebrafish mutants and in vivo imaging studies recently showed that slow-twitch muscle cell migration is an active, cell autonomous migration mediated by the transcription factor Prdm1a (Ono, Yu, Jackson, Parkin, & Ingham, 2015). Slow-twitch muscle fibers in *prdm1a* mutant embryos migrate more slowly and not as far as slow fibers in wild-type embryos (Ono et al., 2015). Elegant experiments using genetic mosaic analysis and time-lapse confocal microscopy indicate that slow-twitch muscle cells actively migrate by extending filopodia along their leading edge (Fig. 2 and Ono et al., 2015). This process is disrupted in *prdm1a* mutant embryos, where slow-twitch muscle cells extend filopodia along both their leading and trailing edges (Ono et al., 2015). Taken together, these data suggest the model that Prdm1a, which is induced by Hedgehog signaling (Baxendale et al., 2004), mediates polarization of slow-twitch muscle fibers and that this polarization enables directed cell migration (Ono et al., 2015). One candidate for this is Akap12, a slow muscle-specific cytoskeletal remodeling protein required for slow muscle migration (Kim, Kim, Jeong, Han, & Kim, 2014) (Fig. 2). In light of these recent data, what role does differential cadherin expression play in this process? Given the compelling evidence that both M- and N-cadherin are required for slow muscle migration, we suggest that these proteins contribute to efficient migration, that their differential expression functions as an additional guidance cue, and that homophilic cell–cell adhesion enables the collective migration of the slow muscle cell cohort.

2.2.3 Fast-Twitch Muscle Development

Live imaging studies identified morphogenetic cell behaviors mediating fast muscle fiber morphogenesis (Snow, Goody, et al., 2008). Similar behaviors have been identified during muscle morphogenesis in the chick embryo (Gros, Scaal, & Marcelle, 2004; Gros, Serralbo, & Marcelle, 2009), suggesting potential conservation of muscle morphogenesis in vertebrates.

First, short fast muscle precursor cells exhibit dynamic protrusive activity in all directions (Snow, Goody, et al., 2008) (Fig. 3). Second, protrusions are only formed in the direction of elongation (Snow, Goody, et al., 2008). During this phase, intercalation/elongation of muscle fibers results in elongation of fast muscle cells. Next is boundary capture, where muscle cells interact with the anterior and posterior myotome boundaries and cease elongating. After boundary capture, cellular shape changes generate more regularly shaped myotubes (myotube formation) (Snow, Goody, et al., 2008). Fusion occurs concurrently with myotube formation. However, fusion is not necessary for myotube formation because some zebrafish mutant strains with defective muscle cell fusion have mononucleate, regularly shaped myotubes (Powell & Wright, 2011). The next phase includes myofibril formation, the generation of strings of sarcomeres, which are the basic units that generate muscle contraction. The earlier studies have provided a rudimentary understanding of cellular mechanisms and have allowed for the identification of some of the molecular mechanisms that drive muscle morphogenesis in zebrafish.

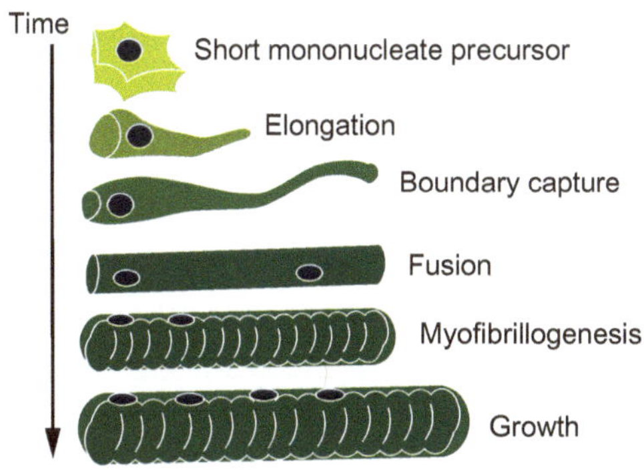

Fig. 3 The cellular morphologies of fast-twitch muscle cells during zebrafish muscle development. Developmental time is represented vertically with earlier developmental time on the top and later developmental time on the bottom. Presumptive fast-twitch muscle fibers are short, extend protrusions in all directions, and contain one nucleus (*light green* cell on top, *gray circle* represents nucleus). Next, cells extend protrusions and elongate. Elongating fast-twitch muscle cells then interact with the extracellular matrix at segment boundaries (i.e., boundary capture) and cease elongating. Fusion generates multinucleated cells (*green* cell, fourth from top). Proteins are organized into repeating contractile units called sarcomeres during myofibrillogenesis (*white outlines in dark green* cells represent sarcomeres, fifth from top). Sarcomeres and interactions between functional muscle fibers and the extracellular matrix at segment boundaries must be actively maintained during muscle use and further growth of zebrafish embryos. Fusion also continually occurs (*dark green* cell, sixth from top).

2.2.4 Fast-Twitch Muscle Cell Elongation

The basement membrane (BM) protein laminin 111, and its receptor, Integrin alpha6beta1 are necessary for efficient and oriented fast muscle cell elongation in zebrafish (Goody et al., 2012; Snow, Goody, et al., 2008). Muscle cell adhesion to the extracellular matrix (ECM) is also critical for fast muscle cell elongation, as laminin beta1 and gamma1 are necessary for timely fast muscle cell elongation (Snow, Goody, et al., 2008). Hedgehog signaling is necessary for the delayed fast-twitch muscle cell elongation that occurs in *laminin beta1* or *gamma1* mutant zebrafish (Peterson & Henry, 2010). Understanding the latent secondary, or "back-up," mechanisms for developmental processes will provide deeper insight into the mechanisms of development.

2.2.5 Boundary Capture and Myotube Formation

Axial skeletal muscle fibers stop elongating when they contact the anterior and posterior segment boundaries. This process is called boundary capture and controls the length of muscle fibers. In zebrafish, boundary capture is mediated by laminin 111 signaling (Snow, Goody, et al., 2008). The ECM glycoprotein Fibronectin (Fn) also plays a role in regulating length of axial skeletal muscle fibers in zebrafish (Snow, Peterson, Khalil, & Henry, 2008). After fast-twitch muscle cells have elongated, they are initially irregularly shaped and mononucleate. The next phase of muscle morphogenesis involves transition from an irregularly shaped cell to a rod-shaped myotube (Snow, Goody, et al., 2008). Increased organization of the actin cytoskeleton correlates with myotube formation. Actin organization occurs prior to the formation of sarcomeres (Mei, Li, & Gui, 2009; Naganawa & Hirata, 2011), but the mechanisms that regulate this distinct cellular process in vertebrates in vivo are unclear. What little is known came from work in zebrafish, which identified the actin regulatory proteins flightless homolog 1 (Naganawa & Hirata, 2011) and myotubularin-related 8 (Mei et al., 2009). A major question is what is the function of this increased actin organization and is this change in actin a necessary step for sarcomere formation? Altogether, the transparency of zebrafish embryos along with the spatial segregation of slow- and fast-twitch muscle fibers has enabled visualization of cellular behaviors during morphogenesis.

3. TRADEMARK TRAITS OF SKELETAL MUSCLE: MUSCLE CELL FUSION AND SARCOMERES

3.1 Muscle Cell Fusion

A defining trait of skeletal muscle cells is multiple nuclei per cell. This multinucleate state is achieved via cell fusion events. Skeletal muscle cells fuse during primary muscle morphogenesis, during hypertrophy (when skeletal muscle cells become bigger in response to exercise), and during repair (in response to damage, injury, or disease). It is important to understand the regulation of vertebrate skeletal muscle cell fusion at the cellular and molecular levels because this information could potentially hold clues to understanding muscle regeneration and to identifying targets for muscle disease therapies. The zebrafish model has been particularly informative in terms of identifying genes and their roles in vertebrate muscle cell fusion.

3.1.1 Vertebrate-Specific Innovations

A landmark study identified two vertebrate-specific cell surface receptors required for fast-twitch muscle fusion in zebrafish: Jamb and Jamc (Powell & Wright, 2011). These proteins are deuterostome-specific Ig superfamily cell surface proteins with two extracellular Ig superfamily domains, a transmembrane domain, and a short cytoplasmic domain. Although fast-twitch fibers are specified and appear morphologically normal in *jamb* or *jamc* mutants, fusion does not occur (Powell & Wright, 2011). Biochemical and genetic mosaic analyses demonstrated that Jamb and Jamc interact and this interaction is required for fusion (Powell & Wright, 2011).

Another landmark study identified the transmembrane protein Tmem8c, also known as Myomaker, as necessary for myoblast fusion and muscle formation in mouse embryos (Millay et al., 2013). Myomaker is the only vertebrate muscle-specific factor identified to date that is both necessary and sufficient to direct myoblast fusion. Although the intracellular C-terminal region of Myomaker is necessary for myoblast fusion and Myomaker exerts its function through the cytoskeleton (Millay et al., 2016, 2013), it is not yet clear how Myomaker interacts with other factors known to play roles in myoblast fusion. Myomaker knockdown via morpholino injection in zebrafish embryos results in short mononucleate presumptive fast muscle cells when analyzed at a developmental stage where fast muscle fibers should be long (Landemaine, Rescan, & Gabillard, 2014). One interpretation of these findings is that Myomaker is required for muscle cell fusion in zebrafish. An alternate interpretation is that Myomaker is required for timely fast-twitch muscle cell elongation. Given that the initial myotome contains mononucleate muscle cells in zebrafish, any putative fusion mutant would be expected to have *long* muscle cells that are mononucleate, as is the case for *jamb* or *jamc* mutants (Powell & Wright, 2011).

The distinction between fast muscle cell elongation vs fusion phenotypes impacts the interpretation of the function of Ckip-1 in muscle cell fusion in zebrafish. Ckip-1 was shown to play a role in mammalian muscle fusion in C2C12 myoblasts (Baas et al., 2012). Like Myomaker, knockdown of Ckip-1 in zebrafish resulted in short mononucleate cells that could result from either fusion or elongation defects (Baas et al., 2012). Future studies should examine whether fast muscle cell elongation recovers in Myomaker- or Ckip-1-deficient zebrafish and, if so, whether long fast-twitch fibers contain fewer nuclei.

3.1.2 Do Proteins From Insects Mediate Cell Fusion in Vertebrates?

Zebrafish orthologs of *Drosophila* Kirre (Kirrel3l), *Drosophila* CDM and Crk family members (Dock1, Dock5, Crk, and Crkl), and of *Drosophila* Sns (Nephrin) have been identified (Moore, Parkin, Bidet, & Ingham, 2007; Sohn et al., 2009; Srinivas, Woo, Leong, & Roy, 2007). Loss of function for many of these factors results in short mononucleate muscle cells in zebrafish that could result from either elongation or fusion defects (Moore et al., 2007; Sohn et al., 2009; Srinivas et al., 2007). By immunohistochemically analyzing phenotypes later in development, it was shown that fast-twitch muscle elongation recovered and there were fewer nuclei in long fast-twitch muscle fibers in Kirrel3l-deficient zebrafish (Srinivas et al., 2007). Therefore, Kirrel3l does appear to play a conserved role in muscle cell fusion. Short fast-twitch muscle cells were also observed upon knockdown of Nephrin in zebrafish (Sohn et al., 2009). Whether muscle cell elongation recovers and whether there were actual fusion defects in these embryos remains to be determined. Taken together, these data suggest the hypothesis that the heterophilic adhesion mediated by Kirrel3l and Nephrin may play dual roles in zebrafish muscle development: initially mediating fast-twitch muscle cell elongation and then promoting fusion. Improved visualization of muscle cell fusion using fluorescent markers will greatly assist with this issue. For example, the *Tg(mylpfa:*H2B-*GFP)* zebrafish line, in which fast muscle cell nuclei fluoresce, should be very useful toward this end (Zhang & Roy, 2016).

3.2 Myofibrillogenesis

A second trademark feature of skeletal muscle cells is the repeating, contractile units called sarcomeres. Movement is generated by the coordinated contraction and relaxation of sarcomeres within muscle cells. Sarcomeres consist of densely packed and highly ordered myofilaments (i.e., actin and myosin) and other proteins that bundle them together, anchor them to the sarcolemma, or are otherwise involved. Myofibrillogenesis is the process by which long strings of sarcomeres are generated in muscle cells. Myofibrillogenesis occurs during primary muscle development as well as during muscle growth and repair. Disruptions of myofibrillogenesis and/or the inability to maintain sarcomeres leads to a subset of myopathies/cardiomyopathies called the myofibrillar myopathies. Thus, understanding how sarcomeres are assembled and how sarcomeres are incorporated into myofibrils are critical issues in muscle development and disease (see Section 6.4).

3.2.1 Sarcomeres as Dynamic Structures

The ability to image myofibrillogenesis in zebrafish has dramatically contributed to our understanding of sarcomere formation and homeostasis. Fluorescence recovery after photobleaching (FRAP) analysis demonstrated that sarcomeric proteins—Actin, Alpha-actinin, FATZ, Myotilin, and

Telethonin—are dynamically exchanged between sarcomeres and a cytoplasmic pool (Sanger, Wang, Holloway, Du, & Sanger, 2009). Live imaging also provided insight into the biomechanics of muscle contractions. Muscle contraction is powered by skeletal myosin transducing ATP into mechanical work. Understanding how myosin lever-arm orientation changes during aging and disease may provide insight into the etiology of how myopathies linked to myosin deleteriously affect muscle function. The application of superresolution microscopy to measuring the orientation of single myosin molecules in zebrafish skeletal muscle revealed differences in the orientations of myosin lever-arms in skeletal vs cardiac muscle (Sun et al., 2014). Applying this technology to different types of muscle diseases will be an important approach to learn more about how myosin mediates muscle function. Additionally, developing methods to utilize superresolution microscopy to simultaneously analyze dynamics of multiple sarcomeric proteins and their conformations during different types of exercise will greatly enhance our understanding of how myofibrils regulate muscle performance and health.

3.2.2 It's All About Connections: Impacts of the Membrane, ECM, and Neuromuscular Junctions on Myofibrillogenesis

Recent insight indicates that myofibrillogenesis is integrated with, and regulated by, adhesion to the sarcolemma and to the ECM (Raeker, Shavit, Dowling, Michele, & Russell, 2014; Weitkunat, Kaya-Çopur, Grill, & Schnorrer, 2014). In *Drosophila*, attachment to the ECM and organization of myosin into myofibrils results in an increase in tension, which induces myofibrillogenesis simultaneously throughout the length of whole muscle cells (Weitkunat et al., 2014). In support of the idea that adhesion to the sarcolemma and the ECM could regulate myofibrillogenesis in vertebrates, it has been shown in zebrafish that the sarcolemma protein Caveolin-3 (Nixon et al., 2005) and the ECM proteins Periostin (Kudo, Amizuka, Araki, Inohaya, & Kudo, 2004), Collagen XV (Pagnon-Minot et al., 2008), and transforming growth factor beta induced (TGFBi) (Kim & Ingham, 2009) are necessary for myofibrillogenesis. Additionally, disrupting cholesterol in cell membranes via statins disrupted myofibrillogenesis in zebrafish (Campos et al., 2015; Huang et al., 2011). It was hypothesized that cholesterol might be important for maintenance of membrane microdomains and potentially for secretion of ECM proteins to the extracellular space (Campos et al., 2015). Thus, studies in the zebrafish support the recent hypothesis, generated from experiments in *Drosophila*, that tension mediated by adhesion to the ECM may mediate myofibrillogenesis. Furthermore, myofibrillogenesis may also depend on the neuromuscular junction (NMJ) because genes encoding multiple acetylcholine receptor (AchR) subunits have been shown to have roles in myofibrillogenesis in zebrafish (Behra et al., 2002; Brennan, Mangoli, Dyer, & Ashworth, 2005; van der Meulen, Schipper, van Leeuwen, & Kranenbarg, 2005). Integrating studies of the sarcolemma, muscle cell–ECM adhesion, and NMJ with myofibrillogenesis certainly adds layers of complexity, but also the opportunity for an integrated physiological understanding of myofibrillogenesis and muscle development and homeostasis. The zebrafish system, with its accessible genetics and live imaging capabilities, is poised to have a tremendous impact on our understanding of myofibrillogenesis and sarcomere maintenance.

3.2.3 Interactions Between Myofibrillogenesis and Muscle Fusion?

Myofibrillogenesis and muscle cell fusion may be more closely linked than has been appreciated. In particular, a recent study links metabolism with both muscle cell fusion and muscle growth (Tixier et al., 2013). In this study, a screen for genes expressed in both zebrafish and *Drosophila* muscle identified the glycolysis gene *pgam2*. Injection of morpholinos against zebrafish *pgam2* resulted in thinner fast muscle fibers with significantly fewer nuclei (Tixier et al., 2013). The observation of thinner fast fibers in zebrafish embryos is highly suggestive of defects in myofibrillogenesis (Kim and Ingham, 2009; Kotani et al., 2015). While the observation of thinner slow-twitch fibers in zebrafish would suggest that myofibrillogenesis could regulate fiber thickness independently of fusion events, the occurrence of thinner fast fibers suggests the possibility that there is crosstalk between the processes of fusion and myofibrillogenesis in regulating fast fiber thickness. Thus, when thinner fast-twitch muscle fibers are observed, we believe investigating whether fusion and/or myofibrillogenesis defects are contributing factors may be a fruitful avenue of investigation. The functional/biochemical consequences and potential translational implications of this thinner fiber phenotype also need to be addressed in future studies.

4. MTJ DEVELOPMENT

4.1 Developmental Defects in the MTJ May Lead to Muscle Disease

Muscles and tendons function as an integrated unit to transduce force to the skeletal system and stabilize joints. Muscle and tendons anchor to each other through receptor complexes that indirectly link the actin cytoskeleton inside muscle cells to the tendon ECM. Cell–ECM adhesions mechanically link muscles to tendons and are required for muscle physiology and function. The main receptor complexes linking muscle cells to their surrounding ECM are the dystrophin–glycoprotein complex and Integrins (Marshall & Crosbie-Watson, 2013).

Many muscle diseases, such as Duchenne, Becker, Merosin-deficient muscular dystrophies, and CMD with integrin deficiency, result from mutations that disrupt adhesion of muscle fibers to their surrounding BM. This weakened link between muscle fibers and their surrounding BM results in increased susceptibility to fiber damage and death during repeated cycles of contraction and relaxation. Data from mouse and zebrafish models of muscular dystrophies implicate MTJ damage and/or fiber detachment from the MTJ as the primary etiology behind muscle dysfunction in some types of muscular dystrophy (Hall et al., 2007; Jacoby et al., 2009; Mayer et al., 1997). It is not known if detachment of muscle fibers from the MTJ BM contributes to human muscular dystrophies because biopsies are excised away from the MTJ to avoid injury to the tendon. However, MRI studies in humans do suggest that muscle damage is more severe closer to the MTJ (Hasegawa et al., 1992; Nagao et al., 1991). Thus, studies of MTJ formation during zebrafish development could provide important clinical insights into how defects in this structure contribute to the onset or progression of muscular dystrophy.

4.2 MTJ Composition During Early Development

The composition of the MTJ ECM changes during MTJ development (Fig. 4). The ECM protein Fn is abundant at initial somite boundaries (Julich, Geisler, & Holley, 2005; Koshida et al., 2005; Trinh & Stainier, 2004), but is downregulated after fast-twitch muscle fibers elongate and attach to the MTJ (Snow & Henry, 2009). Increased organization of the BM protein laminin 111 is observed concomitant with Fn downregulation. These changes occur by approximately 1 day postfertilization (dpf) in zebrafish embryos. At this stage, there are sparse collagen fibrils that are not well organized or anchored to the BM (Charvet, Malbouyres, Pagnon-Minot, Ruggiero, & Le Guellec, 2011). Similar to muscle development in mouse embryos, the predominant laminin isoform in the BM shifts from laminin 111 to laminin 211; this likely occurs between 2 and 3 dpf in zebrafish (Hall et al., 2007). By 3 dpf, the myoseptum has enlarged to

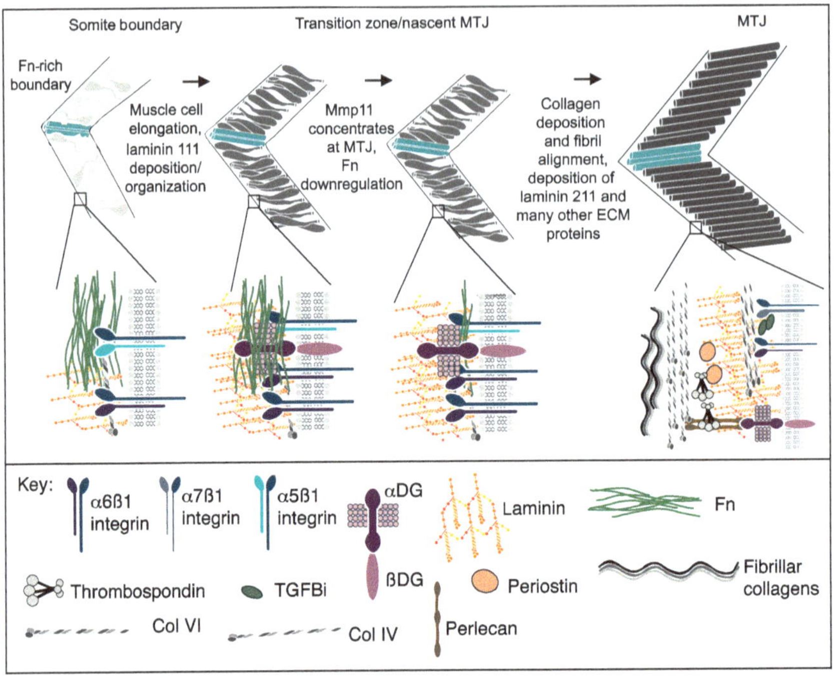

Fig. 4 Molecular changes at the developing zebrafish MTJ. Simplified cartoon of some of the major proteins involved in muscle cell–ECM adhesion at the zebrafish MTJ over developmental time. Developmental time progresses horizontally from left to right. Concurrent with the cellular changes that accompany fast-twitch muscle morphogenesis (cells in *various shades of gray*), the molecular composition of the MTJ ECM changes from Fn-rich somite boundaries, to laminin 111-rich nascent MTJs, to laminin 211-rich mature MTJs (*chevron-shaped lines* represent segment boundaries). *Blue muscle fibers* represent slow-twitch muscle pioneers. Magnifications of the boxed section of each segment boundary cartoon the muscle cell plasma membrane (sarcolemma), transmembrane receptors, and major components of the MTJ ECM (see key). Multiple transmembrane receptor complexes indirectly link the intracellular actin cytoskeleton (not shown) to the ECM. At the developmental stage furthest to the right, the laminin-rich basement membrane (BM) attaches to the collagen-rich interstitial matrix. Mutations in the genes encoding these adhesion components cause many muscle diseases.

500 nm and contains thick collagen fibrils that are aligned either transversely or longitudinally (Charvet et al., 2011).

Many other ECM proteins expressed at the MTJ such as Periostin, Thrombospondin, and multiple collagens (Kudo et al., 2004; Pagnon-Minot et al., 2008; Subramanian & Schilling, 2014; Telfer, Busta, Bonnemann, Feldman, & Dowling, 2010) play a critical role in maintaining the MTJ during repeated cycles of contraction and relaxation, injury, and repair. Zebrafish mutant or morpholino-treated embryos deficient for laminin alpha2, Thrombospondin 4b, or Collagen 22a1 proteins have muscle defects characterized by muscle fibers retracting from the MTJ and the defects are enhanced when muscle contraction is stimulated (either mechanically or electrically) (Charvet et al., 2013; Hall et al., 2007; Subramanian & Schilling, 2014). In these examples, the muscle fibers do not rupture but retract from the MTJ intact. Therefore, these proteins play a role in muscle health by maintaining integrity of muscle cell adhesion to the ECM at the MTJ.

4.3 Mechanisms of Laminin 111 Organization and Regulation

Given the critical role that specific ECM proteins play in MTJ development and homeostasis, it is necessary to determine the mechanisms underlying the dynamic changes in ECM composition during MTJ development. Understanding these mechanisms may provide insight into therapeutic strategies in muscle disease.

Nicotinamide riboside kinase (Nrk)-mediated NAD+ synthesis is conserved from yeast to humans, and members of the Nrk family function to generate nicotinamide adenine dinucleotide (NAD+) (Bieganowski & Brenner, 2004; Tempel et al., 2007). MTJ morphogenesis in Nrk2b-deficient zebrafish embryos can be rescued by providing exogenous NAD+, and it was shown that this metabolite is critical for laminin 111 organization at the MTJ. This is important because some models of CMDs present with disorganized laminin 111 at the MTJ (Goody et al., 2012). NAD+ acts as a small molecule agonist of laminin 111 organization and muscle fiber–MTJ adhesion; exogenous NAD+ supplementation reduced muscle fiber degeneration and improved mobility in zebrafish models of some CMDs (Goody et al., 2012). These data highlight the important contribution that asking basic questions about developmental biology can have with regards to potential therapies. It will be interesting to identify additional components in the Nrk2b pathway. In this regard, laminin 111 organization also appears to be reduced in a zebrafish model of autosomal recessive adolescent onset distal myopathy caused by mutations in *adssl1* (Park et al., 2016). Adssl1 is an adenylosuccinate synthase that is expressed mainly in muscle. Adssl1 may interact with the Nrk2b pathway. Nrk2b is a nicotinamide riboside kinase and thus requires nicotinamide riboside. Human purine nucleotide phosphorylase is required for nicotinamide riboside utilization (Belenky, Christensen, Gazzaniga, Pletnev, & Brenner, 2009) and Adssl1 plays a role in purine biosynthesis (Park et al., 2016). Thus, in the future, it will be interesting to determine the mechanisms of action of Adssl1 and whether Adssl1 interacts with the Nrk2b pathway to promote laminin 111 organization at the MTJ.

4.4 Mechanisms of Fn Regulation

Fn is critically important in multiple developmental contexts (somite segmentation, cell migration, cellular branching) and also for regeneration (reviewed in Bentzinger et al., 2013; Goody & Henry, 2010). Fn adjacent to fast-twitch muscle fibers becomes downregulated during MTJ development (Snow & Henry, 2009) but, until recently, it was unknown how Fn was degraded in vivo. Recent work showed that laminin 111 organization acts as a "checkpoint" for Fn downregulation. Laminin 111 organization potentiates the localization of a matrix metalloproteinase (Mmp11), which is necessary and sufficient for Fn downregulation at the zebrafish fast-twitch muscle MTJ, providing one mechanism by which laminin 111 acts as a signal that results in Fn downregulation (Jenkins, Alrowaished, Goody, Crawford, & Henry, 2016). Thus, understanding how crosstalk between ECM proteins influences cellular outputs during development resulted in the identification of a MMP that may have therapeutic utility in the many conditions where fibrosis contributes to disease progression.

5. FROM DEVELOPMENT TO HOMEOSTASIS

An interesting chapter in a recent *Current Topics in Developmental Biology* volume asks (1) if we should define developmental biology and, if so, (2) what would that definition be (Pradeu et al., 2016)? These questions are especially important given that the impact of development on health, aging, and congenital diseases is becoming more recognized. Here, we discuss evidence for the concept that muscle development does not "end" per se, but, rather, active homeostasis maintains muscle structure even after initial muscle development occurs.

5.1 Homeostasis Part 1: The MTJ and Regulation of Muscle Fiber Length

The axis of time is critical when interpreting phenotypes because sometimes a given phenotype could result from disruptions at multiple stages of development. For example, defects in muscle fiber length can be due to either early boundary defects or later homeostasis defects. As mentioned earlier, elongating muscle fibers are "captured" by the ECM at the MTJ, which is derived from somite boundaries. If somite boundaries do not form properly, then elongating fibers are not captured and abnormally long muscle fibers are observed (Snow, Peterson, et al., 2008). Such cases are examples of early developmental defects. However, in zebrafish muscle disease models, we and others provided examples where somite boundaries form normally, but later in development, muscle fibers become abnormally long (Etard et al., 2010; Goody, Kelly, Lessard, Khalil, & Henry, 2010; Goody et al., 2012; Snow, Goody, et al., 2008) (Fig. 5A). This phenotype shows that maintenance of fiber length and MTJ integrity is an active process, which we refer to as active homeostasis.

Evidence from zebrafish studies elucidated a critical role for laminin 111 in fiber length and MTJ integrity. The laminin chains laminin beta1 and laminin gamma1 as well as an integrin receptor for laminin 111, Integrin alpha6, are required to maintain fiber length/MTJ integrity (Goody et al., 2012; Snow, Goody, et al., 2008). Data also indicate that just the presence of laminin 111 is not sufficient: laminin 111 is present but disrupted in zebrafish deficient for Nrk2b or Apo2a (Etard et al., 2010). In the latter case, Hsp45b binds to Apo2a and its silencing causes a similar MTJ failure phenotype, suggesting that both are required for MTJ integrity (Etard et al., 2010). Laminin 111 is not the only ECM protein required for MTJ maintenance, as zebrafish lacking Periostin also show MTJ failure and abnormally long muscle fibers (Kudo et al., 2004). In contrast, although collagens are critical components of the MTJ, disruption of many of the collagen chains studied thus far (Col15a1, Col6a2, Col6a4a, Col4a4b, Col22a1) does not result in MTJ failure or abnormally long fibers in zebrafish (Charvet et al., 2013; Pagnon-Minot et al., 2008; Ramanoudjame et al., 2015). These data highlight the utility of the zebrafish model in identifying discrete roles for different MTJ ECM proteins during development and homeostasis.

5.2 Homeostasis Part 2: When the Resilience of Muscle Fibers Fails in Dystrophies

Muscle homeostasis involves maintenance of sarcolemma integrity, fiber adhesion to the ECM, ECM integrity, and the satellite cell pool. Muscle use can result in small tears in the sarcolemma (plasma membrane) that need to be repaired. Muscle fibers can also sustain enough damage to the sarcolemma that they die, at which point the muscle resident stem cells (satellite cells) mediate

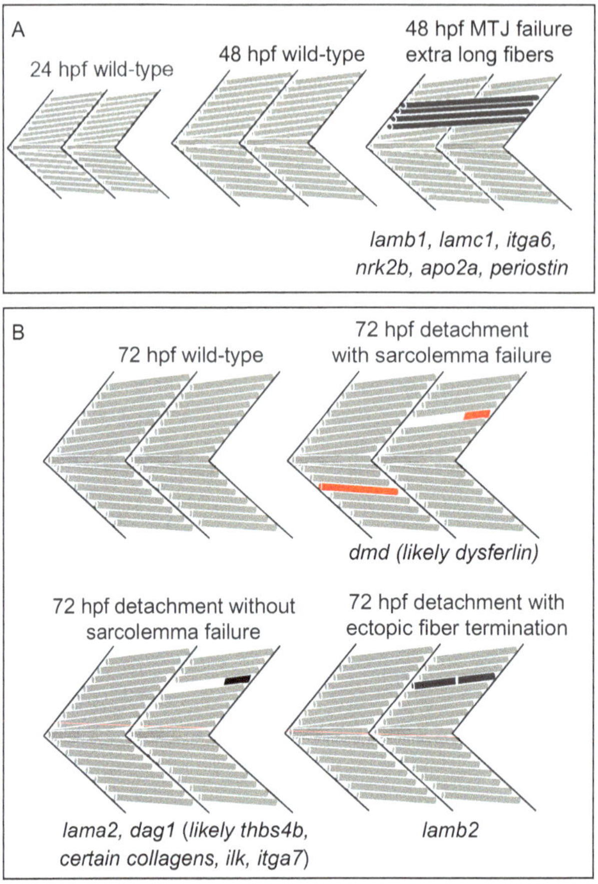

Fig. 5 Schematic of muscle disease phenotypes in zebrafish embryos. Two muscle segments of a side oriented, anterior left, dorsal top zebrafish embryo are diagrammed at the different developmental stages indicated in the figure. Muscle fibers are *gray cylinders* and segment boundaries are *black chevron-shaped lines*. Muscle fibers with disease phenotypes are *dark gray* or *red*. (A) In zebrafish deficient in laminin beta1, laminin gamma1, Integrin alpha6, Nrk2b, Apo2a, or Periostin gene products, MTJs fail and some muscle fibers are abnormally long (*dark gray* fibers). (B) In 72 hpf *dmd* mutant zebrafish, fiber detachment is observed. Detached (*short red fiber*) and even some attached fibers (*long red fiber*) uptake Evans blue dye, showing that *dmd* mutations result in loss of sarcolemma integrity. Fiber detachment occurs independently of loss of sarcolemma integrity in *laminin alpha2* or *dystroglycan* mutant zebrafish (*short dark gray fiber*), suggesting that loss of adhesion to the ECM can also cause fiber detachment. Fibers detach yet remain viable in *laminin beta2* mutant zebrafish by forming an ectopic MTJ (*dark gray fiber* with *white line* in the middle representing an ectopic MTJ).

muscle repair. There is a great deal of interest in understanding muscle repair as it pertains to muscle aging, traumatic muscle injury, and muscle diseases, and zebrafish provides a powerful model for such studies.

Dysferlin is a membrane protein important for membrane resealing. Mutations in human *DYSF* can result in multiple types of myopathies (Bashir et al., 1998; Liu et al., 1998). In mammalian cell culture models, Dysferlin has been associated with sites of damaged muscle cell membranes, suggesting that it might play a role in membrane repair (Lennon et al., 2003). Important questions being addressed using zebrafish are (1) how does Dysferlin potentiate membrane healing in vivo, (2) how does the membrane reseal and where does the membrane patch come from, and (3) can any membrane domain contribute to repair? These questions were addressed by using live imaging to monitor membrane repair in zebrafish embryos (Roostalu & Strähle, 2012). The fact that Dysferlin, but not many other membrane/vesicle markers tested, rapidly relocates to damage suggests the hypothesis that there is a specialized membrane domain, rich in Dysferlin, that participates in membrane repair.

By combining live imaging and genetic approaches in zebrafish, the steps for the highly ordered process of membrane repair involving Annexins and Dysferlin have been uncovered (Roostalu & Strähle, 2012). First, the lipid-binding Annexin A6 accumulates rapidly and might "clog" the membrane. Simultaneously, Dysferlin from a specialized membrane subdomain accumulates at the damage site. Next, Annexin A2a is added to the patch. Finally, Annexin A1a arrives. Thus, these data support the model where a multilayered scaffold forms in a sequential fashion to promote membrane repair (Roostalu & Strähle, 2012). The hope is that this information could be used to develop methodologies for sarcolemma repair to either enhance muscle recovery after injury or potentially reduce degeneration of diseased muscle.

6. NEW INSIGHTS INTO MUSCLE DISEASE FROM ZEBRAFISH

One premise underlying the effort dedicated to understanding muscle development, growth, and repair is that knowledge gained can be leveraged to develop therapies to combat muscle disease. The zebrafish model has made many important contributions to our understanding of muscle disease, particularly with regards to muscular dystrophies and myopathies. Here, we will discuss some recent studies that used the zebrafish model to provide insight into muscle disease.

6.1 Testing Candidate Disease Causing Mutations

Morpholinos can be used to knockdown gene expression to help validate potential human disease causing genes identified by whole-exome sequencing: if injection of morpholinos against the gene in question results in muscle disease then this result bolsters the conclusion that the disease causing gene has been identified. Some recent examples include a distal myopathy caused by mutation in *ADSSL1* (Park et al., 2016) and Native American myopathy in which *Stac3* is mutated (Horstick et al., 2013). Similar approaches identified the causative genes for a nemaline myopathy (Yuen et al., 2014) and a mitochondrial myopathy (Shamseldin, Smith, et al., 2016). This approach was also used to identify a novel lethal congenital contracture syndrome caused by mutation in *ZBTB42* (Patel et al., 2014). Thus, one use of the zebrafish system has been to fairly readily narrow down potential candidates to identify and/or validate mutations responsible for disease phenotypes.

6.2 Identification of Candidate Genes for Muscle Diseases of Unknown Genetic Cause

A recent study investigated how fast skeletal myosin-binding protein C (MyBPC-2) mediates sarcomere length and muscle contraction (Li et al., 2016). The finding that injection of morpholinos against MyBPC-2 results in myopathy may aid clinicians endeavoring to identify the genetic basis for unknown muscular dystrophies.

Genetic screens in zebrafish have also led to identification of genes disrupting muscle integrity. One such mutant, *trage*, had disruption of muscle integrity as assessed by birefringence and impaired swimming behavior (Berger et al., 2014). Positional cloning identified a nonsense mutation in *tropomodulin4* (Berger et al., 2014). Phenotypic characterization showed that cytoplasmic rods, similar to the nemaline rods found in patients with nemaline myopathy, were prevalent in *tmod4/trg* mutants (Berger et al., 2014). Thus, *tmod4/trg* is a new candidate gene for nemaline myopathies of unknown etiology.

6.3 Novel Disease Mechanisms

The cellular pathology underlying disease progression is not well understood in muscular dystrophies and myopathies. This lack of understanding is a significant barrier for therapy development. Thus, identification of the cellular pathology in zebrafish disease models is a high priority because this knowledge could lead to new therapeutic avenues.

One example is the divergent cellular pathology between a Duchenne muscular dystrophy model and a dystroglycanopathy model. Muscle fibers in the zebrafish model of Duchenne muscular dystrophy uptake Evans blue dye, which only infiltrates cells that have damaged membranes. This dye was used to label and track the behavior of muscle fibers with damaged membranes. Evans blue dye-labeled fibers were observed to detach from the BM at the MTJ, and then undergo necrosis/apoptosis (Hall et al., 2007) (Fig. 5B). In contrast, muscle fibers in some zebrafish muscular dystrophy models actually detach from the BM at the MTJ *prior* to being permeable to Evans blue dye (Hall et al., 2007; Jacoby et al., 2009) (Fig. 5B). These data indicate that the cellular pathology of muscular dystrophies varies depending upon the genetic basis for disease. Remarkably, muscle fibers in one zebrafish mutant with muscular dystrophy survive after they detach from the MTJ (Jacoby et al., 2009). Muscle fibers in *softy/lamb2* mutant zebrafish generate and attach to ectopic BMs and thus remain viable (Fig. 5B). *Softy* mutants are unique in that they are homozygous viable and it is hypothesized that these ectopic fiber terminations function to stabilize the damaged myotome (Jacoby et al., 2009). This stabilization could allow more robust regeneration or slow degeneration. Certainly, understanding the mechanisms underlying the recovery from early and severe muscle degeneration is critical because it may provide novel insight into therapeutic approaches.

Another example of zebrafish contributing new knowledge regarding disease mechanisms is autosomal dominant centronuclear myopathy, which is caused by mutations in *dynamin-2* (*DNM2*) (Hanisch et al., 2011; Jeub et al., 2008). Generation of a zebrafish model revealed defects in excitation-contraction coupling that were hypothesized to be caused by defective membrane tubulation (Gibbs et al., 2014). This hypothesis was tested in an in vitro tubulation assay, where the dominant effect of the mutant *DNM2* mRNA on normal tubule formation was confirmed. Thus, this study provides new insight into the cellular mechanisms of disease, and this knowledge can be used as a platform for therapy development.

6.4 Myofibrillar Myopathies

Myofibrillar myopathies are heterogeneous in the muscle groups affected and the timing of onset, but they share progressive muscle weakness and protein aggregation. Zebrafish models have been developed for most known myofibrillar myopathy-causing genes (Table 2). Interestingly, overexpression

Table 2 Zebrafish Models of Muscle Diseases not in Gibbs, Horstick, and Dowling (2013) Review

Myopathy	Gene Product	Method of Gene Manipulation	References
Centronuclear myopathy	Dynamin2	Translation and splice blocking *dnm2a* MOs; overexpression of human *DNM2* R522H or S619L mutants	Gibbs, Davidson, Telfer, Feldman, and Dowling (2014) and Bragato et al. (2016)
	Bin1	Translation and splice blocking *bin1* MOs; overexpression of human *BIN1* K35N or K575* mutants	Smith, Gupta, and Beggs (2014)
X-linked myotubular myopathy (subtype of centronuclear myopathy)	Mtm1, Mtmr12	Translation (*mtm1*, *mtmr12*) and splice (*mtmr12*) blocking MOs; overexpression of human *MTM1*, *MTMR12*	Gupta, Hnia, et al. (2013)
Limb-girdle muscular dystrophy (LGMD)	Popdc1	Splice blocking *popcd1* MOs; overexpression of human *POPDC1* S201F; TALEN-mediated mutagenesis *popdc1 s191f*	Schindler et al. (2016)
	Hnrpdl	Translation blocking *hnrpdl* MOs	Vieira et al. (2014)
LGMD, movement myopathy, congenital muscular dystrophy (CMD) with fatty liver and infant onset cataract	Trappc11	Mutant line. Liver and eye phenotypes previously characterized in zebrafish. Skeletal muscle phenotype not characterized in zebrafish	Liang et al. (2015)
Mitochondrial myopathy	Slc25a42	Translation and splice blocking *slc25a42* MOs; overexpression of human *SLC25A42* N291D mutant	Shamseldin, Smith, et al. (2016)

Distal myopathy	Adssl1	Splice blocking *adssl1* MOs; overexpression of human *ADSSL1* D304N and I350fs mutants	Park et al. (2016)
Ullrich congenital muscular dystrophy (UCMD), Bethlem myopathy	Col6a1	TALEN-mediated mutagenesis of splice site resulting in exon skipping	Radev et al. (2015)
	Col6a1	Splice blocking *col6a1* MOs	Zulian et al. (2014)
Myofibrillar myopathy	FilaminC	*flnc* MOs and mutant line; overexpression of human *FLNC* W2710X mutant	Ruparelia, Oorschot, Ramm, and Bryson-Richardson (2016) and Ruparelia, Zhao, Currie, and Bryson-Richardson (2012)
	Abcrystallina, Abcrystallinb, Desmina, Desminb, Bag3, Dnajb6a, Dnajb6b, Fhl1a, Fhl1b, Filaminca, Filamincb, Myotilin, Plectina, Plectinb, Vcp, Zasp	Translation or splice blocking MOs	Bührdel et al. (2015)
	Desmina	Mutant line from ENU mutagenesis screen; translation and splice blocking *desma* MOs	Ramspacher et al. (2015)
	Bag3	Translation and splice blocking MOs; overexpression of human *BAG3* P209L mutant	Ruparelia, Oorschot, Vaz, Ramm, and Bryson-Richardson (2014)

Continued

Table 2 Zebrafish Models of Muscle Diseases not in Gibbs, Horstick, and Dowling (2013) Review—cont'd

Myopathy	Gene Product	Method of Gene Manipulation	References
Nemaline myopathy	Acta1, Neb	Translation (*actc1b*) and splice (*actc1b*, *neb*) blocking MOs; overexpression of human *ACTA1* D286G mutant	Sztal et al. (2015)
	Lmod3	Splice blocking *lmod3* MOs	Yuen et al. (2014)
	Tmod4	Mutant line from ENU mutagenesis screen; translation blocking *tmod4* MOs	Berger et al. (2014)
	Klhl40, Klhl41	Translation and splice blocking MOs for *klhl41a* and *klhl41b*; Translation blocking MOs for *klhl40a* and *klhl40b*	Gupta, Ravenscroft, et al. (2013), and Ravenscroft et al. (2013)
Myotonia congenita	Clcn1	Overexpression of human *CLCN1* I553F/ H555N and *CLCN1* L844F mutants	Cheng, Tian, Burgunder, Hunziker, and Eng (2014)
Duchenne muscular dystrophy (DMD)	DMD	Mutant lines (*sapje*, *sapje-like*, *dmdpc2*)	Waugh et al. (2014), Li, Andersson-Lendahl, Sejersen, and Arner (2014), Kawahara et al. (2014), Kawahara and Kunkel (2013), Giacomotto et al. (2013), Johnson, Farr, and Maves (2013), Winder, Lipscomb, Angela Parkin, and Juusola (2011), and Bassett et al. (2003)

Stormorken syndrome	Stim1, Orai1	Overexpression of human *STIM1* R304W, *STIM1* D76A, and *ORAI1* P245L mutants. Thrombocytopenia, but not tubular myopathy phenotype investigated in zebrafish	Nesin et al. (2014)
GNE myopathy	Gne	Translation and splice blocking *gne* MOs	Daya et al. (2014)
Myotonic dystrophy	Dmpk 3′UTR	Overexpression of RNA containing 91 uninterrupted CUG repeats	Todd et al. (2014)
Secondary dystroglycanopathy	Gmppb	Splice blocking *gmppb* MOs	Carss et al. (2013)
	B3galnt2	Translation and splice blocking *b3galnt2* MOs	Stevens et al. (2013)
	Pomk	Translation and splice blocking *pomk* MOs; overexpression of human *POMK* Q109* mutant mRNA	Di Costanzo et al. (2014)
Native American myopathy	Stac3	Mutant line; overexpression of zebrafish *stac3* mutant mRNA corresponding to human *STAC3* W284S mutation	Horstick et al. (2013)
Arthrogryposis	Zc4h2	Translation and splice blocking *zc4h2* MOs; overexpression of mouse *Zc4h2* mutant mRNAs corresponding to human *ZC4H2* mutants (Val63Leu, Arg198Gln, Pro201Ser, Arg213Trp)	Hirata et al. (2013)
Arthrogryposis, lethal congenital contracture syndrome (LCCS)	Mybpc1	Translation and splice blocking *mybpc1* MOs, overexpression of zebrafish *Mybpc1* W220R, Y845H mutants (correspond to pathogenic human mutations)	Ha et al. (2013)

Continued

Table 2 Zebrafish Models of Muscle Diseases not in Gibbs, Horstick, and Dowling (2013) Review—cont'd

Myopathy	Gene Product	Method of Gene Manipulation	References
LCCS	Zbtb42	Translation and splice blocking *zbtb42* MOs; overexpression of human *ZBTB42* Arg397His mutant	Patel et al. (2014)
Nutritional myopathy		Vitamins E and C deficiency	Lebold et al. (2013)
Ceroid lipofuscinosis, neuronal, 10	Cathepsin D	Translation blocking *ctsd* MOs	Follo, Ozzano, Montalenti, Santoro, and Isidoro (2013)
Unnamed neuromuscular disease	Golga2	Translation and splice blocking *golga2* MOs	Shamseldin, Bennett, Alfadhel, Gupta, and Alkuraya (2016)
No known human disease	Apobec2	Translation blocking *apobec2* MOs	Etard, Roostalu, and Strähle (2010)
	Atrogin1	Translation and splice blocking *atrogin1* MOs	Bühler et al. (2016)
	Col22a1	Translation blocking *col22a1* MOs, overexpression of human wild-type *COLXXII*	Charvet et al. (2013)
	Mybpc2	MOs	Li, Andersson-Lendahl, Sejersen, and Arner (2016)

*Stop codon.

of mutated human *BAG3* or *FLNC*, despite causing protein aggregation, rescues the fiber phenotype caused by knockdown of the corresponding zebrafish protein, suggesting that these disease causing mutations still result in functional proteins (Ruparelia et al., 2016, 2014). Zebrafish studies have linked impaired autophagy to disrupted myofibrillogenesis and shown that pharmacological activation of autophagy may be a potential therapeutic intervention for myofibrillar myopathies (Ruparelia et al., 2016; Skobo et al., 2014). The importance of autophagy in the dynamic maintenance of sarcomeres and the contractile function of muscle tissue suggests that investigating autophagy in muscle development, regeneration, aging, and other myopathies is an important goal for future studies.

7. CONCLUSIONS AND FUTURE PROSPECTS

One major premise underlying efforts to elucidate how muscle develops during embryogenesis is that understanding normal developmental processes and homeostasis can inform regenerative medicine (traumatic muscle injuries), aging (sarcopenia), and disease (cachexia, muscular dystrophies, myopathies). Studies in zebrafish have made seminal contributions to our understanding of vertebrate muscle specification and morphogenesis. The zebrafish model has also been invaluable for identification of disease causing genes and novel disease mechanisms. What are the major challenges that remain? With regards to muscular dystrophies, there are at least two pressing issues. One is the dramatic phenotypic spectrum of disease progression and the other is therapeutic crossover. One of the challenges in understanding muscular dystrophies and myopathies is that patients present with varying phenotypes, both within a particular type of dystrophy/myopathy, and between different dystrophies/myopathies. The extent to which potential aberrations in muscle development are linked to such variations in the age of onset and speed of progression is not well understood. Use of the zebrafish model has already made significant insights into the genetic basis of multiple muscle diseases, and revealed new cellular pathologies in these diseases. We predict that leveraging the advantages of the zebrafish model will have an enormous impact on understanding the molecular genetics underlying the dramatic phenotypic spectrum of muscle diseases. It is difficult to imagine another vertebrate model where it is feasible to quantify multiple phenotypic traits in large populations through time, followed by genomic analyses to identify biomarkers and loci that interact with disease progression. Thus, the zebrafish model, with its facile genetics and in vivo cell biology, stands poised to contribute tremendously to our understanding of muscle diseases. The question of whether there is any potential therapeutic crossover between treatments for Duchenne/Becker muscular dystrophy and other types of muscular dystrophies has yet to be answered. This is an important question because although therapeutic strides are being made for the most common type of muscular dystrophy, Duchenne/Becker muscular dystrophy, treatment for most other types of muscle diseases lags behind. Therefore, in the future, it is critical to determine whether differences in cellular pathology affect treatment options. This is an area where zebrafish models of muscle disease have high potential to contribute to pathologic mechanisms and questions regarding therapeutic crossover.

ACKNOWLEDGMENTS

We would like to acknowledge our colleague's work that was not highlighted in the main text of this review. References for much of this work can be found in Table 2 as well as the References section. This work was supported by the Graduate School of Biomedical Sciences and Engineering at the University of Maine, March of Dimes Award number #1-FY14-284 to C.A.H., and National Institute of Child Health and Human Development Award numbers 5RO3HD077545 and R15HD088217 to C.A.H.

REFERENCES

Baas, D., Caussanel-Boude, S., Guiraud, A., Calhabeu, F., Delaune, E., Pilot, F., et al. (2012). CKIP-1 regulates mammalian and zebrafish myoblast fusion. *Journal of Cell Science*, *125*, 3790–3800.

Bashir, R., Britton, S., Strachan, T., Keers, S., Vafiadaki, E., Lako, M., et al. (1998). A gene related to *Caenorhabditis elegans* spermatogenesis factor fer-1 is mutated in limb-girdle muscular dystrophy type 2B. *Nature Genetics*, *20*, 37–42.

Bassett, D. I., Bryson-Richardson, R. J., Daggett, D. F., Gautier, P., Keenan, D. G., & Currie, P. D. (2003). Dystrophin is required for the formation of stable muscle attachments in the zebrafish embryo. *Development*, *130*, 5851–5860.

Baxendale, S., Davison, C., Muxworthy, C., Wolff, C., Ingham, P. W., & Roy, S. (2004). The B-cell maturation factor Blimp-1 specifies vertebrate slow-twitch muscle fiber identity in response to Hedgehog signaling. *Nature Genetics*, *36*, 88–93.

Behra, M., Cousin, X., Bertrand, C., Vonesch, J.-L., Biellmann, D., Chatonnet, A., et al. (2002). Acetylcholinesterase is required for neuronal and muscular development in the zebrafish embryo. *Nature Neuroscience*, *5*, 111–118.

Belenky, P., Christensen, K. C., Gazzaniga, F., Pletnev, A. A., & Brenner, C. (2009). Nicotinamide riboside and nicotinic acid riboside salvage in fungi and mammals. Quantitative basis for Urh1 and purine nucleoside phosphorylase function in NAD+ metabolism. *The Journal of Biological Chemistry*, *284*, 158–164.

Bentzinger, C. F., Wang, Y. X., Maltzahn von, J., Soleimani, V. D., Yin, H., & Rudnicki, M. A. (2013). Fibronectin regulates Wnt7a signaling and satellite cell expansion. *Cell Stem Cell*, *12*, 75–87.

Berger, J., Tarakci, H., Berger, S., Li, M., Hall, T. E., Arner, A., et al. (2014). Loss of Tropomodulin4 in the zebrafish mutant träge causes cytoplasmic rod formation and muscle weakness reminiscent of nemaline myopathy. *Disease Models & Mechanisms*, *7*, 1407–1415.

Bieganowski, P., & Brenner, C. (2004). Discoveries of nicotinamide riboside as a nutrient and conserved NRK genes establish a Preiss-Handler independent route to NAD+ in fungi and humans. *Cell*, *117*, 495–502.

Bragato, C., Gaudenzi, G., Blasevich, F., Pavesi, G., Maggi, L., Giunta, M., et al. (2016). Zebrafish as a model to investigate dynamin 2-related diseases. *Scientific Reports*, *6*, 20466.

Brennan, C., Mangoli, M., Dyer, C. E. F., & Ashworth, R. (2005). Acetylcholine and calcium signalling regulates muscle fibre formation in the zebrafish embryo. *Journal of Cell Science*, *118*, 5181–5190.

Buckingham, M., Bajard, L., Chang, T., Daubas, P., Hadchouel, J., Meilhac, S., et al. (2003). The formation of skeletal muscle: From somite to limb. *Journal of Anatomy*, *202*, 59–68.

Bühler, A., Kustermann, M., Bummer, T., Rottbauer, W., Sandri, M., & Just, S. (2016). Atrogin-1 deficiency leads to myopathy and heart failure in zebrafish. *International Journal of Molecular Sciences*, *17*, E187.

Bührdel, J. B., Hirth, S., Kessler, M., Westphal, S., Forster, M., Manta, L., et al. (2015). In vivo characterization of human myofibrillar myopathy genes in zebrafish. *Biochemical and Biophysical Research Communications*, *461*, 217–223.

Campos, L. M., Rios, E. A., Midlej, V., Atella, G. C., Herculano-Houzel, S., Benchimol, M., et al. (2015). Structural analysis of alterations in zebrafish muscle differentiation induced by simvastatin and their recovery with cholesterol. *The Journal of Histochemistry and Cytochemistry*, *63*, 427–437.

Cann, G. M., Lee, J. W., & Stockdale, F. E. (1999). Sonic hedgehog enhances somite cell viability and formation of primary slow muscle fibers in avian segmented mesoderm. *Anatomy and Embryology*, *200*, 239–252.

Carss, K. J., Stevens, E., Foley, A. R., Cirak, S., Riemersma, M., Torelli, S., et al. (2013). Mutations in GDP-mannose pyrophosphorylase B cause congenital and limb-girdle muscular dystrophies associated with hypoglycosylation of α-dystroglycan. *American Journal of Human Genetics*, *93*, 29–41.

Charvet, B., Guiraud, A., Malbouyres, M., Zwolanek, D., Guillon, E., Bretaud, S., et al. (2013). Knockdown of col22a1 gene in zebrafish induces a muscular dystrophy by disruption of the myotendinous junction. *Development*, *140*, 4602–4613.

Charvet, B., Malbouyres, M., Pagnon-Minot, A., Ruggiero, F., & Le Guellec, D. (2011). Development of the zebrafish myoseptum with emphasis on the myotendinous junction. *Cell and Tissue Research*, *346*, 439–449.

Cheng, W., Tian, J., Burgunder, J. M., Hunziker, W., & Eng, H. L. (2014). Myotonia congenita-associated mutations in chloride channel-1 affect zebrafish body wave swimming kinematics. *PLoS One*, *9*, e103445.

Ciciliot, S., Rossi, A. C., Dyar, K. A., Blaauw, B., & Schiaffino, S. (2013). Muscle type and fiber type specificity in muscle wasting. *The International Journal of Biochemistry & Cell Biology*, *45*, 2191–2199.

Cortes, F., Daggett, D., Bryson-Richardson, R. J., Neyt, C., Maule, J., Gautier, P., et al. (2003). Cadherin-mediated differential cell adhesion controls slow muscle cell migration in the developing zebrafish myotome. *Developmental Cell*, *5*, 865–876.

Currie, P. D., & Ingham, P. W. (1996). Induction of a specific muscle cell type by a hedgehog-like protein in zebrafish. *Nature*, *382*, 452–455.

Daggett, D. F., Domingo, C. R., Currie, P. D., & Amacher, S. L. (2007). Control of morphogenetic cell movements in the early zebrafish myotome. *Developmental Biology*, *309*, 169–179.

Daya, A., Vatine, G. D., Becker-Cohen, M., Tal-Goldberg, T., Friedmann, A., Gothilf, Y., et al. (2014). Gne depletion during zebrafish development impairs skeletal muscle structure and function. *Human Molecular Genetics*, *22*, 3349–3361.

Devakanmalai, G. S., Zumrut, H. E., & Ozbudak, E. M. (2013). Cited3 activates Mef2c to control muscle cell differentiation and survival. *Biology Open*, *2*, 505–514.

Devoto, S. H., Melançon, E., Eisen, J. S., & Westerfield, M. (1996). Identification of separate slow and fast muscle precursor cells in vivo, prior to somite formation. *Development*, *122*, 3371–3380.

Di Costanzo, S., Balasubramanian, A., Pond, H. L., Rozkalne, A., Pantaleoni, C., Saredi, S., et al. (2014). POMK mutations disrupt muscle development leading to a spectrum of neuromuscular presentations. *Human Molecular Genetics*, *23*, 5781–5792.

Etard, C., Roostalu, U., & Strähle, U. (2010). Lack of Apobec2-related proteins causes a dystrophic muscle phenotype in zebrafish embryos. *The Journal of Cell Biology*, *189*, 527–539.

Feng, X., Adiarte, E. G., & Devoto, S. H. (2006). Hedgehog acts directly on the zebrafish dermomyotome to promote myogenic differentiation. *Developmental Biology*, *300*, 736–746.

Follo, C., Ozzano, M., Montalenti, C., Santoro, M. M., & Isidoro, C. (2013). Knockdown of cathepsin D in zebrafish fertilized eggs determines congenital myopathy. *Bioscience Reports*, *33*, e00034.

Giacomotto, J., Brouilly, N., Walter, L., Mariol, M. C., Berger, J., Ségalat, L., et al. (2013). Chemical genetics unveils a key role of mitochondrial dynamics, cytochrome c release and IP3R activity in muscular dystrophy. *Human Molecular Genetics*, *22*, 4562–4578.

Gibbs, E. M., Davidson, A. E., Telfer, W. R., Feldman, E. L., & Dowling, J. J. (2014). The myopathy-causing mutation DNM2-S619L leads to defective tubulation in vitro and in developing zebrafish. *Disease Models & Mechanisms*, 7, 157–161.

Gibbs, E. M., Horstick, E. J., & Dowling, J. J. (2013). Swimming into prominence: the zebrafish as a valuable tool for studying human myopathies and muscular dystrophies. *FEBS Journal*, *280*(17), 4187–4197. http://doi.org/10.1111/febs.12412.

Goody, M. F., & Henry, C. A. (2010). Dynamic interactions between cells and their extracellular matrix mediate embryonic development. *Molecular Reproduction and Development*, 77, 475–488.

Goody, M. F., Kelly, M. W., Lessard, K. N., Khalil, A., & Henry, C. A. (2010). Nrk2b-mediated NAD+ production regulates cell adhesion and is required for muscle morphogenesis in vivo. *Developmental Biology*, *344*, 809–826.

Goody, M. F., Kelly, M. W., Reynolds, C. J., Khalil, A., Crawford, B. D., & Henry, C. A. (2012). NAD+ biosynthesis ameliorates a zebrafish model of muscular dystrophy. *PLoS Biology*, *10*, e1001409.

Gros, J., Scaal, M., & Marcelle, C. (2004). A two-step mechanism for myotome formation in chick. *Developmental Cell*, *6*, 875–882.

Gros, J., Serralbo, O., & Marcelle, C. (2009). WNT11 acts as a directional cue to organize the elongation of early muscle fibres. *Nature*, *457*, 589–593.

Gupta, V. A., Hnia, K., Smith, L. L., Gundry, S. R., McIntire, J. E., Shimazu, J., et al. (2013). Loss of catalytically inactive lipid phosphatase myotubularin-related protein 12 impairs myotubularin stability and promotes centronuclear myopathy in zebrafish. *PLoS Genetics*, *9*, e1003583.

Gupta, V. A., Ravenscroft, G., Shaheen, R., Todd, E. J., Swanson, L. C., Shiina, M., et al. (2013). Identification of KLHL41 mutations implicates BTB-kelch-mediated ubiquitination as an alternate pathway to myofibrillar disruption in nemaline myopathy. *American Journal of Human Genetics*, *93*, 1108–1117.

Ha, K., Buchan, J. G., Alvarado, D. M., McCall, K., Vydyanath, A., Luther, P. K., et al. (2013). MYBPC1 mutations impair skeletal muscle function in zebrafish models of arthrogryposis. *Human Molecular Genetics*, *22*, 4967–4977.

Hall, T. E., Bryson-Richardson, R. J., Berger, S., Jacoby, A. S., Cole, N. J., Hollway, G. E., et al. (2007). The zebrafish candyfloss mutant implicates extracellular matrix adhesion failure in laminin alpha2-deficient congenital muscular dystrophy. *Proceedings of the National Academy of Sciences of the United States of America*, *104*, 7092–7097.

Hammond, C. L., Hinits, Y., Osborn, D. P., Minchin, J. E., Tettamanti, G., & Hughes, S. M. (2007). Signals and myogenic regulatory factors restrict pax3 and pax7 expression to dermomyotome-like tissue in zebrafish. *Developmental Biology*, *302*, 504–521.

Hanisch, F., Müller, T., Dietz, A., Bitoun, M., Kress, W., Weis, J., et al. (2011). Phenotype variability and histopathological findings in centronuclear myopathy due to DNM2 mutations. *Journal of Neurology*, *258*, 1085–1090.

Hasegawa, T., Matsumura, K., Hashimoto, T., Ikehira, H., Fukuda, H., & Tateno, Y. (1992). Intramuscular degeneration process in Duchenne muscular dystrophy—Investigation by longitudinal MR imaging of the skeletal muscles. *Rinshō Shinkeigaku*, *32*, 333–335.

Henry, C. A., & Amacher, S. L. (2004). Zebrafish slow muscle cell migration induces a wave of fast muscle morphogenesis. *Developmental Cell*, *7*, 917–923.

Hirata, H., Nanda, I., van Riesen, A., McMichael, G., Hu, H., Hambrock, M., et al. (2013). ZC4H2 mutations are associated with arthrogryposis multiplex congenita and intellectual disability through impairment of central and peripheral synaptic plasticity. *American Journal of Human Genetics*, *92*, 681–695.

Hollway, G. E., Bryson-Richardson, R. J., Berger, S., Cole, N. J., Hall, T. E., & Currie, P. D. (2007). Whole-somite rotation generates muscle progenitor cell compartments in the developing zebrafish embryo. *Developmental Cell*, *12*, 207–219.

Horstick, E. J., Linsley, J. W., Dowling, J. J., Hauser, M. A., McDonald, K. K., Ashley-Koch, A., et al. (2013). Stac3 is a component of the excitation-contraction coupling machinery and mutated in Native American myopathy. *Nature Communications*, *4*, 1952.

Hu, J. K., McGlinn, E., Harfe, B. D., Kardon, G., & Tabin, C. J. (2012). Autonomous and nonautonomous roles of Hedgehog signaling in regulating limb muscle formation. *Genes & Development*, *26*, 2088–2102.

Huang, S. H., Hsiao, C. D., Lin, D. S., Chow, C. Y., Chang, C. J., & Liau, I. (2011). Imaging of zebrafish in vivo with second-harmonic generation reveals shortened sarcomeres associated with myopathy induced by statin. *PLoS One*, *6*, e24764.

Jackson, H. E., & Ingham, P. W. (2013). Control of muscle fibre-type diversity during embryonic development: The zebrafish paradigm. *Mechanisms of Development*, *130*, 447–457.

Jacoby, A. S., Busch-Nentwich, E., Bryson-Richardson, R. J., Hall, T. E., Berger, J., Berger, S., et al. (2009). The zebrafish dystrophic mutant softy maintains muscle fibre viability despite basement membrane rupture and muscle detachment. *Development*, *136*, 3367–3376.

Jenkins, M. H., Alrowaished, S. S., Goody, M. F., Crawford, B. D., & Henry, C. A. (2016). Laminin and Matrix metalloproteinase 11 regulate Fibronectin levels in the zebrafish myotendinous junction. *Skeletal Muscle*, *6*, 18.

Jeub, M., Bitoun, M., Guicheney, P., Kappes-Horn, K., Strach, K., Druschky, K. F., et al. (2008). Dynamin 2-related centronuclear myopathy: Clinical, histological and genetic aspects of further patients and review of the literature. *Clinical Neuropathology*, *27*, 430–438.

Johnson, N. M., Farr, G. H., 3rd, & Maves, L. (2013). The HDAC inhibitor TSA ameliorates a zebrafish model of duchenne muscular dystrophy. *PLoS Currents*, *5*. ecurrents.md.8273cf41db10e2d15dd3ab827cb4b027.

Julich, D., Geisler, R., & Holley, S. A. (2005). Integrinalpha5 and delta/notch signaling have complementary spatiotemporal requirements during zebrafish somitogenesis. *Developmental Cell*, *8*, 575–586.

Kawahara, G., Gasperini, M. J., Myers, J. A., Widrick, J. J., Eran, A., Serafini, P. R., et al. (2014). Dystrophic muscle improvement in zebrafish via increased heme oxygenase signaling. *Human Molecular Genetics, 23*, 1869–1878.

Kawahara, G., & Kunkel, L. M. (2013). Zebrafish based small molecule screens for novel DMD drugs. *Drug Discovery Today: Technologies, 10*, e91–e96.

Ketley, A., Warren, A., Holmes, E., Gering, M., Aboobaker, A. A., & Brook, J. D. (2013). The miR-30 microRNA family targets smoothened to regulate hedgehog signalling in zebrafish early muscle development. *PLoS One, 8*, e65170.

Kim, H. R., & Ingham, P. W. (2009). The extracellular matrix protein TGFBI promotes myofibril bundling and muscle fibre growth in the zebrafish embryo. *Developmental Dynamics, 238*, 56–65.

Kim, H.-H., Kim, J.-G., Jeong, J., Han, S.-Y., & Kim, K.-W. (2014). Akap12 is essential for the morphogenesis of muscles involved in zebrafish locomotion. *Differentiation, 88*, 106–116.

Koshida, S., Kishimoto, Y., Ustumi, H., Shimizu, T., Furutani-Seiki, M., Kondoh, H., et al. (2005). Integrinalpha5-dependent fibronectin accumulation for maintenance of somite boundaries in zebrafish embryos. *Developmental Cell, 8*, 587–598.

Kotani, Y., Morito, D., Yamazaki, S., Ogino, K., Kawakami, K., Takashima, S., et al. (2015). Neuromuscular regulation in zebrafish by a large AAA+ ATPase/ubiquitin ligase mysterin/RNF213. *Scientific Reports, 4*, 16161.

Kudo, H., Amizuka, N., Araki, K., Inohaya, K., & Kudo, A. (2004). Zebrafish periostin is required for the adhesion of muscle fiber bundles to the myoseptum and for the differentiation of muscle fibers. *Developmental Biology, 267*, 473–487.

Landemaine, A., Rescan, P.-Y., & Gabillard, J.-C. (2014). Myomaker mediates fusion of fast myocytes in zebrafish embryos. *Biochemical and Biophysical Research Communications, 451*, 480–484.

Lebold, K. M., Löhr, C. V., Barton, C. L., Miller, G. W., Labut, E. M., Tanguay, R. L., et al. (2013). Chronic vitamin E deficiency promotes vitamin C deficiency in zebrafish leading to degenerative myopathy and impaired swimming behavior. *Comparative Biochemistry and Physiology Part C: Toxicology & Pharmacology, 157*, 382–389.

Lennon, N. J., Kho, A., Bacskai, B. J., Perlmutter, S. L., Hyman, B. T., & Brown, R. H., Jr. (2003). Dysferlin interacts with annexins A1 and A2 and mediates sarcolemmal wound-healing. *The Journal of Biological Chemistry, 278*, 50466–50473.

Li, M., Andersson-Lendahl, M., Sejersen, T., & Arner, A. (2014). Muscle dysfunction and structural defects of dystrophin-null sapje mutant zebrafish larvae are rescued by ataluren treatment. *The FASEB Journal, 28*, 1593–1599.

Li, M., Andersson-Lendahl, M., Sejersen, T., & Arner, A. (2016). Knockdown of fast skeletal myosin-binding protein C in zebrafish results in a severe skeletal myopathy. *The Journal of General Physiology, 147*, 309–322.

Li, X., Blagden, C. S., Bildsoe, H., Bonnin, M. A., Duprez, D., & Hughes, S. M. (2004). Hedgehog can drive terminal differentiation of amniote slow skeletal muscle. *BMC Developmental Biology, 6*, 9.

Liang, W. C., Zhu, W., Mitsuhashi, S., Noguchi, S., Sacher, M., Ogawa, M., et al. (2015). Congenital muscular dystrophy with fatty liver and infantile-onset cataract caused by TRAPPC11 mutations: Broadening of the phenotype. *Skeletal Muscle, 5*, 29.

Lin, C. Y., Chen, J. S., Loo, M. R., Hsiao, C. C., Chang, W. Y., & Tsai, H. J. (2013). MicroRNA-3906 regulates fast muscle differentiation through modulating the target gene homer-1b in zebrafish embryos. *PLoS One, 8*, e70187.

Liu, J., Aoki, M., Illa, I., Wu, C., Fardeau, M., Angelini, C., et al. (1998). Dysferlin, a novel skeletal muscle gene, is mutated in Miyoshi myopathy and limb girdle muscular dystrophy. *Nature Genetics, 20*, 31–36.

Ljubicic, V., Burt, M., & Jasmin, B. J. (2014). The therapeutic potential of skeletal muscle plasticity in Duchenne muscular dystrophy: Phenotypic modifiers as pharmacological targets. *The FASEB Journal*, *28*, 548–568.

Long, J. H., Adcock, B., & Root, R. G. (2002). Force transmission via axial tendons in undulating fish: A dynamic analysis. *Comparative Biochemistry and Physiology. Part A, Molecular & Integrative Physiology*, *133*, 911–929.

Marshall, J. L., & Crosbie-Watson, R. H. (2013). Sarcospan: A small protein with large potential for Duchenne muscular dystrophy. *Skeletal Muscle*, *3*(1), 1.

Mayer, U., Saher, G., Fässler, R., Bornemann, A., Echtermeyer, F., von der Mark, H., et al. (1997). Absence of integrin alpha 7 causes a novel form of muscular dystrophy. *Nature Genetics*, *17*, 318–323.

Mei, J., Li, Z., & Gui, J. F. (2009). Cooperation of Mtmr8 with PI3K regulates actin filament modeling and muscle development in zebrafish. *PLoS One*, *4*, e4979.

Millay, D. P., Gamage, D. G., Quinn, M. E., Min, Y. L., Mitani, Y., Bassel-Duby, R., et al. (2016). Structure-function analysis of myomaker domains required for myoblast fusion. *Proceedings of the National Academy of Sciences of the United States of America*, *113*, 2116–2121.

Millay, D. P., O'Rourke, J. R., Sutherland, L. B., Bezprozvannaya, S., Shelton, J. M., Bassel-Duby, R., et al. (2013). Myomaker is a membrane activator of myoblast fusion and muscle formation. *Nature*, *499*, 301–305.

Moore, C. A., Parkin, C. A., Bidet, Y., & Ingham, P. W. (2007). A role for the Myoblast city homologues Dock1 and Dock5 and the adaptor proteins Crk and Crk-like in zebrafish myoblast fusion. *Development*, *134*, 3145–3153.

Murayama, E., Sarris, M., Redd, M., Le Guyader, D., Vivier, C., Horsley, W., et al. (2015). NACA deficiency reveals the crucial role of somite-derived stromal cells in haematopoietic niche formation. *Nature Communications*, *28*, 8375.

Naganawa, Y., & Hirata, H. (2011). Developmental transition of touch response from slow muscle-mediated coilings to fast muscle-mediated burst swimming in zebrafish. *Developmental Biology*, *355*, 194–204.

Nagao, H., Morimoto, T., Sano, N., Takahashi, M., Nagai, H., Tawa, R., et al. (1991). Magnetic resonance imaging of skeletal muscle in patients with Duchenne muscular dystrophy—Serial axial and sagittal section studies. *No to Hattatsu*, *23*, 39–43.

Nesin, V., Wiley, G., Kousi, M., Ong, E. C., Lehmann, T., Nicholl, D. J., et al. (2014). Activating mutations in STIM1 and ORAI1 cause overlapping syndromes of tubular myopathy and congenital miosis. *Proceedings of the National Academy of Sciences of the United States of America*, *111*, 4197–4202.

Nguyen, P. D., Hollway, G. E., Sonntag, C., Miles, L. B., Hall, T. E., Berger, S., et al. (2014). Haematopoietic stem cell induction by somite-derived endothelial cells controlled by meox1. *Nature*, *512*, 314–318.

Nixon, S. J., Wegner, J., Ferguson, C., Méry, P.-F., Hancock, J. F., Currie, P. D., et al. (2005). Zebrafish as a model for caveolin-associated muscle disease; caveolin-3 is required for myofibril organization and muscle cell patterning. *Human Molecular Genetics*, *14*, 1727–1743.

O'Brien, J. H., Hernandez-Lagunas, L., Artinger, K. B., & Ford, H. L. (2014). MicroRNA-30a regulates zebrafish myogenesis through targeting the transcription factor Six1. *Journal of Cell Science*, *127*, 2291–2301.

Ono, Y., Yu, W., Jackson, H. E., Parkin, C. A., & Ingham, P. W. (2015). Adaxial cell migration in the zebrafish embryo is an active cell autonomous property that requires the Prdm1a transcription factor. *Differentiation*, *89*, 77–86.

Pagnon-Minot, A., Malbouyres, M., Haftek-Terreau, Z., Kim, H. R., Sasaki, T., Thisse, C., et al. (2008). Collagen XV, a novel factor in zebrafish notochord differentiation and muscle development. *Developmental Biology*, *316*, 21–35.

Park, H. J., Hong, Y. B., Choi, Y.-C., Lee, J., Kim, E. J., Lee, J.-S., et al. (2016). ADSSL1 mutation relevant to autosomal recessive adolescent onset distal myopathy. *Annals of Neurology*, *79*, 231–243.

Patel, N., Smith, L. L., Faqeih, E., Mohamed, J., Gupta, V. A., & Alkuraya, F. S. (2014). ZBTB42 mutation defines a novel lethal congenital contracture syndrome (LCCS6). *Human Molecular Genetics*, *23*, 6584–6593.

Peterson, M. T., & Henry, C. A. (2010). Hedgehog signaling and laminin play unique and synergistic roles in muscle development. *Developmental Dynamics*, *239*, 905–913.

Powell, G. T., & Wright, G. J. (2011). Jamb and jamc are essential for vertebrate myocyte fusion. *PLoS Biology*, *9*, e1001216.

Pradeu, T., Laplane, L., Prvot, K., Hoquet, T., Reynaud, V., Fusco, G., et al. (2016). Defining "Development." *Current Topics in Developmental Biology*, *117*, 171–183.

Qiu, J., Fan, X., Wang, Y., Jin, H., Song, Y., Han, Y., et al. (2016). Embryonic hematopoiesis in vertebrate somites gives rise to definitive hematopoietic stem cells. *Journal of Molecular Cell Biology*, *8*, 288–301. http://dx.doi.org/10.1093/jmcb/mjw024.

Radev, Z., Hermel, J. M., Elipot, Y., Bretaud, S., Arnould, S., Duchateau, P., et al. (2015). A TALEN-Exon skipping design for a Bethlem Myopathy model in zebrafish. *PLoS One*, *10*, e0133986.

Raeker, M. O., Shavit, J. A., Dowling, J. J., Michele, D. E., & Russell, M. W. (2014). Membrane-myofibril cross-talk in myofibrillogenesis and in muscular dystrophy pathogenesis: Lessons from the zebrafish. *Frontiers in Physiology*, *5*, 14.

Ramanoudjame, L., Rocancourt, C., Laine, J., Klein, A., Joassard, L., Gartioux, C., et al. (2015). Two novel COLVI long chains in zebrafish that are essential for muscle development. *Human Molecular Genetics*, *24*, 6624–6639.

Ramspacher, C., Steed, E., Boselli, F., Ferreira, R., Faggianelli, N., Roth, S., et al. (2015). Developmental alterations in heart biomechanics and skeletal muscle function in desmin mutants suggest an early pathological root for desminopathies. *Cell Reports*, *11*, 1564–1576.

Ravenscroft, G., Miyatake, S., Lehtokari, V. L., Todd, E. J., Vornanen, P., Yau, K. S., et al. (2013). Mutations in KLHL40 are a frequent cause of severe autosomal-recessive nemaline myopathy. *American Journal of Human Genetics*, *93*, 6–18.

Roostalu, U., & Strähle, U. (2012). In vivo imaging of molecular interactions at damaged sarcolemma. *Developmental Cell*, *22*, 515–529.

Ruparelia, A. A., Oorschot, V., Ramm, G., & Bryson-Richardson, R. J. (2016). FLNC myofibrillar myopathy results from impaired autophagy and protein insufficiency. *Human Molecular Genetics*, *25*, 2131–2142. http://dx.doi.org/10.1093/hmg/ddw080.

Ruparelia, A. A., Oorschot, V., Vaz, R., Ramm, G., & Bryson-Richardson, R. J. (2014). Zebrafish models of BAG3 myofibrillar myopathy suggest a toxic gain of function leading to BAG3 insufficiency. *Acta Neuropathologica*, *128*, 821–833.

Ruparelia, A. A., Zhao, M., Currie, P. D., & Bryson-Richardson, R. J. (2012). Characterization and investigation of zebrafish models of filamin-related myofibrillar myopathy. *Human Molecular Genetics*, *21*, 4073–4083.

Sanger, J. W., Wang, J., Holloway, B., Du, A., & Sanger, J. M. (2009). Myofibrillogenesis in skeletal muscle cells in zebrafish. *Cell Motility and the Cytoskeleton*, *66*, 556–566.

Schiaffino, S., & Reggiani, C. (2011). Fiber types in mammalian skeletal muscles. *Physiological Reviews*, *91*, 1447–1531.

Schindler, R. F., Scotton, C., Zhang, J., Passarelli, C., Ortiz-Bonnin, B., Simrick, S., et al. (2016). POPDC1(S201F) causes muscular dystrophy and arrhythmia by affecting protein trafficking. *The Journal of Clinical Investigation*, *126*, 239–253.

Shamseldin, H. E., Bennett, A. H., Alfadhel, M., Gupta, V., & Alkuraya, F. S. (2016). GOLGA2, encoding a master regulator of golgi apparatus, is mutated in a patient with a neuromuscular disorder. *Human Genetics*, *135*, 245–251.

Shamseldin, H. E., Smith, L. L., Kentab, A., Alkhalidi, H., Summers, B., Alsedairy, H., et al. (2016). Mutation of the mitochondrial carrier SLC25A42 causes a novel form of mitochondrial myopathy in humans. *Human Genetics*, *135*, 21–30.

Skobo, T., Benato, F., Grumati, P., Meneghetti, G., Cianfanelli, V., Castagnaro, S., et al. (2014). Zebrafish ambra1a and ambra1b knockdown impairs skeletal muscle development. *PLoS One*, *9*, e99210.

Smith, L. L., Gupta, V. A., & Beggs, A. H. (2014). Bridging integrator 1 (Bin1) deficiency in zebrafish results in centronuclear myopathy. *Human Molecular Genetics*, *23*, 3566–3578.

Snow, C. J., Goody, M., Kelly, M. W., Oster, E. C., Jones, R., Khalil, A., et al. (2008). Time-lapse analysis and mathematical characterization elucidate novel mechanisms underlying muscle morphogenesis. *PLoS Genetics*, *4*, e1000219.

Snow, C. J., & Henry, C. A. (2009). Dynamic formation of microenvironments at the myotendinous junction correlates with muscle fiber morphogenesis in zebrafish. *Gene Expression Patterns*, *9*, 37–42.

Snow, C. J., Peterson, M. T., Khalil, A., & Henry, C. A. (2008). Muscle development is disrupted in zebrafish embryos deficient for fibronectin. *Developmental Dynamics*, *237*, 2542–2553.

Sohn, R. L., Huang, P., Kawahara, G., Mitchell, M., Guyon, J., Kalluri, R., et al. (2009). A role for nephrin, a renal protein, in vertebrate skeletal muscle cell fusion. *Proceedings of the National Academy of Sciences of the United States of America*, *106*, 9274–9279.

Srinivas, B. P., Woo, J., Leong, W. Y., & Roy, S. (2007). A conserved molecular pathway mediates myoblast fusion in insects and vertebrates. *Nature Genetics*, *39*, 781–786.

Stellabotte, F., Dobbs-McAuliffe, B., Fernández, D. A., Feng, X., & Devoto, S. H. (2007). Dynamic somite cell rearrangements lead to distinct waves of myotome growth. *Development*, *134*, 1253–1257.

Stevens, E., Carss, K. J., Cirak, S., Foley, A. R., Torelli, S., Willer, T., et al. (2013). Mutations in B3GALNT2 cause congenital muscular dystrophy and hypoglycosylation of α-dystroglycan. *American Journal of Human Genetics*, *92*, 354–365.

Subramanian, A., & Schilling, T. F. (2014). Thrombospondin-4 controls matrix assembly during development and repair of myotendinous junctions. *eLife*, *3*, e02372.

Sun, X., Ekker, S. C., Shelden, E. A., Takubo, N., Wang, Y., & Burghardt, T. P. (2014). In vivo orientation of single myosin lever arms in zebrafish skeletal muscle. *Biophysical Journal*, *107*, 1403–1414.

Sztal, T. E., Zhao, M., Williams, C., Oorschot, V., Parslow, A. C., Giousoh, A., et al. (2015). Zebrafish models for nemaline myopathy reveal a spectrum of nemaline bodies contributing to reduced muscle function. *Acta Neuropathologica*, *130*, 389–406.

Talbot, J., & Maves, L. (2016). Skeletal muscle fiber type: Using insights from muscle developmental biology to dissect targets for susceptibility and resistance to muscle disease. *Wiley Interdisciplinary Reviews. Developmental Biology*, *5*, 518–534.

Telfer, W. R., Busta, A. S., Bonnemann, C. G., Feldman, E. L., & Dowling, J. J. (2010). Zebrafish models of collagen VI-related myopathies. *Human Molecular Genetics*, *19*, 2433–2444.

Tempel, W., Rabeh, W. M., Bogan, K. L., Belenky, P., Wojcik, M., Seidle, H. F., et al. (2007). Nicotinamide riboside kinase structures reveal new pathways to NAD+. *PLoS Biology*, *5*, e263.

Tixier, V., Bataille, L., Etard, C., Jagla, T., Weger, M., Daponte, J. P., et al. (2013). Glycolysis supports embryonic muscle growth by promoting myoblast fusion. *Proceedings of the National Academy of Sciences of the United States of America*, *110*, 18982–18987.

Todd, P. K., Ackall, F. Y., Hur, J., Sharma, K., Paulson, H. L., & Dowling, J. J. (2014). Transcriptional changes and developmental abnormalities in a zebrafish model of myotonic dystrophy type 1. *Disease Models & Mechanisms*, *7*, 143–155.

Trinh, L. A., & Stainier, D. Y. R. (2004). Fibronectin regulates epithelial organization during myocardial migration in zebrafish. *Developmental Cell*, *6*, 371–382.

Tu, C. F., Tsao, K. C., Lee, S. J., & Yang, R. B. (2014). SCUBE3 (signal peptide-CUB-EGF domain-containing protein 3) modulates fibroblast growth factor signaling during fast muscle development. *The Journal of Biological Chemistry*, *289*, 18928–18942.

van der Meulen, T., Schipper, H., van Leeuwen, J. L., & Kranenbarg, S. (2005). Effects of decreased muscle activity on developing axial musculature in nicb107 mutant zebrafish (Danio rerio). *The Journal of Experimental Biology*, *208*, 3675–3687.

Vieira, N. M., Naslavsky, M. S., Licinio, L., Kok, F., Schlesinger, D., Vainzof, M., et al. (2014). A defect in the RNA-processing protein HNRPDL causes limb-girdle muscular dystrophy 1G (LGMD1G). *Human Molecular Genetics*, *23*, 4103–4110.

Waugh, T. A., Horstick, E., Hur, J., Jackson, S. W., Davidson, A. E., Li, X., et al. (2014). Fluoxetine prevents dystrophic changes in a zebrafish model of Duchenne muscular dystrophy. *Human Molecular Genetics*, *23*, 4651–4662.

Weitkunat, M., Kaya-Çopur, A., Grill, S. W., & Schnorrer, F. (2014). Tension and force-resistant attachment are essential for myofibrillogenesis in drosophila flight muscle. *Current Biology*, *24*, 705–716.

Winder, S. J., Lipscomb, L., Angela Parkin, C., & Juusola, M. (2011). The proteasomal inhibitor MG132 prevents muscular dystrophy in zebrafish. *PLoS Currents*, *3*, RRN1286.

Wolff, C., Roy, S., & Ingham, P. W. (2003). Multiple muscle cell identities induced by distinct levels and timing of hedgehog activity in the zebrafish embryo. *Current Biology*, *13*, 1169–1181.

Yao, Z., Farr, G. H., 3rd, Tapscott, S. J., & Maves, L. (2013). Pbx and Prdm1a transcription factors differentially regulate subsets of the fast skeletal muscle program in zebrafish. *Biology Open*, *2*, 546–555.

Yin, C., & Solnica-Krezel, L. (2007). Convergence and extension movements affect dynamic notochord-somite interactions essential for zebrafish slow muscle morphogenesis. *Developmental Dynamics*, *236*, 2742–2756.

Yuen, M., Sandaradura, S. A., Dowling, J. J., Kostyukova, A. S., Moroz, N., Quinlan, K. G., et al. (2014). Leiomodin-3 dysfunction results in thin filament disorganization and nemaline myopathy. *The Journal of Clinical Investigation*, *124*, 4693–4708.

Zhang, W., & Roy, S. (2016). The zebrafish fast myosin light chain mylpfa: H2B-GFP transgene is a useful tool for in vivo imaging of myocyte fusion in the vertebrate embryo. *Gene Expression Patterns*, *20*, 106–110.

Zulian, A., Rizzo, E., Schiavone, M., Palma, E., Tagliavini, F., Blaauw, B., et al. (2014). NIM811, a cyclophilin inhibitor without immunosuppressive activity, is beneficial in collagen VI congenital muscular dystrophy models. *Human Molecular Genetics*, *23*, 5353–5363.

CHAPTER SEVEN

Zebrafish Pancreas Development and Regeneration: Fishing for Diabetes Therapies

Victoria E. Prince*[,1], Ryan M. Anderson[†], Gokhan Dalgin*
*The University of Chicago, Chicago, IL, United States
[†]Indiana University School of Medicine, Indianapolis, IN, United States
[1]Corresponding author: e-mail address: vprince@uchicago.edu

Contents

Abstract

The zebrafish pancreas shares its basic organization and cell types with the mammalian pancreas. In addition, the developmental pathways that lead to the establishment of the pancreatic islets of Langherhans are generally conserved from fish to mammals. Zebrafish provides a powerful tool to probe the mechanisms controlling establishment of the pancreatic endocrine cell types from early embryonic progenitor cells, as well as the regeneration of endocrine cells after damage. This knowledge is, in turn, applicable to refining protocols to generate renewable sources of human pancreatic islet cells that are critical for regulation of blood sugar levels. Here, we review how previous and

Current Topics in Developmental Biology, Volume 124
ISSN 0070-2153
http://dx.doi.org/10.1016/bs.ctdb.2016.10.005

ongoing studies in zebrafish and beyond are influencing the understanding of molecular mechanisms underlying various forms of diabetes and efforts to develop cell-based approaches to cure this increasingly widespread disease.

1. INTRODUCTION TO THE PANCREAS

Diseases of the human pancreas, such as the growing Western epidemic of diabetes, have devastating consequences. The desire to better understand and indeed cure such diseases drives a great deal of biomedical research on the pancreas. In addition, the developmental mechanisms that build this complex organ present an intrinsically fascinating problem for study. We discuss how studies in zebrafish embryos, larvae, and adults are informing our understanding of pancreas development and regeneration. These studies provide mechanistic understanding of normal processes as well as insight into the pathophysiology of diabetes and how the disease can potentially be cured. The influence of findings from zebrafish is far reaching, as lessons learned from this system have provided a valuable foundation for efforts to produce renewable sources of human pancreatic cells.

1.1 Diabetes Mellitus

Diabetes mellitus is a collection of related metabolic disorders characterized by the common disruption of normal blood glucose homeostasis (DeFronzo, Ferrannini, Alberti, & Zimmet, 2015). Type 1 diabetes (T1D), affecting 5–10% of all diabetic patients, is primarily an autoimmune disease in which the insulin-secreting beta cells are destroyed. The far more prevalent Type 2 diabetes (T2D) is typically a later onset disease, with both genetic and environmental components; the disease initiates with peripheral insulin resistance, progressing to beta cell stress, dysfunction, and ultimately loss of functional beta cells. Less prevalent forms of diabetes are caused by mutations in single genes, causing permanent forms of neonatal diabetes and the "MODYs" (maturity-onset diabetes of the young). In 2014 it was estimated that over 22 million Americans suffered diabetes of all forms, a fourfold increase since 1980 (Center for Disease Control), correlating with increased incidence of obesity. Here, we review how pancreas development occurs in zebrafish, noting examples where zebrafish has been used to model particular classes of diabetes. We focus on how the hormone-producing pancreatic islet cell types form, as gaining an understanding of this process holds promise for recapitulating the process in vitro to enable cell replacement therapy, or for manipulating the islets to regenerate in diabetes.

1.2 Pancreas Structure and Development

The pancreas is composed of two distinct yet colocalized compartments: the exocrine compartment comprises acinar cells producing digestive enzymes and the ducts through which these flow, and the endocrine compartment made up of the "islets of Langerhans," which secrete hormones that regulate blood glucose levels. The pancreatic islets are critical for survival, which becomes dependent on regulated exogenous supply of insulin hormone in their absence. In humans the endocrine islets comprise insulin-secreting beta cells, antagonistically acting glucagon-secreting alpha cells, somatostatin-secreting delta cells, pancreatic polypeptide-secreting cells, and a small number of ghrelin-secreting epsilon cells. The fundamental composition and organization of the zebrafish pancreatic islets (Fig. 1) is very similar (reviewed by Kinkel & Prince, 2009); although zebrafish PP cells have not been described.

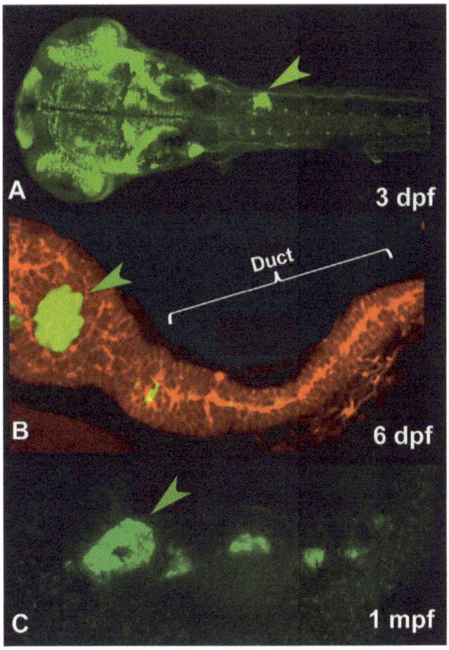

Fig. 1 Development of the zebrafish pancreas. (A) A zebrafish larva at 3 days postfertilization (dpf) shows pancreatic endocrine cells labeled with *TgBAC(Neurod1:EGFP)*, which also labels neurons in the spinal cord, brain, and associated sensory structures. Anterior is to left. (B) In a 6 dpf larva, pancreatic endocrine cells are labeled with *TgBAC(Neurod1:EGFP)*, epithelial cells, including the duct, are labeled with Zn8 antibody (*red*). (C) A 1-month postfertilization juvenile fish, showing formation of secondary islets; image acquired through the ventral skin of a live larva. *Arrowheads* indicate the principal islet in each panel.

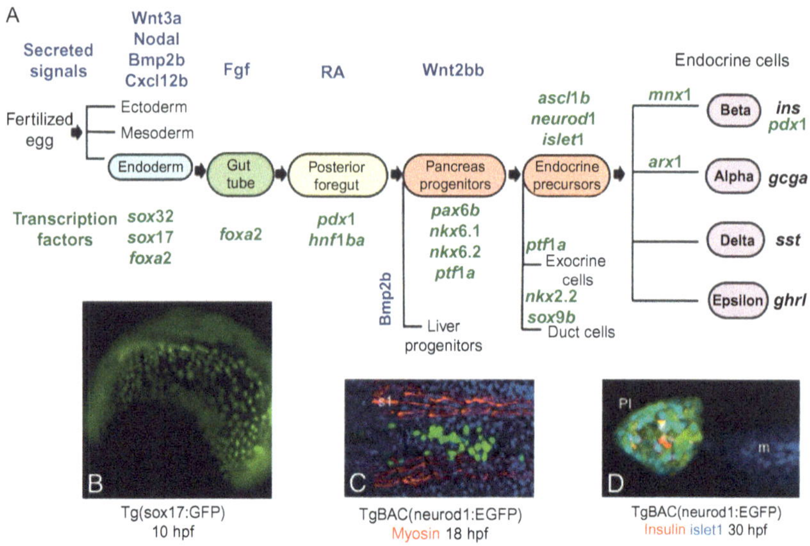

Fig. 2 Zebrafish endocrine pancreas cell lineages. (A) Key secreted signals involved in cell fate decisions, and transcription factors that mark various cell types, are indicated (see text for additional information). (B) Lateral view, anterior to left, 10 hpf *Tg(sox17:GFP)* late gastrula; dispersed GFP-positive endodermal cells are in the process of converging toward the midline where they will form the gut tube. (C) Dorsal view, anterior to left, 18 hpf *TgBAC(Neurod1:EGFP)* specimen, antimyosin antibody (*red*) labels muscle of the somites, s1 = somite 1. Neurod-expressing pancreatic progenitors (*green*) lie in the midline adjacent to somites 1–4, nuclei are labeled with TO-PRO. (D) Dorsal view, anterior to left, 30 hpf *TgBAC(Neurod1:EGFP)* specimen. Neurod-expressing endocrine cells of the principal islet (*green*: PI), antiinsulin antibody (*red*) labels beta cells, antiislet1 antibody (*blue*) labels endocrine cells, and adjacent mesenchyme (m).

The pancreatic cell lineages, their relationships, and key molecules that function during their development in the zebrafish, are schematized in Fig. 2.

The initial regionalization of endoderm cells along the primary body axis, which is necessary to establish the pancreatic field (or "anlage"), is dependent on secreted signals, discussed in more detail later. In both mammals and zebrafish the pancreas forms from both dorsal and ventral pancreatic buds, which emerge from a domain of Pdx1-expressing foregut endoderm (Herrera, Nepote, & Delacour, 2002; Yee, Yusuff, & Pack, 2001). In zebrafish embryos, the early forming dorsal bud forms a small islet by 24 hours postfertilization (hpf) (Biemar et al., 2001). A second, more anteriorly located ventral bud forms by 32 hpf and contributes acinar, ductal, and a second wave of endocrine cell types (Field, Dong, Beis, & Stainier, 2003). Morphological reorganization of the gut brings the dorsal and ventral bud tissues together at

52 hpf, enlarging the principal islet (Field et al., 2003). Additional, secondary islet cells emerge along the length of the intrapancreatic duct (IPD) as larval development continues (Fig. 1). The organization of the mature zebrafish pancreas, with multiple islets distributed throughout the organ is reminiscent of the organization of the mammalian pancreas.

One interesting variation in vertebrate endocrine pancreas development is the order in which the first cell types form. Specifically, in the zebrafish principal islet insulin-expressing cells are the earliest endocrine cell type detected, followed by glucagon-expressing cells; this is also the order of cell type appearance found in human development (reviewed by Jennings, Berry, Strutt, Gerrard, & Hanley, 2015). By contrast, in the mouse this pattern is reversed: glucagon-expressing cells appear before insulin-expressing cells. Thus, in this regard zebrafish islet development seems more similar to human than does the mouse.

2. GENE-FUNCTION ANALYSIS STREAMLINED IN ZEBRAFISH

Genome wide association studies, transcriptome profiling, and human genetic conditions that cause diabetes have identified candidate genes that potentially impact pancreatic function, although testing the relevance of these genes requires in vivo models. Experiments in mice have proven utility in this regard, but zebrafish is becoming an increasingly appealing option, in part due to modern CRISPR/Cas9 genome-editing tools (Varshney, Sood, & Burgess, 2015), but also because of the well-stocked toolkit of available transgenic lines that mark specific cell types or manipulate their function (Table 1), as well as defined markers and assays for pancreatic form and function that allow effective tests of pancreatic gene function in zebrafish. In this section, we describe how these tools are providing new insights into pancreas development.

2.1 The Many Uses of Transgenic Zebrafish

Transgenic zebrafish lines used for studying pancreas development and regeneration are listed in Table 1. As with similar tables in this book, this list will become obsolete as soon as published; we therefore advise readers to also make use of the regularly updated information at www.zfin.org. Some lines we showcase conveniently label a specific differentiated cell type,

Table 1 Transgenes and Markers for Zebrafish Pancreas Development and Regeneration

Cell Type	Transgenic Markers and Cre Drivers	Gene Marker	Antibody
Beta cells	*Tg(-1.2ins:EGFP)* (Xu et al., 2010 #78)	*ins* (Milewski, Duguay, Chan, & Steiner, 1998)	Insulin
	Tg(ins:dsRed)^{m1081Tg} (Shin et al., 2008 #305)		
	Tg(ins:GFP)^{jh3Tg} (Pisharath, Rhee, Swanson, Leach, & Parsons, 2007)		Anti C-peptide antibody (Eames, Kinkel, Rajan, Prince, & Philipson, 2013)
	Tg(ins:GFP)^{zf5Tg} (Huang, Vogel, Liu, Melton, & Lin, 2001)		
	Tg(ins:kaede)^{jh6Tg} (Pisharath et al., 2007)		
	Tg3(ins:kaede)^{s949Tg} (Andersson et al., 2012)		
	Tg(ins:mCherry)^{jh2Tg} (Pisharath et al., 2007 #220)		
	Tg(ins:Cre)^{s924Tg} (Hesselson, Anderson, Beinat, & Stainier, 2009)		
Delta cells	*Tg(sst2:GFP)*^{gz18Tg} (Li, Wen, Peng, Korzh, & Gong, 2009)	*sst2* (Argenton, Zecchin, & Bortolussi, 1999)	Somatostatin
	Tg(sst2:RFP)^{gz19Tg} (Li et al., 2009)		
	Tg(−2.5sst2:GFP)^{jh20Tg} (Wang, Rovira, Yusuff, & Parsons, 2011)		
	Tg(sst2:Cre)^{s963Tg} (Ye, Robertson, Hesselson, Stainier, & Anderson, 2015)		
Alpha cells	*Tg(gcga:GFP)*^{ia1Tg} (Zecchin et al., 2007)	*gcga* (Argenton et al., 1999)	Glucagon
	Tg(gcga:Cre)^{s962Tg} (Ye et al., 2015)		
Beta cell progenitors	*Tg(mnx1:0.6hsp70:GFP)*^{os26} (Dalgin et al., 2011)	*mnx1* (Wendik, Maier, & Meyer, 2004)	
	Tg(mnx1:GFP)^{ml2TG} (Arkhipova et al., 2012)		

Alpha cell progenitors		*arx1* (Djiotsa et al., 2012)	
Endocrine progenitors	*TgBAC(Neurod1:EGFP)*^{nl1} (Obholzer et al., 2008)	*neurod1* (Korzh, Sleptsova, Liao, He, & Gong, 1998) *acsl1* (Flasse, Stern, et al., 2013)	Neurod1 (Kani et al., 2010)
	Tg(ascl1b:eGFP-2A-creERT2) (Ghaye et al., 2015)	*islet1* (Korzh, Edlund, & Thor, 1993) *nkx6.1* (Binot et al., 2010)	
	TgBAC(nkx6.1:eGFP) (Ghaye et al., 2015)	*nkx6.2* (Binot et al., 2010)	
Pancreatic duct	*Tg(-3.5nkx2.2a:GFP)*^{ia3Tg} (duct:GFP) (Pauls, Zecchin, Tiso, Bortolussi, & Argenton, 2007)	*nkx2.2* (Pauls et al., 2007)	2F11 (Abcam ab71286)
			Zn8 (ZIRC)
Exocrine pancreas	*Tg(ela3l:EGFP)* (Wan et al., 2006)	*ptf1a* (Zecchin et al., 2004)	CarboxypeptidaseA
	Tg(fabp10:dsRed; ela3l:GFP)^{gz15Tg} (Farooq et al., 2008)	*ela3l* (Wan et al., 2006)	
	Tg(ptf1a:EGFP)^{jh1Tg} (Godinho et al., 2005)	*trypsin* (Biemar et al., 2001)	
	Tg(−5.5ptf1a:DsRed)^{ia6Tg} (Leung, Klopper, Grill, Harris, & Norden, 2011)		
	TgBAC(ptf1a:Cre-ERT2)^{mk201Tg} (Wang, Park, Parsons, & Leach, 2015)		
	Tg(ela3l:Cre)^{s932Tg} (Hesselson, Anderson, & Stainier, 2011)		
Pancreatic progenitors	*Tg(−6.5pdx1:GFP)* (Huang et al., 2001)	*pdx1* (Milewski et al., 1998)	Pax6a/b (Verbruggen et al., 2010)
		pax6b (Biemar et al., 2001)	
		hnf1ba (vhnf1, tcf2) (Sun & Hopkins, 2001)	

Continued

Table 1 Transgenes and Markers for Zebrafish Pancreas Development and Regeneration—cont'd

Cell Type	Transgenic Markers and Cre Drivers	Gene Marker	Antibody
Endoderm	*Tg(sox17:GFP)*^{s870Tg} (Sakaguchi, Kikuchi, Kuroiwa, Takeda, & Stainier, 2006)	*sox17* (Alexander & Stainier, 1999)	
	Tg(−5.0sox17:EGFP)^{zf99} (Mizoguchi, Verkade, Heath, Kuroiwa, & Kikuchi, 2008)	*sox32* (Alexander & Stainier, 1999)	
	Tg(sox17:DsRed)^{s930Tg} (Chung & Stainier, 2008)	*foxa2 (hnf3b/axial)* (Chang, Blader, Fischer, Ingham, & Strahle, 1997)	
	Tg(Xla.Eef1a1:GFP)^{s852Tg} (gut:GFP) (Field et al., 2003)		
	Tg(EPV.TP1-Mmu.Hbb:hist2h2l-mCherry)^{s939Tg} (Delous et al., 2012)		
Notch-responsive cells	*Tg(Tp1glob:hmgb1-mCherry)*^{jh32} (nuclear mCherry) (Parsons et al., 2009)		
	Tg(Tp1bglob:eGFP)$^{um14\ Tp1}$ (cytoplasmic eGFP) (Parsons et al., 2009)		
	Tg(Tp1glob:H2BmCherry)^{S939}, (nuclear mCherry; long half-life) (Ninov, Borius, & Stainier, 2012)		
	Tg(Tp1glob:VenusPEST)$^{S940\ Tp1}$, VenusPEST (destabilized protein) (Ninov et al., 2012)		
	Tg(EPV.Tp1-Ocu.Hbb2:CreERT2)^{jh12Tg} (Wang et al., 2011) (Cre driver line)		
	Tg(EPV.Tp1-Mmu.Hbb:CreERT2,cryaa:mCherry)^{s959Tg} (Ninov et al., 2013) (Cre driver line)		
	(Tp1 = six concatemerized Notch-responsive elements from the Epstein Barr Virus terminal protein 1 (TP1) gene, containing in total 12 Rbp-Jk binding sites (Parsons et al., 2009))		

CFTR fusion lines	*TgBAC(cftr-GFP)*^{pd1041} (Navis, Marjoram, & Bagnat, 2013)	
	TgBAC(cftr-RFP)^{pd104} (Navis et al., 2013)	
Cell cycle markers	*Tg(ins:zFucci-G1)* (Ninov et al., 2013)	
	Tg(ins:zFucci-S/G2/M) (Ninov et al., 2013)	
Lox responder lines	*Tg(bactin2:loxP-RFP-loxP-GFP;cryaa:CFP)*^{s928Tg} (Kikuchi et al., 2010)	
	Tg(ubi:loxP-GFP-loxP-RFP;cryaa:CFP)^{cz1701Tg}, also known as *Tg(ubiquitin:switch)* (Mosimann et al., 2011)	
	Tg(hsp70l:loxP-RFP-loxP-H2B-GFP;cryaa:CFP)^{s923Tg} (Hesselson et al., 2011 #60)	
	Tg(ins:loxP-RFP-loxP-H2B-GFP;cryaa:CFP)^{s934Tg} (Hesselson et al., 2011)	
Beta cell injury lines	*Tg(ins:CFP-NTR)*^{s892Tg} (Curado et al., 2007)	
	Tg(ins:flag-NTR)^{s950Tg} (Andersson et al., 2012)	
	Tg(ins:NTR-mCherry)^{jh5Tg} (Pisharath et al., 2007)	
	Tg(ins:LOXP-BFP-LOXPCdi.Tox,ins:Hsa.HIST1H2BJ-GFP)^{s954Tg} (Ninov et al., 2013)	
Gluconeogenesis markers	*Tg(pck1:Venus,cryaa:mcherryn)*^{s953} (Gut et al., 2013)	*pck1* (Jurczyk et al., 2011)

while others provide a read out of cell function or signaling state, or even induce specific cell type death to facilitate the study of pancreas regeneration.

The first relevant transgenic line was *Tg(insulin:GFP)* (Huang et al., 2001), which provides a convenient fluorescent marker of beta cells, and similar transgenes were subsequently established that label other endocrine cell types. Additional transgenes label cell populations at specific developmental stages. For example, all endoderm is labeled by *Tg(sox17:GFP)* from gastrulation stages through the fusion of pancreatic buds, providing an easy assay of early endoderm morphology and behavior, and also allowing a straightforward approach to separate endoderm cells using FACS. We used this powerful sorting approach to compare the endodermal transcriptomes of control vs retinoic acid (RA) manipulated specimens, to uncover targets of RA signaling (Dalgin et al., 2011). Transgenes that label precursor populations are also useful: for example, *TgBAC(Neurod1:EGFP)* labels all endocrine pancreas precursors (Dalgin & Prince, 2015; Dalgin et al., 2011) and *Tg(mnx1:0.6hsp70:GFP)* labels not only beta cell precursors, but in addition, due to perdurance of stable GFP, their differentiated beta cell derivatives (Dalgin et al., 2011), thus providing a useful approach to track the fates of beta cell precursors.

Fluorescent transgene approaches can also be deployed to read out activity of signaling pathways. A Notch reporter line was used to establish that Notch-responsive cells in the ventral pancreatic bud give rise to endocrine cells (Parsons et al., 2009). Other transgene-based approaches have taken advantage of the differential half lives of fluorescent reporters, for example, to study the dynamics of Notch signaling (Ninov et al., 2012). This approach revealed that high levels of Notch signaling promote quiescence, lower levels promote amplification of the progenitor pool by stimulating mitosis, and sustained downregulation triggers cell cycle exit and endocrine differentiation. Fluorescent molecule stability has also been exploited to measure cell divisions: dilution of fluorescence intensity resulting from multiple rounds of cell division was used to distinguish the proliferative capacity of different populations of pancreatic progenitors during zebrafish development. This revealed that dorsal bud-derived (DBD) endocrine cells are largely quiescent, and thus retain H2B-RFP label expressed from mRNA microinjected at the zygote stage, whereas ventral bud-derived (VBD) cells continue to divide until later stages and thus the label is diluted and lost (Hesselson et al., 2009). This method provides an effective method to distinguish between dorsal vs ventral bud-derived endocrine cells that together populate the principal islet. This same study reported the "HOT-Cre" method, where H2B-GFP transgene is spatially confined to beta cells using insulin regulatory elements and temporally controlled by heat shock. Coupled with EDU labeling, this approach showed that while H2B-GFP-labeled DBD cells did not enter into cell division, the VBD cells went through divisions with an average cell cycle length of 9.2 days, presumably contributing to expanding beta cell mass (Hesselson et al., 2009).

A more direct method to assay the cell cycle is the "FUCCI" system, which provides a real-time readout of cell cycle dynamics using two fluorophores (Sakaue-Sawano et al., 2008). FUCCI components have been placed under control of the insulin promoter to assess proliferation of beta cells. This approach was used to screen drugs that promote beta cell replication during regeneration (Tsuji et al., 2014) and nutrient overload (Ninov et al., 2013). Remarkably, in 15-day postfertilization (dpf) zebrafish larvae, beta cells proliferate rapidly in response to a high calorie diet, and progenitors in the IPD are also stimulated to differentiate into new beta cells (neogenesis) (Ninov et al., 2013). These experiments show the promise of using zebrafish to explore the relationship between diet and beta cell differentiation and/or replication, which is particularly important since the compensatory beta cell expansion normally associated with weight gain fails in diabetic humans. In addition, there is much interest in developing approaches to stimulate proliferative responses from human beta cells (Bouwens & Rooman, 2005).

Transgenic approaches based on Cre-Lox, or similar recombination systems, can permanently label cells at a specific point in their developmental history and thus provide lineage-tracing information. A recent sophisticated example of this approach used a bicistronic BAC-based transgene *Tg(ascl1b:eGFP-2A-creER*^{T2}*)* to label zebrafish cells expressing the transcription factor gene *ascl1b* (Ghaye et al., 2015). In this case, the *ascl1b* regulatory sequence drives expression of both GFP and Cre, with the two proteins cleaved posttranslationally at the 2A peptide. In this system Cre expression is tamoxifen dependent, enabling temporal control of Cre-based recombination by addition of 4-hydroxytamoxifen. Fish carrying the *Tg(ascl1b:eGFP-2A-creER*^{T2}*)* transgene were crossed with fish carrying a reporter transgene, such as *Tg(ubiquitin:switch)*, enabling recombination at LoxP sites to label cells of interest. By applying tamoxifen at different stages, this approach confirmed that endocrine cells derived from both the DB and VB originate from Ascl1b-positive progenitors, providing a deeper understanding of the gene regulatory networks that underlie endocrine cell fates.

In a final example, cells expressing the nitroreductase transgene can be cleanly killed by adding a prodrug into the water (Curado et al., 2007; Curado, Stainier, & Anderson, 2008; Pisharath et al., 2007). This approach has allowed efficient beta cell deletion by expressing nitroreductase under the control of insulin regulatory elements. In Section 4, we discuss how this system has provided a key tool to study pancreas regeneration.

2.2 Transcription Factors in Pancreas Development and Disease

Much research on pancreas development has focused on transcription factors, and reviews of the transcriptional networks driving pancreatic development have been previously published (Dassaye, Naidoo, & Cerf, 2016; Kinkel & Prince, 2009; Pan & Wright, 2011; Tiso, Moro, & Argenton, 2009). Here, we focus on recent findings and areas where our own work has had impact. In each case the precise roles of specific transcription factors are not fully conserved between mammals and zebrafish, revealing a complex set of strategies that have evolved to form the pancreas. Despite this, findings from zebrafish can suggest hypotheses broadly applicable to all vertebrate systems and thus enrich research relevant to human disease.

2.2.1 Basic Helix-Loop-Helix Factors as Key Drivers of Pancreas Development

Endocrine pancreas cells in the mouse develop from precursors that transiently express the basic helix-loop-helix (bHLH) domain transcription factor Ngn3 (Gu, Dubauskaite, & Melton, 2002; Mellitzer et al., 2004; Schonhoff, Giel-Moloney, & Leiter, 2004) and mice lacking Ngn3 fail to produce any endocrine cells (Gradwohl, Dierich, LeMeur, & Guillemot, 2000). Further, RA signals are required to maintain the pancreatic progenitors and to promote the commitment of these progenitors to Ngn3-positive endocrine precursors (Ostrom et al., 2008). Ngn3 initiates the endocrine development program by inducing an epithelial to mesenchymal transition in the pancreatic foregut epithelium (Rukstalis & Habener, 2007), and by activating several downstream target genes including the MODY6 gene *Neurod1* (Huang et al., 2000). Recently, CRISPR/Cas9 genome-editing experiments in human ES cells have shown that NGN3 also plays a vital role in human endocrine pancreas development (McGrath, Watson, Ingram, Helmrath, & Wells, 2015), although a small number of endocrine cells are able to form through NGN3-independent mechanisms (Zhu et al., 2016). By contrast, zebrafish endocrine precursors do not express Ngn3, or any other Neurogenin-family member, and zebrafish mutant for Ngn3 do not display endocrine pancreas defects (Flasse, Pirson, et al., 2013).

In zebrafish, it is proposed that two bHLH transcription factors, Ascl1b and Neurod1, work together to perform the function ascribed to mammalian Ngn3. Ascl1b is expressed in zebrafish precursor cells, and lineage tracing revealed that both DBD and VBD endocrine cells derive from Ascl1b-expressing progenitors (Ghaye et al., 2015). By contrast, Neurod1 expression is maintained in differentiated endocrine cells (Dalgin & Prince, 2015; Dalgin et al., 2011; Flasse, Pirson, et al., 2013). Knockdown of both Ascl1b and Neurod1 results in complete loss of endocrine hormone expression in zebrafish (Flasse, Pirson, et al., 2013), showing that cooperation between these two genes is required for endocrine pancreas development. Intriguingly, our work has shown that IPD cells are sensitive to Neurod1 levels, where high and low levels of Neurod1 are required for alpha and beta cell differentiation, respectively (Dalgin & Prince, 2015). An important consideration is that our finding relied on dose–response experiments using morpholinos; although these findings were confirmed by CRISPR mutagenesis, morpholino reagents remain important for zebrafish research. Gaining a better understanding of how bHLH transcription factors are activated in the IPD, and the mechanisms that allow precise control of expression levels, may ultimately be instrumental for optimizing in vitro differentiation of endocrine precursors to beta cells.

2.2.2 Motor Neuron and Pancreas Homeobox 1 (mnx1)

Motor neuron and pancreas homeobox 1 (Mnx1) (previously Hb9) plays an important role in zebrafish endocrine cell fate decisions. Morpholino knockdown of this transcription factor not only depletes beta cells, but concomitantly increases the number of alpha cells (Dalgin et al., 2011). Conveniently, expression of the *Tg(mnx1:0.6hsp70:GFP)* transgene is unaffected by a morpholino targeting the 5′UTR of the endogenous *mnx1* gene, allowing expression of the stable Mnx1:GFP reporter to provide a cell fate marker. This approach revealed that in the absence of functional Mnx1 protein, GFP-expressing cells that would normally have become beta cells instead switched fate to produce alpha cells. Interestingly, in mice mutant for Mnx1, as in zebrafish, there is a loss of beta cells, but in this case the loss is accompanied by an increase in number of delta cells (Li, Arber, Jessell, & Edlund, 1999). Thus, the role of Mnx1 in establishing endocrine cell ratios is conserved, whereas the details of specific cell types under regulation vary.

2.3 Outside the Nucleus: Insights Into the Roles of Secreted and Cell Surface Signals

There has been significant research in multiple systems investigating the signaling molecules that function in regionalization, specification, commitment, and differentiation of the pancreas. Some of this work is referred to in Section 3, and this topic has also been effectively reviewed (Pan & Wright, 2011). Here, we provide updates on relevant recent work in zebrafish that has provided new insights into the roles of signaling molecules and downstream pathways.

2.3.1 Early Endoderm Development

Before specification of the pancreatic anlage, the endoderm germ layer must be established. In addition to signaling events, a complex series of cell movements during gastrulation produce the germ layers. Two groups (Mizoguchi et al., 2008; Nair & Schilling, 2008) reported a vital role for chemokine signaling in coordinating the movements of zebrafish mesoderm and endoderm. The mesodermally expressed chemokine Cxcl12b and its endodermally expressed receptor Cxcr4 work in an integrin-dependent manner to "tether" the endoderm to migrating mesoderm. In addition, chemokine signaling is necessary for the morphology, particularly polarized filopodial processes, of migrating endoderm cells. When chemokine signaling is deficient, endoderm cells migrate too far anteriorly and the gut bifurcates causing "viscera bifida" with a duplicated pancreas and liver, as cells fail to reach the midline in time to fuse. This is one of many examples where signaling from one germ layer to another is key to establishment of pancreatic fates.

2.3.2 Establishment and Patterning of the Hepatopancreatic Anlage

The liver and pancreas anlagen have been postulated to derive from a common hepatopancreatic domain within the foregut endoderm at somitogenesis stages. To test this hypothesis, individual cell fates were followed in postgastrulation zebrafish embryos (Chung, Shin, & Stainier, 2008), confirming the existence of a hepatopancreatic progenitor domain. Within the monolayer of foregut endoderm the most medial cells contribute to endocrine pancreas, those lying just 2 cell-widths lateral contribute to liver, and the intermediate zone contributes exocrine pancreas. This pattern of cell fates correlates with a gradient of Pdx1 expression, where the highest Pdx1 expression correlates with endocrine fates.

Chung et al. (2008) also revealed a role for BMP signaling in controlling the decision between liver vs pancreas. Bmp2b signaling from the lateral plate mesoderm, at about 14.5 hpf, promotes liver fates in the immediately adjacent endodermal cells, whereas more medial cells differentiate toward Pdx1-positive pancreatic fates. By manipulating BMP signaling the investigators showed that Bmp2b promotes liver at the expense of pancreas. Thus, Bmp levels at these early developmental stages define the size of the liver vs pancreatic anlagen. In a follow-up study, the same authors (Chung, Andersson, Row, Kimelman, & Stainier, 2010) probed more deeply into the roles of BMP signaling at different stages by genetically manipulating individual endoderm cells using cell transplantation. This study confirmed that active BMP signaling is incompatible with beta cell differentiation, whether at early stages in the dorsal bud or later in the ventral bud. A recent study (Xu, Cui, Del Campo, & Shin, 2016) has identified the *four and a half LIM domains 1b* (*fhl1b*) gene as a novel Bmp2b target. Fhl1b functions downstream of Bmp2b to promote liver fates, and its expression is incompatible with endocrine fates. Interestingly, this study also showed that *fhl1b* must be downregulated for endocrine cells to regenerate. Thus, FHL1b may prove to be an important player in negatively regulating both development and regeneration of pancreatic endocrine cells.

Understanding the initial establishment of the broader hepatopancreatic domain, which includes progenitors for gall bladder and extrapancreatic ducts, as well as liver and pancreas, has been a somewhat intractable problem. However, a recent zebrafish study has revealed that the transcription factor encoding gene *hnf1ba*, and Wnt signaling work synergistically in hepatopancreatic development (Lancman et al., 2013). A heterozygous mutation of human *HNF1b* underlies MODY5, suggesting that partial loss of function underlies pathology. Loss of *Hnf1b* function in animal models causes profound endoderm defects, providing an ineffective disease model. By contrast, the zebrafish $hnf1ba^{s430}$ mutant allele is a hypomorph, which does not cause major disruptions in endoderm patterning and instead leads to pancreas hypoplasia and loss of beta cells, highly reminiscent of MODY5 pathology. This zebrafish allele has thus uncovered a later role for an *hnf1b* gene in ventral pancreas development and provided an effective model of MODY5. Importantly, the $hnf1ba^{S430}$ hypomorph has allowed interactions between Hnf1b and Wnt2bb signaling to be tested, revealing synergistic functions. The majority of double $hnf1ba^{S430}/wnt2bb^{s404}$ mutants have a complete loss of hepatocytes, confirming that these genes play synergistic roles in liver development. In addition, although $wnt2bb^{s404}$ mutants show no signs of pancreatic defects, the double mutants have severe pancreatic agenesis/hypoplasia, with about 6% of double mutants missing the entire ventrally derived hepatopancreatic system. Together, these examples illustrate how the embryology and genetics of the zebrafish are providing important insights into how the initial hepatopancreatic anlage develops.

3. DEVELOPING RENEWABLE SOURCES OF BETA CELLS FOR CELL-BASED DIABETES THERAPIES

Both T1D and advanced T2D are characterized by beta cell loss, requiring regular monitoring of blood glucose levels and injection of insulin to restore homeostasis. If glucose homeostasis is not carefully controlled serious clinical problems endanger the lives of diabetic patients (Fullerton et al., 2014). A cure, as opposed to a treatment, for diabetes, can only be achieved by restoring functional beta cells. It has long been known that transplantation of a healthy substitute pancreas can restore function (Kelly, Lillehei, Merkel, Idezuki, & Goetz, 1967), and more recently the Edmonton Protocol reported that transfer of isolated islets into the portal vein of patients can similarly restore insulin secretion (Shapiro et al., 2000). In both cases, the limited number of suitable donors coupled with the necessity for constant potent immunosuppression means that only a small subset of patients are appropriate to benefit (Bromberg, Kaplan, Halloran, & Robertson, 2007). However, if renewable sources of beta cells that can avoid immunorejection become available, they should provide an effective cell-based diabetes therapy.

3.1 Zebrafish Developmental Studies Influenced Early Stem Cell Protocols

Stem cells represent an attractive source to replace dysfunctional or diseased cells, for instance, failing beta cells could be replaced with stem cell-derived cells that accurately regulate glucose homeostasis. Further, the ability to induce human pluripotent stem cells (iPSCs) (Takahashi & Yamanaka, 2006; Yabe et al., 2016) has opened up the exciting possibility of generating patient-specific beta cells for transplantation therapies. Although adult humans show variation in pancreatic islet size and number (Bonner-Weir, Sullivan, & Weir, 2015; Kilimnik, Jo, Periwal, Zielinski, & Hara, 2012), it is estimated that a healthy human adult has nearly 10 billion beta cells: 3–4 million islets, with ~4000 endocrine cells in each islet, of which ~60% are beta cells (Bonner-Weir et al., 2015; Ionescu-Tirgoviste et al., 2015). Producing this quantity of beta cells in vitro represents a significant challenge, and much recent effort has gone toward enhancing efficiency of differentiation protocols. The functional characterization of stem cell-derived beta cells is also extremely important, as full function relies on properly regulated insulin secretion. Here, we summarize how work to identify the key factors driving zebrafish pancreas development has informed protocols to differentiate human embryonic or iPS cells into beta cells in culture.

In 2006, a seminal study (D'Amour et al., 2006) reported a protocol to shepherd human embryonic stem (ES) cells through developmental stages equivalent to those taken by developing embryos, leading the cells along a progressively narrowing pathway to the specification of insulin-positive cells. Importantly, this work used developmental knowledge in two ways: (1) to suggest candidate secreted molecules able to direct cells toward the next stage and (2) to provide key molecular markers to verify the step-wise evolution of cell fate. This general concept, of pushing cells through progressive developmental stages toward increasingly more limited fates, is schematized in Fig. 3. This general scheme is, as one would expect, very similar to the process of normal pancreatic development (Fig. 2).

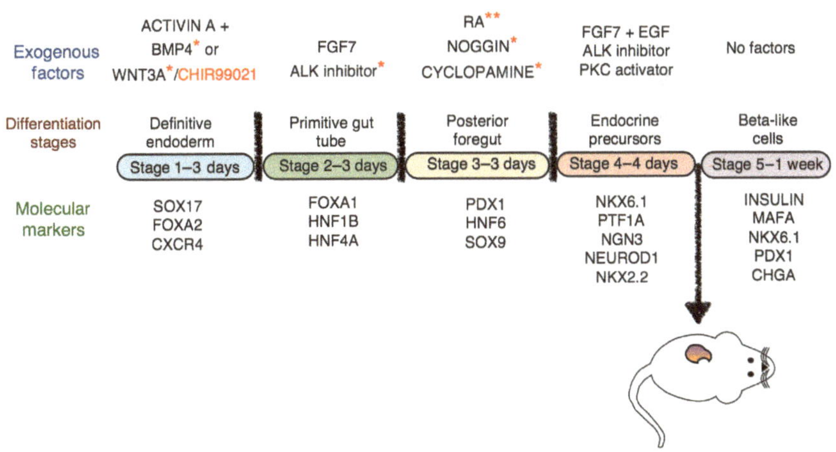

Fig. 3 Overview of beta cell differentiation protocols. Human ES or iPS cells are treated with exogenous factors to push them through the differentiation process, mimicking developmental stages as indicated. Alternative factors used in some protocols are shown in *red/gray*. Signaling factors first uncovered in zebrafish are indicated by **; signaling factors confirmed in zebrafish are indicated by *. Molecular markers allow the differentiation stage to be monitored. Maturation of beta cells is significantly improved by transferring stage 4 progenitors under the kidney capsule of a mouse, providing an appropriate in vivo environment to foster formation of functional beta cells (see text).

Zebrafish developmental biology research played an important role in establishment of the D'Amour protocol: Zebrafish studies first revealed the requirement for RA in selecting pancreatic precursor fates from the broader foregut endoderm field (Stafford & Prince, 2002). Zebrafish mutant alleles in Aldh2 (Begemann, Schilling, Rauch, Geisler, & Ingham, 2001; Grandel et al., 2002), the aldehyde dehydrogenase that catalyzes production of RA, were key to enabling this work. Aldh2 mutant embryos, as well as specimens treated with pharmacological inhibitors of RA synthesis, lacked both pancreatic progenitors and differentiated pancreatic cells types. Reciprocally, RA treatment drove a dramatic expansion of the pancreatic field in zebrafish embryos (Stafford & Prince, 2002). The role for RA signaling in this early step of pancreas development proved common to amphibians, avians, and mammals (Chen et al., 2004; Martin et al., 2005; Molotkov, Molotkova, & Duester, 2005; Stafford, Hornbruch, Mueller, & Prince, 2004). Together, these studies revealed that RA plays a key role in the development of pancreatic islet cells, leading to its inclusion in the early D'Amour protocol (2006); RA is now a typical element of most protocols to differentiate pancreatic endocrine cells from stem cells (Fig. 3).

3.2 Advances in Stem Cell Protocols

While the D'Amour protocol represented a major step forward, multiple challenges confounded the production of transplantable beta cells. Paramount was the generation of many dysfunctional polyhormonal cells, expressing both glucagon and insulin, a situation similar to that observed in human fetal endocrine pancreas (D'Amour et al., 2006). Further, the cell population resulting from this initial five-stage protocol was unable to appropriately secrete insulin in response to glucose. A follow-up study (Kroon et al., 2008) showed that differentiation of human ES cells through a slightly modified version of the first four stages of the D'Amour protocol, followed by implantation of cell aggregates under the kidney capsule of mice, allowed the cells to complete their differentiation in vivo to result in cells expressing only a single hormone (Fig. 3). These human cells responded appropriately to glucose and regulated blood sugar homeostasis when the host mice were rendered diabetic. This represented a major step forward, but the resultant mixed cell population had a worrying propensity to produce teratomas, and only relatively small numbers of ES cell-derived beta cells were produced.

Building on these initial studies, many laboratories have published protocols that promote the production of insulin-expressing beta-like cells from stem cells (Bruin et al., 2013; D'Amour et al., 2006; Kroon et al., 2008; McGrath et al., 2015; Pagliuca et al., 2014; Rezania et al., 2013; Toyoda et al., 2015). A common theme is the attempt to recapitulate the molecular steps governing development of endocrine pancreas cell types (Figs. 2 and 3). Below we discuss how studies of development in zebrafish and other models have continued to influence efforts to efficiently and accurately produce beta cells in vitro for cell replacement therapies.

3.2.1 Building Definitive Endoderm

The initial step toward producing pancreatic endocrine cells is the induction of definitive endoderm (DE), which during development occurs early in gastrulation. In most differentiation protocols, this is achieved using the transforming growth factor β (TGFβ) superfamily member ACTIVIN A (related to NODAL) (D'Amour et al., 2005; Kubo et al., 2004) (Fig. 2). In mouse, high levels of Nodal are necessary and sufficient to induce endoderm from bipotential mesendodermal cells, while lower doses induce mesoderm (Collignon, Varlet, & Robertson, 1996; Conlon et al., 1994; Zhou, Sasaki, Lowe, Hogan, & Kuehn, 1993). This dose-dependent role for Nodal is broadly conserved across vertebrate models, including zebrafish (Chen & Schier, 2001; Schier, Neuhauss, Helde, Talbot, & Driever, 1997; Thisse, Wright, & Thisse, 2000). During zebrafish gastrulation, cells expressing high levels of Nodal are the first to involute, with those fated to become endoderm becoming arranged in a head-anterior to tail-posterior (AP) array and forming a sparse monolayer atop the yolk (Warga & Nusslein-Volhard, 1999). At this stage, *sox17* and *foxA2* (pan-endoderm markers) commence expression in the endoderm monolayer (Alexander & Stainier, 1999; Strahle, Blader, Henrique, & Ingham, 1993). Typically, human cell in vitro differentiation protocols monitor production of DE by robust expression of the homologous human *Sox17* and *FoxA2* genes (Fig. 3). Interestingly, forced expression of SOX17 in ES cells is sufficient to induce DE (Seguin, Draper, Nagy, & Rossant, 2008).

In addition to Nodal, normal gastrulation also depends on Wnt signaling. Indeed, dorsally expressed Wnt functions as a positive regulator of Nodal in zebrafish (Fan et al., 2007). Combining WNT3A (or Wnt agonist CHIR99021) with ACTIVIN A treatment in the initial stages of human ES cell differentiation improves efficiency of DE formation

(Kroon et al., 2008; McGrath et al., 2015; Rezania et al., 2013). During zebrafish gastrulation, Nodal induces downstream BMP (bone morphogenetic protein) signaling, which is in turn necessary for proper AP patterning of the endoderm (Poulain, Furthauer, Thisse, Thisse, & Lepage, 2006; Tiso, Filippi, Pauls, Bortolussi, & Argenton, 2002). A dorsal–ventral gradient of Bmp signaling instructs initial AP patterning (Tiso et al., 2002); however, as discussed in Section 2.3.2, Bmp signals act subsequently to promote liver fates at the expense of pancreas. In accord with this biphasic developmental role, BMP signaling similarly modulates endoderm formation at two different stages in vitro. In the first stage, low levels of BMP promote DE formation, with multiple studies finding that addition of a low concentration of BMP4 to ACTIVIN A and WNT3A improves DE formation (McGrath et al., 2015; Nostro et al., 2011; Russ et al., 2015; Teo et al., 2012). At the second stage, BMP signaling must be blocked for differentiation of endocrine cells to occur (Loh et al., 2014). These results with BMP nicely illustrate how developmental studies can reveal subtleties of time-dependent signaling to improve differentiation protocols.

3.2.2 From Endoderm to Endocrine Fates

After DE has been produced, it becomes regionalized into specific AP domains, including the posterior foregut from which pancreas derives. The foregut endoderm then gives rise to the dorsal and ventral buds of the pancreas. Developmental studies have again uncovered signaling molecules that promote these subsequent steps, with adjacent mesoderm tissue frequently acting as the source of signals. These studies have been well reviewed, and we refer the reader to (Pan & Wright, 2011) for a detailed treatment of this topic. Recently, it has become clear that cell transitions and marker gene expression must occur in the appropriate sequence, and one approach to isolating specific pools of precursors is to flow-sort cells based on cell surface markers (Kelly et al., 2011). For example, cells must pass through a PDX1(+);NKX6.1(+) progenitor stage if they are to ultimately produce metabolically active insulin-secreting cells (Rezania et al., 2013).

The mesenchyme surrounding the pancreatic epithelium is a source of secreted signals including FGF10 and TGFβ, and these signals are required for proper endocrine cell fate specification and morphogenesis (Bhushan et al., 2001; Miralles, Battelino, Czernichow, & Scharfmann, 1998; Sanvito et al., 1994). In addition, Notch signaling plays a role in self-renewal vs differentiation of endocrine progenitors (reviewed by Afelik & Jensen,

2013). One role for FGF10 is to block premature pancreatic epithelial cell differentiation by maintaining Notch activity in endocrine progenitors (Hart, Papadopoulou, & Edlund, 2003; Norgaard, Jensen, & Jensen, 2003). Studies in both mouse and zebrafish have demonstrated that prematurely forcing Notch downregulation promotes endocrine cell differentiation (Apelqvist et al., 1999; Jensen et al., 2000; Ninov et al., 2012; Parsons et al., 2009). These data suggest that inactivation of Notch signaling could similarly promote endocrine cell production in cultured cells. Interestingly, the single report exploring treatment of cultured endocrine precursors with Notch (Rezania et al., 2011) found that most cells matured as alpha not beta cells, implying a changing competence of progenitors and suggesting we need a deeper understanding of Notch signaling before we can exploit it to promote in vitro beta cell production.

After insulin expression commences, the beta cells must mature to become functional. In mammals, the maturation correlates with changing profiles of Maf gene expression (Hang & Stein, 2011; Riedel et al., 2012). To date, there has been minimal work in the zebrafish on *maf* class genes (Hesselson et al., 2009), and it is unclear whether mechanisms controlling the final stages of beta cell maturation are well conserved across vertebrates. Nevertheless, gaining an in depth understanding of the maturation process, and the appropriate markers to assay for it, will likely be key to optimizing efficient production of functional beta cells.

3.2.3 Challenges and Future Directions: Improving Cell-Based Therapies

One challenge in generating functional beta cells in large quantities is the high cost of the relevant cytokines and growth factors. Synthetic molecules provide a cost-effective alternative, and zebrafish provides an excellent platform for testing candidate molecules in vivo or screening for new ones (MacRae & Peterson, 2015; Rennekamp & Peterson, 2015). The small size and transparency of zebrafish larvae, coupled with the ease of obtaining large numbers of specimens, allow rapid in vivo screening, with transgenic tools providing a fluorescent read out for specific cell types. This approach has been successfully used to identify molecules relevant to beta cell production. For instance, an initial manual screen of FDA-approved compounds identified six "hits" that promote beta cell differentiation in zebrafish (Rovira et al., 2011). The Stanier group used a cell cycle reporter to identify 20 compounds that promote beta cell proliferation, and two drug classes identified in this screen also promoted regeneration after beta cell ablation (Tsuji et al., 2014). Recently, a high-throughput approach ("automated reporter quantification in vivo"; ARQiv) enabled over half a million zebrafish larvae to be screened for drugs inducing endocrine cell differentiation or beta cell proliferation (Wang, Rajpurohit, et al., 2015). This approach identified 11 drugs that induce endocrine differentiation and 15 that stimulate beta cell proliferation, with two compounds showing activity in both assays.

Reassuringly, these three chemical screens, performed on different scales, showed significant overlap in results. Interestingly, most of the hits from these screens are neuromodulators, pointing to the nervous system playing an important role in pancreas development (Borden, Houtz, Leach, & Kuruvilla, 2013). Indeed, two studies identified a role for serotonergic signaling in beta cell replication (Tsuji et al., 2014; Wang, Rajpurohit, et al., 2015). Another pathway identified in all three screens was RA signaling, which as mentioned earlier was initially shown to promote pancreatic cell fates in zebrafish. Intriguingly, at later stages RA signaling must be suppressed for "secondary" beta cells to differentiate from the IPD (Rovira et al., 2011), yet RA signaling also promotes beta cell proliferation (Wang, Rajpurohit, et al., 2015), suggesting that this molecule like others has developmental stage-dependent roles. Wang, Rajpurohit, et al. (2015), with their high-throughput approach, also identified a new pathway of interest: NF-κB signaling is active in pancreatic progenitors and inhibiting this pathway promotes endocrine cell differentiation. In human pancreatic cancers NF-κB promotes expression of endocrine progenitor marker SOX9 (Sun et al., 2013), suggesting NF-κB may be important in human pancreas development. Candidate molecules identified in these zebrafish chemical screens are yet to be tested more broadly in mammalian models of pancreas development, or most importantly in human ES cell in vitro differentiation protocols, but these results highlight an exciting new, clinically relevant, role for zebrafish.

Over the last decade, stem cell differentiation protocols have become significantly more effective at generating beta-like cells; however, three main challenges remain: efficiently producing sufficient quantities of beta cells, ensuring the beta cells are fully mature and hence functional, and eluding the immune system. Research into these areas continues apace, and animal models, including the zebrafish, will continue to provide key information.

4. PANCREAS REGENERATION AS A THERAPEUTIC TOOL

A cure for diabetes will depend upon the restoration of a sufficient mass of functional beta cells. While there has been significant progress in developing approaches to provide exogenous beta cells to patients, many challenges remain, including invasive surgical procedures, allograft rejection or life-long immune suppression, cost, and most importantly availability of functionally competent glucose-responsive beta cells. Stimulation of endogenous repair mechanisms to regenerate lost beta cells in situ, from the diabetic patient's own tissues, would circumvent many of these difficulties. Such repair mechanisms could involve recapitulation of developmental processes, or distinct mechanisms evolved to maintain homeostasis and regeneration of the mature pancreas. Currently, the full regenerative capacity of the human endocrine pancreas is unknown, but there are tantalizing clues that regeneration of human beta cells occurs. For example, 11–40% of long-term T1D patients produce measurable C-peptide (cleaved from the insulin molecule during processing) throughout their lives (Rother & Harlan, 2004) and such patients retain some beta cells (Baiu, Merriam, & Odorico, 2011; Wu, Yang, Chen, & Xu, 2015). In pancreata of T2D patients, many insulin/glucagon double-positive cells are found, and these may represent alpha cells caught during conversion into beta cells, similar to alpha to beta cell transdifferentiation described in animal models (Chung, Hao, Piran, Keinan, & Levine, 2010; Thorel et al., 2010; Ye et al., 2015). Together, these observations suggest that the human pancreas may have regenerative potential. To date, zebrafish experiments focused on pancreas regeneration have validated mechanisms hypothesized or discovered in mice and have also revealed novel pathways that regulate islet repair (Gemberling, Bailey, Hyde, & Poss, 2013).

4.1 Pancreas Injury Models

A variety of injury models (surgical, chemical, or genetic) have been developed to study both the pathogenesis of diabetes and the potential mechanisms of beta cell regeneration. These models have shown that regeneration of beta cells may proceed by multiple modes, including proliferation of surviving/preexisting beta cells, de novo differentiation of resident stem cells (neogenesis), and transdifferentiation of other nonbeta pancreatic cell types. The major injury models used in zebrafish are summarized below.

4.1.1 Surgical Injury of the Pancreas

A straightforward model of pancreas damage is surgical resection. After a 90% pancreatectomy in adult rats, significant regeneration of both endocrine and exocrine tissues occurs as the result of both proliferation of differentiated tissues and expansion and differentiation of duct tissue into new pancreatic lobes, i.e., neogenesis (Bonner-Weir, Baxter, Schuppin, & Smith, 1993; Bonner-Weir et al., 2010; Bonner-Weir, Trent, & Weir, 1983; Brockenbrough, Weir, & Bonner-Weir, 1988). In humans, there is conflicting evidence for pancreas regeneration after pancreatectomy. While there are clinical reports of regeneration of the entire pancreas, unaccompanied by diabetes (Schlegel et al., 2000), most clinical studies provide little evidence for regeneration of pancreas after 50–60% resection to treat cancer (Tsiotos, Barry, Johnson, & Sarr, 1999). By contrast, the adult zebrafish shows a robust regenerative response to surgical pancreatectomy. In pancreatectomized fish PCNA-positive proliferating cells are associated with the pancreatic ducts, and existing beta cells have been found to proliferate (Delaspre et al., 2015; Moss et al., 2009).

4.1.2 Chemical Destruction of Beta Cells

Streptozotocin (STZ) is a common reagent used to injure the pancreas in rodents. STZ is also effective at beta cell ablation in adult fish, but relatively high concentrations are needed both in zebrafish (Moss et al., 2009; Olsen, Sarras, & Intine, 2010; Sarras, Leontovich, Olsen, & Intine, 2013) and in another teleost fish model, Tilapia (Wright, Abraham, Dickson, Yang, & Morrison, 1999). Use of STZ in zebrafish embryos or larvae has not been reported, nor was this approach effective in our hands (RMA, unpublished). Interestingly, human beta cells are very resistant to STZ (Yang & Wright, 2002), suggesting that fish beta cells may be more similar to human beta cells than to those of mice in this regard.

4.1.3 Genetic and Hybrid Chemical-Genetic Beta Cell Destruction Models

Several genetic models of beta cell destruction have been developed in the mouse and zebrafish, and these can be divided into two classes: (1) purely genetic models and (2) inducible hybrid chemical-genetic models.

4.1.3.1 Genetic Models

A spontaneous genetic rodent model that closely recapitulates aspects of human diabetes is the Akita mouse, which carries an autosomal dominant allele consisting of a missense mutation in the *Ins2* gene. The Akita mutation (C96Y) blocks proinsulin protein folding to impair insulin processing and secretion, resulting in ER stress pathway activation and beta cell apoptosis (Oyadomari et al., 2002). In humans, multiple heterozygous dominant mutations in the insulin gene have been described that lead to permanent neonatal diabetes; these mutations were shown to block proper protein folding and processing in cultured cells (Rajan et al., 2010), similar to the Akita phenotype. For one of these, the C34G mutation, we developed a zebrafish model in which the human mutant form of insulin is misexpressed as a transgene in zebrafish beta cells (Eames et al., 2013). Although these transgenic fish accumulated excess insulin in the secretory pathway, suggesting partial recapitulation of the phenotype observed in cell culture, and similarities with the Akita phenotype, neither beta cell mass nor glucose homeostasis was affected. Nonetheless, islet architecture was altered, resembling regenerating islets of other zebrafish models (Curado et al., 2007; Eames et al., 2013; Moss et al., 2009) (see later). This remodeling is consistent with the possibility of enhanced regeneration and/or homeostasis in the zebrafish, relative to the mouse, that suffices to maintain normal glucose metabolism.

4.1.3.2 Hybrid Chemical-Genetic Models

Specific and inducible destruction of beta cells can be mediated by the expression of toxin genes under the control of pancreas-specific regulatory sequences. In mice, ablation systems using diphtheria toxin (DT) have been used most frequently (Nir, Melton, & Dor, 2007). Comparable genetic beta cell ablation has been carried out in zebrafish using two approaches: constitutive ablation with DT (Ninov et al., 2013) and conditional ablation mediated by nitroreductase (Curado et al., 2007; Pisharath et al., 2007). Both approaches use zebrafish insulin regulatory sequences to drive toxin gene expression in beta cells.

The constitutive DT-based approach was used to eliminate beta cells from the zebrafish islet throughout embryonic and larval development (Ninov et al., 2013). The larvae were hyperglycemic, growth retarded, and showed both proliferation and remodeling of ductal tissues. This was accompanied by excessive and precocious differentiation of endocrine cells, suggesting an active "regenerative" response or developmental regulation. Conditional, nitroreductase-based beta cell ablation, has been more widely employed in the zebrafish. This relies on beta cell expression of the *Escherichia coli* nitroreductase gene (*nfsb*, encoding NTR), which is innocuous under normal physiological conditions. When the NTR substrate (metronidazole; MTZ) is introduced, this prodrug rapidly induces apoptosis (reviewed in Curado et al., 2008), with evidence of beta cell injury within 3 h (RMA, unpublished observations). In zebrafish embryos and larvae, total elimination of beta cells occurs in under 24 h, altering islet structure and

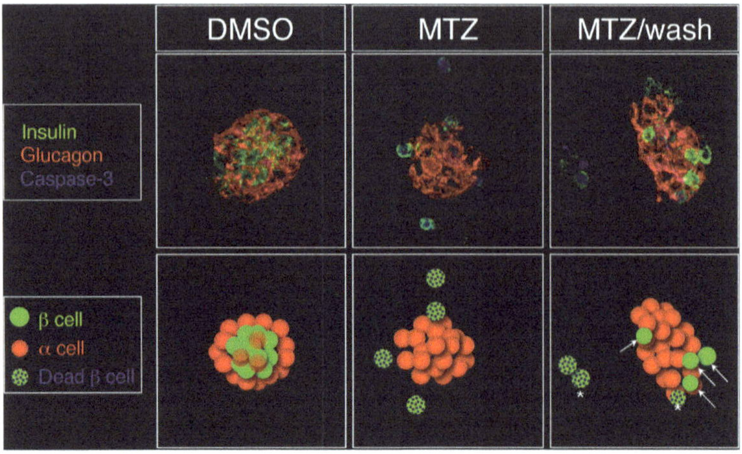

Fig. 4 Regeneration of the zebrafish pancreas after beta cell ablation. *Upper panels* show antibody-labeled principal islets of *Tg(ins:CFP-NTR)* larvae, following treatment commencing at 60 hpf with DMSO carrier for 48 h (control), MTZ for 48 h (ablation), or MTZ for 24 h followed by washing for 24 h (recovery). *Lower panels* show schematized versions of the upper panels. Beta cells are labeled with antiinsulin antibody (regenerated beta cells are indicated with *arrows* on the schematics), alpha cells are labeled with antiglucagon antibody, and dying cells (*) with caspase-3. MTZ efficiently kills NTR-expressing cells within 24 h, within 24 h of recovery new beta cells have formed. Part of this figure previously appeared in modified form in Curado et al. (2008).

elevating free glucose levels (Andersson, 2014; Curado et al., 2007; Pisharath et al., 2007). Importantly, after MTZ is washed out, beta cell mass and function rapidly reconstitute (Andersson, 2014; Ye et al., 2015) (Fig. 4). In adult fish, elevated blood glucose and absence of beta cells are observed 3 days after MTZ treatment, but recover by 2 weeks (Delaspre et al., 2015; Moss et al., 2009). In both larvae and adults, increased proliferation was observed at the periphery of the islet during regeneration (Moss et al., 2009; Pisharath et al., 2007), and increased neogenesis was observed in the pancreatic ducts (Wang et al., 2011). In addition, conversion of alpha cells into beta cells has been documented in larvae (Ye et al., 2015).

4.2 Cellular and Molecular Pathways of Beta Cell Regeneration Uncovered in Zebrafish

4.2.1 Transdifferentiation and Plasticity

Cell plasticity describes the ability of some cells to take on the characteristics of other cells in an organism. The conversion of differentiated nonbeta cell types in the pancreas into beta cells through transdifferentiation has the potential to restore glucose homeostasis. Importantly, nonbeta cells are still present in normal or even increased numbers in the diabetic pancreas, and therefore represent a potential source of replacement beta cells.

Zebrafish studies have revealed a role for the maintenance of DNA methyltransferase *dnmt1* in negatively regulating beta cell regeneration (Anderson et al., 2009). Dnmt1 propagates cell-specific methylation patterns during cell replication, and DNA methylation helps restrict and stabilize cell fates. While mouse *Dnmt1* mutants die early in gestation, zebrafish mutants survive to juvenile stages and development of beta and alpha cells is unaffected. However, after NTR/MTZ-mediated ablation, beta cell regeneration in zebrafish *dnmt1*-deficient specimens is increased by 50–100%. This may be due to a "loosening" of cell fate restriction as a consequence of passive DNA hypomethylation in the *dnmt1* mutants, although this remains to be tested by lineage tracing. Interestingly, decitabine, an inhibitor of DNMT1, promotes zebrafish beta cell regeneration in a small molecule screen, independently confirming a role for this pathway in suppressing beta cell regeneration (Wang, Rajpurohit, et al., 2015). In mouse, beta cell-specific deletion of *Dnmt1* resulted in derepression of *Arx* expression, and beta to alpha cell conversion (Dhawan, Georgia, Tschen, Fan, & Bhushan, 2011), further defining a role for DNA methylation in endocrine fate maintenance. Regeneration studies in mouse *Dnmt* mutants have yet to be conducted, but this is a promising avenue for increasing our understanding of plasticity.

Transdifferentiation of alpha to beta cells was first demonstrated in the mouse, becoming significant under conditions of near-total beta cell ablation (Chung, Hao, et al., 2010; Thorel et al., 2010); it is unknown whether this mechanism occurs in humans. Nevertheless, in diabetic patients there have been descriptions of glucagon/insulin double hormone-positive cells, providing hints that transdifferentiation may be conserved (Butler et al., 2003). In zebrafish, conversion of alpha into beta cells was seen following ablation of all beta cells using the NTR/MTZ system, consistent with conservation of transdifferentiation repair across vertebrates. Importantly, this was the first study to identify an endogenous regulator of endocrine transdifferentiation (Ye et al., 2015): products of the glucagon gene (glucagon or glucagon-like peptide 1; Glp1) are essential for the conversion.

Another important zebrafish finding that has provided insight into plasticity of the pancreas is the discovery that Ptf1a, a key pancreas progenitor and acinar cell transcription factor, is required for maintenance of acinar cell fate in the differentiated pancreas (Hesselson et al., 2011). While mouse mutant studies had shown that Ptf1a is essential for the formation of pancreas tissues (Burlison, Long, Fujitani, Wright, & Magnuson, 2008; Kawaguchi et al., 2002), the isolation of a zebrafish hypomorphic allele showed that high levels of Ptf1a expression drive exocrine differentiation, while lower levels are essential for endocrine differentiation (Dong, Provost, Leach, & Stainier, 2008). It was therefore hypothesized that reducing Ptf1a levels in acinar cells might promote endocrine fate. This model was supported by studies where Ptf1a activity was blocked during larval development using a repressor transgene, which resulted in conversion of exocrine cells into beta cells (Hesselson et al., 2011). Further, recent lineage tracing in developing zebrafish embryos confirmed that Ptf1a-expressing cells are multipotent progenitors, and that experimental reduction of Ptf1a levels decreases the proportion of cells that take on an exocrine fate (Wang, Park, et al., 2015). These findings open up the exciting possibility that reduction of PTF1a levels in the human pancreas could similarly promote transdifferentiation of exocrine cells to the beta cell fate.

4.2.2 Ducts and Centroacinar Cells as Sources of New Beta Cells

During develoment, new endocrine cells are formed from the intrapancreatic ductal epithelium (Section 1.2). Whether the duct can serve as source of new endocrine cells to repopulate adult islets in mammals is controversial (Bonner-Weir et al., 2010; Inada et al., 2008; Kopp et al., 2011; Solar et al., 2009; Xiao et al., 2013; Xu et al., 2008). In zebrafish (Delous et al., 2012; Manfroid et al., 2012), as in mice (Furuyama et al., 2011), the transcription factor *sox9b* is expressed in the hepatopancreatic ductal system where it is crucial for ductal morphogenesis and differentiation and marks differentiating beta cells. Mutant analysis has shown that *sox9b* is essential for the production of the late forming (VBD) beta cells that arise from the ducts during larval development as well as after beta cell ablation (Delous et al., 2012; Manfroid et al., 2012). Whether *sox9b* expression persists in adult ducts or duct-associated progenitors, and whether this regulates adult beta cell regeneration remain to be determined.

Centroacinar cells (CACs) are specialized duct cells, present at the interface of ducts and acinar cells, that have long been hypothesized to act as endocrine cell progenitors in the mature pancreas (reviewed by Beer, Parsons, & Rovira, 2016). Using lineage tracing with a Notch signaling-responsive promoter, Wang et al. (2011) defined the CAC population in both the developing and adult zebrafish. Using both NTR/MTZ beta cell ablation and pancreatectomy, it was definitively demonstrated both that the

CAC population gives rise to new beta cells, and that this population is regulated by Notch signaling (Delaspre et al., 2015). Zebrafish CACs also express *nkx6.1*, as do the ductal progenitors in embryos and larvae (Ghaye et al., 2015). Furthermore, lineage tracing with a TgBAC(*nkx6.1:GFP*) transgenic line has provided robust evidence that the CACs act as endocrine progenitors during homeostasis and regeneration in the adult fish (Ghaye et al., 2015); however, it should be noted that these approaches do not rule out the possibility that other progenitors also produce beta cells.

The Notch-responsive CAC progenitors also express CFTR, the cystic fibrosis transmembrane conductance regulator. Mutations in CFTR cause Cystic Fibrosis (CF), a disease of fluid transport, which leads to mucus build up in multiple organs including lungs and pancreas. A zebrafish mutant in the *cftr* gene (Navis et al., 2013) begins to show growth retardation and a dramatic lost of pancreatic acinar tissues during larval stages; in surviving adults, the acinar tissue is largely absent, replaced by fibrotic tissue, the pancreatic ducts are mucus-filled and dilated, and the islets are small and disorganized as in CF patients (Navis & Bagnat, 2015). A BAC transgene driving a GFP fusion protein (Table 1; Delaspre et al., 2015) revealed that CFTR protein is present in the Notch-responsive CAC progenitor cells of 5 dpf larvae, localized to thin extensions that line the inner lumen of the pancreatic duct. In humans, *cftr*$^{-/-}$ specimens show a reduced number of CACs and a related reduction in potential to produce secondary islets (Delaspre et al., 2015). Together, these results suggest that the *cftr*$^{-/-}$ mutant zebrafish will prove a useful CF model, with potential to provide new insights into the pathophysiology of the disease and act as a drug-screening tool.

Zebrafish has been effectively used to screen for drugs that promote neogenesis during regeneration (Andersson, 2014; Andersson et al., 2012; Tsuji et al., 2014). Andersson et al. (2012) screened more than 7000 compounds in regenerating *Tg(ins:CFP-NTR)* larvae and found several modulators of adenosine signaling, indicating that this pathway promotes beta cell recovery. This approach was validated by a second study in which the adenosine pathway was independently uncovered in a screen for beta cell mitogens performed in rat cell cultures, and confirmed in murine and porcine beta cells (Annes et al., 2012). Importantly, many of the compounds found to promote beta cell proliferation also boost regeneration in beta cell-ablated *Tg(ins:CFP-NTR)* larvae (Tsuji et al., 2014); active compounds included glucocorticoids, retinoids, and modulators of serotonin signaling.

In summary, zebrafish is providing new information on mechanisms of pancreas repair and regeneration. The system provides a powerful genetic model to test candidate gene functions and provides tools to delve into molecular mechanisms. Zebrafish is also becoming an increasingly useful high-throughput screening system to identify novel pathways that promote regeneration. These new pathways, as well as new details of well-studied pathways that regulate beta cell regeneration, can now be validated in mammalian models and their potential to be translated into human therapies explored.

5. PANCREAS AS A METABOLIC REGULATOR: STUDYING METABOLIC DISEASE IN ZEBRAFISH

Zebrafish can be used to study metabolic diseases including diabetes (reviewed by Kimmel & Meyer, 2016; Seth, Stemple, & Barroso, 2013). Investigations of diabetes require reliable measurement of blood glucose levels. Although larval zebrafish are too small for blood to be collected, measurement of whole organism free glucose provides a reliable proxy (Jurczyk et al., 2011); this approach has enabled analysis of the metabolic consequences of disrupting gene function (e.g., Dalgin & Prince, 2015). In addition, a read out of gluconeogenesis has been developed for larval fish, using the *pck1* gene to drive a fluorescent reporter (Gut et al., 2013). In adult fish, we established techniques for blood glucose sampling, and for glucose tolerance testing via intraperitoneal injection (Eames, Philipson, Prince, & Kinkel, 2010; Kinkel, Eames, Philipson, & Prince, 2010). More recently, it was reported that blood can be collected on multiple occasions from a single adult fish (Zang, Shimada, Nishimura, Tanaka, & Nishimura, 2013), enabling longitudinal studies. While the small size of the zebrafish makes such approaches technically challenging, they are worthwhile as they enable metabolic studies to be combined with the power of zebrafish genetics.

Zebrafish has been used to probe how metabolic feedback regulates neogenesis to maintain a proportional beta cell mass (Maddison & Chen, 2012; Ninov et al., 2013). A dramatic expansion of the beta cell mass derived from the pancreatic ducts occurs during the transition from yolk-based nutrition to active larval feeding. The progenitor cell population experiences feedback from beta cells, as chronic beta cell ablation causes rapid depletion of duct progenitors (Ninov et al., 2013). Ninov and colleagues also showed that mTOR signaling in the progenitors responds to the nutritional state of the embryo. In separate studies, it was observed that ductal differentiation is tied to energy availability: overnutrition by feeding larvae on chicken egg yolk or glucose drives an increase in beta cell mass (Maddison & Chen, 2012). Glucose treatment regulates the production of ductal beta cells during regeneration, but has no impact on transdifferentiation (Ye et al., 2015). Lastly, progenitor cells may respond directly to insulin or other secreted factors released from the differentiated beta cell mass, modulating further beta cell differentiation (Li, Maddison, Page-McCaw, & Chen, 2014; Ye et al., 2015).

6. FUTURE DIRECTIONS

With ever more sophisticated genetic manipulations and tools becoming available in the zebrafish, we expect to see continued studies of candidate gene functions during pancreas development and regeneration. The zebrafish is proving a powerful screening tool for drugs and small molecules, and we predict many more such studies of relevance to diabetes in coming years. We also expect zebrafish to become a useful metabolomics model. Metabolomics describes the identification and quantification of metabolites produced by the organism (Fiehn, 2001; Villas-Boas, Rasmussen, & Lane, 2005) and can bridge the gap between genotype and phenotype (Fiehn, 2002; Kaddurah-Daouk, Kristal, & Weinshilboum, 2008). Metabolite profiling in zebrafish has recently been reported, for example, metabolomics was used to interrogate a zebrafish disease model for "alcoholic fatty liver" (Jang et al., 2012). The approach has also uncovered metabolic changes that occur during normal development (Huang et al., 2001). In our own recent studies (GD and VEP, manuscript in prep.), we have compared metabolomic parameters between normal and hyperglycemic zebrafish embryos. We are also taking a broader, systems biology approach, comparing metabolomic and transcriptomic data from normal and hyperglycemic specimens to identify new key pathways relevant to both normal physiology and diabetes.

ACKNOWLEDGMENTS

We thank our colleagues Dr. Mary Kinkel and Dr. Michael Parsons for critical reading of an earlier version of the manuscript and excellent suggestions for its improvement. Relevant studies in the Prince lab have been supported by the JDRF, NIHDK064973, and P&F award from the University of Chicago Diabetes Research Center (P30 DK020595) to G.D. The Anderson lab is supported by The March of Dimes Foundation.

REFERENCES

Afelik, S., & Jensen, J. (2013). Notch signaling in the pancreas: Patterning and cell fate specification. *Wiley Interdisciplinary Reviews. Developmental Biology*, *2*, 531–544.

Alexander, J., & Stainier, D. Y. (1999). A molecular pathway leading to endoderm formation in zebrafish. *Current Biology*, *9*, 1147–1157.

Anderson, R. M., Bosch, J. A., Goll, M. G., Hesselson, D., Dong, P. D., Shin, D., et al. (2009). Loss of Dnmt1 catalytic activity reveals multiple roles for DNA methylation during pancreas development and regeneration. *Developmental Biology*, *334*, 213–223.

Andersson, O. (2014). Role of adenosine signalling and metabolism in beta-cell regeneration. *Experimental Cell Research*, *321*, 3–10.

Andersson, O., Adams, B. A., Yoo, D., Ellis, G. C., Gut, P., Anderson, R. M., et al. (2012). Adenosine signaling promotes regeneration of pancreatic beta cells in vivo. *Cell Metabolism*, *15*, 885–894.

Annes, J. P., Ryu, J. H., Lam, K., Carolan, P. J., Utz, K., Hollister-Lock, J., et al. (2012). Adenosine kinase inhibition selectively promotes rodent and porcine islet beta-cell replication. *Proceedings of the National Academy of Sciences of the United States of America*, *109*, 3915–3920.

Apelqvist, A., Li, H., Sommer, L., Beatus, P., Anderson, D. J., Honjo, T., et al. (1999). Notch signalling controls pancreatic cell differentiation. *Nature*, *400*, 877–881.

Argenton, F., Zecchin, E., & Bortolussi, M. (1999). Early appearance of pancreatic hormone-expressing cells in the zebrafish embryo. *Mechanisms of Development*, *87*, 217–221.

Arkhipova, V., Wendik, B., Devos, N., Ek, O., Peers, B., & Meyer, D. (2012). Characterization and regulation of the hb9/mnx1 beta-cell progenitor specific enhancer in zebrafish. *Developmental Biology*, *365*, 290–302.

Baiu, D., Merriam, F., & Odorico, J. (2011). Potential pathways to restore beta-cell mass: Pluripotent stem cells, reprogramming, and endogenous regeneration. *Current Diabetes Reports*, *11*, 392–401.

Beer, R. L., Parsons, M. J., & Rovira, M. (2016). Centroacinar cells: At the center of pancreas regeneration. *Developmental Biology*, *413*, 8–15.

Begemann, G., Schilling, T. F., Rauch, G. J., Geisler, R., & Ingham, P. W. (2001). The zebrafish neckless mutation reveals a requirement for raldh2 in mesodermal signals that pattern the hindbrain. *Development*, *128*, 3081–3094.

Bhushan, A., Itoh, N., Kato, S., Thiery, J. P., Czernichow, P., Bellusci, S., et al. (2001). Fgf10 is essential for maintaining the proliferative capacity of epithelial progenitor cells during early pancreatic organogenesis. *Development*, *128*, 5109–5117.

Biemar, F., Argenton, F., Schmidtke, R., Epperlein, S., Peers, B., & Driever, W. (2001). Pancreas development in zebrafish: Early dispersed appearance of endocrine hormone expressing cells and their convergence to form the definitive islet. *Developmental Biology*, *230*, 189–203.

Binot, A. C., Manfroid, I., Flasse, L., Winandy, M., Motte, P., Martial, J. A., et al. (2010). Nkx6.1 and nkx6.2 regulate alpha- and beta-cell formation in zebrafish by acting on pancreatic endocrine progenitor cells. *Developmental Biology*, *340*, 397–407.

Bonner-Weir, S., Baxter, L. A., Schuppin, G. T., & Smith, F. E. (1993). A second pathway for regeneration of adult exocrine and endocrine pancreas. A possible recapitulation of embryonic development. *Diabetes*, *42*, 1715–1720.

Bonner-Weir, S., Li, W. C., Ouziel-Yahalom, L., Guo, L., Weir, G. C., & Sharma, A. (2010). Beta-cell growth and regeneration: Replication is only part of the story. *Diabetes*, *59*, 2340–2348.

Bonner-Weir, S., Sullivan, B. A., & Weir, G. C. (2015). Human islet morphology revisited: Human and rodent islets are not so different after all. *The Journal of Histochemistry and Cytochemistry*, *63*, 604–612.

Bonner-Weir, S., Trent, D. F., & Weir, G. C. (1983). Partial pancreatectomy in the rat and subsequent defect in glucose-induced insulin release. *The Journal of Clinical Investigation*, *71*, 1544–1553.

Borden, P., Houtz, J., Leach, S. D., & Kuruvilla, R. (2013). Sympathetic innervation during development is necessary for pancreatic islet architecture and functional maturation. *Cell Reports*, *4*, 287–301.

Bouwens, L., & Rooman, I. (2005). Regulation of pancreatic beta-cell mass. *Physiological Reviews*, *85*, 1255–1270.

Brockenbrough, J. S., Weir, G. C., & Bonner-Weir, S. (1988). Discordance of exocrine and endocrine growth after 90% pancreatectomy in rats. *Diabetes*, *37*, 232–236.

Bromberg, J. S., Kaplan, B., Halloran, P. F., & Robertson, R. P. (2007). The islet transplant experiment: Time for a reassessment. *American Journal of Transplantation*, 7, 2217–2218.

Bruin, J. E., Rezania, A., Xu, J., Narayan, K., Fox, J. K., O'Neil, J. J., et al. (2013). Maturation and function of human embryonic stem cell-derived pancreatic progenitors in macroencapsulation devices following transplant into mice. *Diabetologia*, *56*, 1987–1998.

Burlison, J. S., Long, Q., Fujitani, Y., Wright, C. V., & Magnuson, M. A. (2008). Pdx-1 and Ptf1a concurrently determine fate specification of pancreatic multipotent progenitor cells. *Developmental Biology*, *316*, 74–86.

Butler, A. E., Janson, J., Bonner-Weir, S., Ritzel, R., Rizza, R. A., & Butler, P. C. (2003). Beta-cell deficit and increased beta-cell apoptosis in humans with type 2 diabetes. *Diabetes*, *52*, 102–110.

Chang, B. E., Blader, P., Fischer, N., Ingham, P. W., & Strahle, U. (1997). Axial (HNF3beta) and retinoic acid receptors are regulators of the zebrafish sonic hedgehog promoter. *The EMBO Journal*, *16*, 3955–3964.

Chen, Y., Pan, F. C., Brandes, N., Afelik, S., Solter, M., & Pieler, T. (2004). Retinoic acid signaling is essential for pancreas development and promotes endocrine at the expense of exocrine cell differentiation in Xenopus. *Developmental Biology*, *271*, 144–160.

Chen, Y., & Schier, A. F. (2001). The zebrafish Nodal signal Squint functions as a morphogen. *Nature*, *411*, 607–610.

Chung, W. S., Andersson, O., Row, R., Kimelman, D., & Stainier, D. Y. (2010). Suppression of Alk8-mediated Bmp signaling cell-autonomously induces pancreatic beta-cells in zebrafish. *Proceedings of the National Academy of Sciences of the United States of America*, *107*, 1142–1147.

Chung, C. H., Hao, E., Piran, R., Keinan, E., & Levine, F. (2010). Pancreatic beta-cell neogenesis by direct conversion from mature alpha-cells. *Stem Cells*, *28*, 1630–1638.

Chung, W. S., Shin, C. H., & Stainier, D. Y. (2008). Bmp2 signaling regulates the hepatic versus pancreatic fate decision. *Developmental Cell*, *15*, 738–748.

Chung, W. S., & Stainier, D. Y. (2008). Intra-endodermal interactions are required for pancreatic beta cell induction. *Developmental Cell*, *14*, 582–593.

Collignon, J., Varlet, I., & Robertson, E. J. (1996). Relationship between asymmetric nodal expression and the direction of embryonic turning. *Nature*, *381*, 155–158.

Conlon, F. L., Lyons, K. M., Takaesu, N., Barth, K. S., Kispert, A., Herrmann, B., et al. (1994). A primary requirement for nodal in the formation and maintenance of the primitive streak in the mouse. *Development*, *120*, 1919–1928.

Curado, S., Anderson, R. M., Jungblut, B., Mumm, J., Schroeter, E., & Stainier, D. Y. (2007). Conditional targeted cell ablation in zebrafish: A new tool for regeneration studies. *Developmental Dynamics*, *236*, 1025–1035.

Curado, S., Stainier, D. Y., & Anderson, R. M. (2008). Nitroreductase-mediated cell/tissue ablation in zebrafish: A spatially and temporally controlled ablation method with applications in developmental and regeneration studies. *Nature Protocols*, *3*, 948–954.

Dalgin, G., & Prince, V. E. (2015). Differential levels of Neurod establish zebrafish endocrine pancreas cell fates. *Developmental Biology*, *402*, 81–97.

Dalgin, G., Ward, A. B., Hao le, T., Beattie, C. E., Nechiporuk, A., & Prince, V. E. (2011). Zebrafish mnx1 controls cell fate choice in the developing endocrine pancreas. *Development*, *138*, 4597–4608.

D'Amour, K. A., Agulnick, A. D., Eliazer, S., Kelly, O. G., Kroon, E., & Baetge, E. E. (2005). Efficient differentiation of human embryonic stem cells to definitive endoderm. *Nature Biotechnology*, *23*, 1534–1541.

D'Amour, K. A., Bang, A. G., Eliazer, S., Kelly, O. G., Agulnick, A. D., Smart, N. G., et al. (2006). Production of pancreatic hormone-expressing endocrine cells from human embryonic stem cells. *Nature Biotechnology*, *24*, 1392–1401.

Dassaye, R., Naidoo, S., & Cerf, M. E. (2016). Transcription factor regulation of pancreatic organogenesis, differentiation and maturation. *Islets*, *8*, 13–34.

DeFronzo, R. A., Ferrannini, E., Alberti, K. G. M. M., & Zimmet, P. (2015). International textbook of diabetes mellitus (4th ed.). Chichester, West Sussex; Hoboken, NJ: John Wiley & Sons Inc.

Delaspre, F., Beer, R. L., Rovira, M., Huang, W., Wang, G., Gee, S., et al. (2015). Centroacinar cells are progenitors that contribute to endocrine pancreas regeneration. *Diabetes*, *64*, 3499–3509.

Delous, M., Yin, C., Shin, D., Ninov, N., Debrito Carten, J., Pan, L., et al. (2012). Sox9b is a key regulator of pancreaticobiliary ductal system development. *PLoS Genetics*, *8*, e1002754.

Dhawan, S., Georgia, S., Tschen, S. I., Fan, G., & Bhushan, A. (2011). Pancreatic beta cell identity is maintained by DNA methylation-mediated repression of Arx. *Developmental Cell*, *20*, 419–429.

Djiotsa, J., Verbruggen, V., Giacomotto, J., Ishibashi, M., Manning, E., Rinkwitz, S., et al. (2012). Pax4 is not essential for beta-cell differentiation in zebrafish embryos but modulates alpha-cell generation by repressing arx gene expression. *BMC Developmental Biology*, *12*, 37.

Dong, P. D., Provost, E., Leach, S. D., & Stainier, D. Y. (2008). Graded levels of Ptf1a differentially regulate endocrine and exocrine fates in the developing pancreas. *Genes & Development*, *22*, 1445–1450.

Eames, S. C., Kinkel, M. D., Rajan, S., Prince, V. E., & Philipson, L. H. (2013). Transgenic zebrafish model of the C43G human insulin gene mutation. *Journal of Diabetes Investigation*, *4*, 157–167.

Eames, S. C., Philipson, L. H., Prince, V. E., & Kinkel, M. D. (2010). Blood sugar measurement in zebrafish reveals dynamics of glucose homeostasis. *Zebrafish*, 7, 205–213.

Fan, X., Hagos, E. G., Xu, B., Sias, C., Kawakami, K., Burdine, R. D., et al. (2007). Nodal signals mediate interactions between the extra-embryonic and embryonic tissues in zebrafish. *Developmental Biology*, *310*, 363–378.

Farooq, M., Sulochana, K. N., Pan, X., To, J., Sheng, D., Gong, Z., et al. (2008). Histone deacetylase 3 (hdac3) is specifically required for liver development in zebrafish. *Developmental Biology*, *317*, 336–353.

Fiehn, O. (2001). Combining genomics, metabolome analysis, and biochemical modelling to understand metabolic networks. *Comparative and Functional Genomics*, *2*, 155–168.

Fiehn, O. (2002). Metabolomics—The link between genotypes and phenotypes. *Plant Molecular Biology*, *48*, 155–171.

Field, H. A., Dong, P. D., Beis, D., & Stainier, D. Y. (2003). Formation of the digestive system in zebrafish. II. Pancreas morphogenesis. *Developmental Biology*, *261*, 197–208.

Flasse, L. C., Pirson, J. L., Stern, D. G., Von Berg, V., Manfroid, I., Peers, B., et al. (2013). Ascl1b and Neurod1, instead of Neurog3, control pancreatic endocrine cell fate in zebrafish. *BMC Biology*, *11*, 78.

Flasse, L. C., Stern, D. G., Pirson, J. L., Manfroid, I., Peers, B., & Voz, M. L. (2013). The bHLH transcription factor Ascl1a is essential for the specification of the intestinal secretory cells and mediates Notch signaling in the zebrafish intestine. *Developmental Biology*, *376*, 187–197.

Fullerton, B., Jeitler, K., Seitz, M., Horvath, K., Berghold, A., & Siebenhofer, A. (2014). Intensive glucose control versus conventional glucose control for type 1 diabetes mellitus. *The Cochrane Database of Systematic Reviews*, *2*, CD009122.

Furuyama, K., Kawaguchi, Y., Akiyama, H., Horiguchi, M., Kodama, S., Kuhara, T., et al. (2011). Continuous cell supply from a Sox9-expressing progenitor zone in adult liver, exocrine pancreas and intestine. *Nature Genetics*, *43*, 34–41.

Gemberling, M., Bailey, T. J., Hyde, D. R., & Poss, K. D. (2013). The zebrafish as a model for complex tissue regeneration. *Trends in Genetics*, *29*, 611–620.

Ghaye, A. P., Bergemann, D., Tarifeno-Saldivia, E., Flasse, L. C., Von Berg, V., Peers, B., et al. (2015). Progenitor potential of nkx6.1-expressing cells throughout zebrafish life and during beta cell regeneration. *BMC Biology*, *13*, 70.

Godinho, L., Mumm, J. S., Williams, P. R., Schroeter, E. H., Koerber, A., Park, S. W., et al. (2005). Targeting of amacrine cell neurites to appropriate synaptic laminae in the developing zebrafish retina. *Development*, *132*, 5069–5079.

Gradwohl, G., Dierich, A., LeMeur, M., & Guillemot, F. (2000). neurogenin3 is required for the development of the four endocrine cell lineages of the pancreas. *Proceedings of the National Academy of Sciences of the United States of America*, *97*, 1607–1611.

Grandel, H., Lun, K., Rauch, G. J., Rhinn, M., Piotrowski, T., Houart, C., et al. (2002). Retinoic acid signalling in the zebrafish embryo is necessary during pre-segmentation stages to pattern the anterior-posterior axis of the CNS and to induce a pectoral fin bud. *Development*, *129*, 2851–2865.

Gu, G., Dubauskaite, J., & Melton, D. A. (2002). Direct evidence for the pancreatic lineage: NGN3+ cells are islet progenitors and are distinct from duct progenitors. *Development*, *129*, 2447–2457.

Gut, P., Baeza-Raja, B., Andersson, O., Hasenkamp, L., Hsiao, J., Hesselson, D., et al. (2013). Whole-organism screening for gluconeogenesis identifies activators of fasting metabolism. *Nature Chemical Biology*, *9*, 97–104.

Hang, Y., & Stein, R. (2011). MafA and MafB activity in pancreatic beta cells. *Trends in Endocrinology and Metabolism*, *22*, 364–373.

Hart, A., Papadopoulou, S., & Edlund, H. (2003). Fgf10 maintains notch activation, stimulates proliferation, and blocks differentiation of pancreatic epithelial cells. *Developmental Dynamics*, *228*, 185–193.

Herrera, P. L., Nepote, V., & Delacour, A. (2002). Pancreatic cell lineage analyses in mice. *Endocrine*, *19*, 267–278.

Hesselson, D., Anderson, R. M., Beinat, M., & Stainier, D. Y. (2009). Distinct populations of quiescent and proliferative pancreatic beta-cells identified by HOTcre mediated labeling. *Proceedings of the National Academy of Sciences of the United States of America*, *106*, 14896–14901.

Hesselson, D., Anderson, R. M., & Stainier, D. Y. (2011). Suppression of Ptf1a activity induces acinar-to-endocrine conversion. *Current Biology*, *21*, 712–717.

Huang, H. P., Liu, M., El-Hodiri, H. M., Chu, K., Jamrich, M., & Tsai, M. J. (2000). Regulation of the pancreatic islet-specific gene BETA2 (neuroD) by neurogenin 3. *Molecular and Cellular Biology*, *20*, 3292–3307.

Huang, H., Vogel, S. S., Liu, N., Melton, D. A., & Lin, S. (2001). Analysis of pancreatic development in living transgenic zebrafish embryos. *Molecular and Cellular Endocrinology*, *177*, 117–124.

Inada, A., Nienaber, C., Katsuta, H., Fujitani, Y., Levine, J., Morita, R., et al. (2008). Carbonic anhydrase II-positive pancreatic cells are progenitors for both endocrine and exocrine pancreas after birth. *Proceedings of the National Academy of Sciences of the United States of America*, *105*, 19915–19919.

Ionescu-Tirgoviste, C., Gagniuc, P. A., Gubceac, E., Mardare, L., Popescu, I., Dima, S., et al. (2015). A 3D map of the islet routes throughout the healthy human pancreas. *Scientific Reports*, *5*, 14634.

Jang, Z. H., Chung, H. C., Ahn, Y. G., Kwon, Y. K., Kim, J. S., Ryu, J. H., et al. (2012). Metabolic profiling of an alcoholic fatty liver in zebrafish (Danio rerio). *Molecular BioSystems*, *8*, 2001–2009.

Jennings, R. E., Berry, A. A., Strutt, J. P., Gerrard, D. T., & Hanley, N. A. (2015). Human pancreas development. *Development*, *142*, 3126–3137.

Jensen, J., Heller, R. S., Funder-Nielsen, T., Pedersen, E. E., Lindsell, C., Weinmaster, G., et al. (2000). Independent development of pancreatic alpha- and beta-cells from neurogenin3-expressing precursors: A role for the notch pathway in repression of premature differentiation. *Diabetes*, *49*, 163–176.

Jurczyk, A., Roy, N., Bajwa, R., Gut, P., Lipson, K., Yang, C., et al. (2011). Dynamic glucoregulation and mammalian-like responses to metabolic and developmental disruption in zebrafish. *General and Comparative Endocrinology*, *170*, 334–345.

Kaddurah-Daouk, R., Kristal, B. S., & Weinshilboum, R. M. (2008). Metabolomics: A global biochemical approach to drug response and disease. *Annual Review of Pharmacology and Toxicology*, *48*, 653–683.

Kani, S., Bae, Y. K., Shimizu, T., Tanabe, K., Satou, C., Parsons, M. J., et al. (2010). Proneural gene-linked neurogenesis in zebrafish cerebellum. *Developmental Biology*, *343*, 1–17.

Kawaguchi, Y., Cooper, B., Gannon, M., Ray, M., MacDonald, R. J., & Wright, C. V. (2002). The role of the transcriptional regulator Ptf1a in converting intestinal to pancreatic progenitors. *Nature Genetics*, *32*, 128–134.

Kelly, O. G., Chan, M. Y., Martinson, L. A., Kadoya, K., Ostertag, T. M., Ross, K. G., et al. (2011). Cell-surface markers for the isolation of pancreatic cell types derived from human embryonic stem cells. *Nature Biotechnology*, *29*, 750–756.

Kelly, W. D., Lillehei, R. C., Merkel, F. K., Idezuki, Y., & Goetz, F. C. (1967). Allotransplantation of the pancreas and duodenum along with the kidney in diabetic nephropathy. *Surgery*, *61*, 827–837.

Kikuchi, K., Holdway, J. E., Werdich, A. A., Anderson, R. M., Fang, Y., Egnaczyk, G. F., et al. (2010). Primary contribution to zebrafish heart regeneration by gata4(+) cardiomyocytes. *Nature*, *464*, 601–605.

Kilimnik, G., Jo, J., Periwal, V., Zielinski, M. C., & Hara, M. (2012). Quantification of islet size and architecture. *Islets*, *4*, 167–172.

Kimmel, R. A., & Meyer, D. (2016). Zebrafish pancreas as a model for development and disease. *Methods in Cell Biology*, *134*, 431–461.

Kinkel, M. D., Eames, S. C., Philipson, L. H., & Prince, V. E. (2010). Intraperitoneal injection into adult zebrafish. *Journal of Visualized Experiments*, (42), e2126. http://dx.doi.org/10.3791/2126(note this is an online publication of videos).

Kinkel, M. D., & Prince, V. E. (2009). On the diabetic menu: Zebrafish as a model for pancreas development and function. *Bioessays*, *31*, 139–152.

Kopp, J. L., Dubois, C. L., Schaffer, A. E., Hao, E., Shih, H. P., Seymour, P. A., et al. (2011). Sox9+ ductal cells are multipotent progenitors throughout development but do not produce new endocrine cells in the normal or injured adult pancreas. *Development*, *138*, 653–665.

Korzh, V., Edlund, T., & Thor, S. (1993). Zebrafish primary neurons initiate expression of the LIM homeodomain protein Isl-1 at the end of gastrulation. *Development*, *118*, 417–425.

Korzh, V., Sleptsova, I., Liao, J., He, J., & Gong, Z. (1998). Expression of zebrafish bHLH genes ngn1 and nrd defines distinct stages of neural differentiation. *Developmental Dynamics*, *213*, 92–104.

Kroon, E., Martinson, L. A., Kadoya, K., Bang, A. G., Kelly, O. G., Eliazer, S., et al. (2008). Pancreatic endoderm derived from human embryonic stem cells generates glucose-responsive insulin-secreting cells in vivo. *Nature Biotechnology*, *26*, 443–452.

Kubo, A., Shinozaki, K., Shannon, J. M., Kouskoff, V., Kennedy, M., Woo, S., et al. (2004). Development of definitive endoderm from embryonic stem cells in culture. *Development*, *131*, 1651–1662.

Lancman, J. J., Zvenigorodsky, N., Gates, K. P., Zhang, D., Solomon, K., Humphrey, R. K., et al. (2013). Specification of hepatopancreas progenitors in zebrafish by hnf1ba and wnt2bb. *Development, 140*, 2669–2679.

Leung, L., Klopper, A. V., Grill, S. W., Harris, W. A., & Norden, C. (2011). Apical migration of nuclei during G2 is a prerequisite for all nuclear motion in zebrafish neuroepithelia. *Development, 138*, 5003–5013.

Li, H., Arber, S., Jessell, T. M., & Edlund, H. (1999). Selective agenesis of the dorsal pancreas in mice lacking homeobox gene Hlxb9. *Nature Genetics, 23*, 67–70.

Li, M., Maddison, L. A., Page-McCaw, P., & Chen, W. (2014). Overnutrition induces beta-cell differentiation through prolonged activation of beta-cells in zebrafish larvae. *American Journal of Physiology. Endocrinology and Metabolism, 306*, E799–E807.

Li, Z., Wen, C., Peng, J., Korzh, V., & Gong, Z. (2009). Generation of living color transgenic zebrafish to trace somatostatin-expressing cells and endocrine pancreas organization. *Differentiation, 77*, 128–134.

Loh, K. M., Ang, L. T., Zhang, J., Kumar, V., Ang, J., Auyeong, J. Q., et al. (2014). Efficient endoderm induction from human pluripotent stem cells by logically directing signals controlling lineage bifurcations. *Cell Stem Cell, 14*, 237–252.

MacRae, C. A., & Peterson, R. T. (2015). Zebrafish as tools for drug discovery. *Nature Reviews. Drug Discovery, 14*, 721–731.

Maddison, L. A., & Chen, W. (2012). Nutrient excess stimulates beta-cell neogenesis in zebrafish. *Diabetes, 61*, 2517–2524.

Manfroid, I., Ghaye, A., Naye, F., Detry, N., Palm, S., Pan, L., et al. (2012). Zebrafish sox9b is crucial for hepatopancreatic duct development and pancreatic endocrine cell regeneration. *Developmental Biology, 366*, 268–278.

Martin, M., Gallego-Llamas, J., Ribes, V., Kedinger, M., Niederreither, K., Chambon, P., et al. (2005). Dorsal pancreas agenesis in retinoic acid-deficient Raldh2 mutant mice. *Developmental Biology, 284*, 399–411.

McGrath, P. S., Watson, C. L., Ingram, C., Helmrath, M. A., & Wells, J. M. (2015). The basic helix-loop-helix transcription factor NEUROG3 is required for development of the human endocrine pancreas. *Diabetes, 64*, 2497–2505.

Mellitzer, G., Martin, M., Sidhoum-Jenny, M., Orvain, C., Barths, J., Seymour, P. A., et al. (2004). Pancreatic islet progenitor cells in neurogenin 3-yellow fluorescent protein knock-add-on mice. *Molecular Endocrinology, 18*, 2765–2776.

Milewski, W. M., Duguay, S. J., Chan, S. J., & Steiner, D. F. (1998). Conservation of PDX-1 structure, function, and expression in zebrafish. *Endocrinology, 139*, 1440–1449.

Miralles, F., Battelino, T., Czernichow, P., & Scharfmann, R. (1998). TGF-beta plays a key role in morphogenesis of the pancreatic islets of Langerhans by controlling the activity of the matrix metalloproteinase MMP-2. *The Journal of Cell Biology, 143*, 827–836.

Mizoguchi, T., Verkade, H., Heath, J. K., Kuroiwa, A., & Kikuchi, Y. (2008). Sdf1/Cxcr4 signaling controls the dorsal migration of endodermal cells during zebrafish gastrulation. *Development, 135*, 2521–2529.

Molotkov, A., Molotkova, N., & Duester, G. (2005). Retinoic acid generated by Raldh2 in mesoderm is required for mouse dorsal endodermal pancreas development. *Developmental Dynamics, 232*, 950–957.

Mosimann, C., Kaufman, C. K., Li, P., Pugach, E. K., Tamplin, O. J., & Zon, L. I. (2011). Ubiquitous transgene expression and Cre-based recombination driven by the ubiquitin promoter in zebrafish. *Development, 138*, 169–177.

Moss, J. B., Koustubhan, P., Greenman, M., Parsons, M. J., Walter, I., & Moss, L. G. (2009). Regeneration of the pancreas in adult zebrafish. *Diabetes, 58*, 1844–1851.

Nair, S., & Schilling, T. F. (2008). Chemokine signaling controls endodermal migration during zebrafish gastrulation. *Science, 322*, 89–92.

Navis, A., & Bagnat, M. (2015). Loss of cftr function leads to pancreatic destruction in larval zebrafish. *Developmental Biology*, *399*, 237–248.
Navis, A., Marjoram, L., & Bagnat, M. (2013). Cftr controls lumen expansion and function of Kupffer's vesicle in zebrafish. *Development*, *140*, 1703–1712.
Ninov, N., Borius, M., & Stainier, D. Y. (2012). Different levels of Notch signaling regulate quiescence, renewal and differentiation in pancreatic endocrine progenitors. *Development*, *139*, 1557–1567.
Ninov, N., Hesselson, D., Gut, P., Zhou, A., Fidelin, K., & Stainier, D. Y. (2013). Metabolic regulation of cellular plasticity in the pancreas. *Current Biology*, *23*, 1242–1250.
Nir, T., Melton, D. A., & Dor, Y. (2007). Recovery from diabetes in mice by beta cell regeneration. *The Journal of Clinical Investigation*, *117*, 2553–2561.
Norgaard, G. A., Jensen, J. N., & Jensen, J. (2003). FGF10 signaling maintains the pancreatic progenitor cell state revealing a novel role of Notch in organ development. *Developmental Biology*, *264*, 323–338.
Nostro, M. C., Sarangi, F., Ogawa, S., Holtzinger, A., Corneo, B., Li, X., et al. (2011). Stage-specific signaling through TGFbeta family members and WNT regulates patterning and pancreatic specification of human pluripotent stem cells. *Development*, *138*, 861–871.
Obholzer, N., Wolfson, S., Trapani, J. G., Mo, W., Nechiporuk, A., Busch-Nentwich, E., et al. (2008). Vesicular glutamate transporter 3 is required for synaptic transmission in zebrafish hair cells. *The Journal of Neuroscience*, *28*, 2110–2118.
Olsen, A. S., Sarras, M. P., Jr., & Intine, R. V. (2010). Limb regeneration is impaired in an adult zebrafish model of diabetes mellitus. *Wound Repair and Regeneration*, *18*, 532–542.
Ostrom, M., Loffler, K. A., Edfalk, S., Selander, L., Dahl, U., Ricordi, C., et al. (2008). Retinoic acid promotes the generation of pancreatic endocrine progenitor cells and their further differentiation into beta-cells. *PLoS One*, *3*, e2841.
Oyadomari, S., Koizumi, A., Takeda, K., Gotoh, T., Akira, S., Araki, E., et al. (2002). Targeted disruption of the Chop gene delays endoplasmic reticulum stress-mediated diabetes. *The Journal of Clinical Investigation*, *109*, 525–532.
Pagliuca, F. W., Millman, J. R., Gurtler, M., Segel, M., Van Dervort, A., Ryu, J. H., et al. (2014). Generation of functional human pancreatic beta cells in vitro. *Cell*, *159*, 428–439.
Pan, F. C., & Wright, C. (2011). Pancreas organogenesis: From bud to plexus to gland. *Developmental Dynamics*, *240*, 530–565.
Parsons, M. J., Pisharath, H., Yusuff, S., Moore, J. C., Siekmann, A. F., Lawson, N., et al. (2009). Notch-responsive cells initiate the secondary transition in larval zebrafish pancreas. *Mechanisms of Development*, *126*, 898–912.
Pauls, S., Zecchin, E., Tiso, N., Bortolussi, M., & Argenton, F. (2007). Function and regulation of zebrafish nkx2.2a during development of pancreatic islet and ducts. *Developmental Biology*, *304*, 875–890.
Pisharath, H., Rhee, J. M., Swanson, M. A., Leach, S. D., & Parsons, M. J. (2007). Targeted ablation of beta cells in the embryonic zebrafish pancreas using *E. coli* nitroreductase. *Mechanisms of Development*, *124*, 218–229.
Poulain, M., Furthauer, M., Thisse, B., Thisse, C., & Lepage, T. (2006). Zebrafish endoderm formation is regulated by combinatorial Nodal, FGF and BMP signalling. *Development*, *133*, 2189–2200.
Rajan, S., Eames, S. C., Park, S. Y., Labno, C., Bell, G. I., Prince, V. E., et al. (2010). In vitro processing and secretion of mutant insulin proteins that cause permanent neonatal diabetes. *American Journal of Physiology. Endocrinology and Metabolism*, *298*, E403–E410.
Rennekamp, A. J., & Peterson, R. T. (2015). 15 years of zebrafish chemical screening. *Current Opinion in Chemical Biology*, *24*, 58–70.
Rezania, A., Bruin, J. E., Xu, J., Narayan, K., Fox, J. K., O'Neil, J. J., et al. (2013). Enrichment of human embryonic stem cell-derived NKX6.1-expressing pancreatic progenitor

cells accelerates the maturation of insulin-secreting cells in vivo. *Stem Cells*, *31*, 2432–2442.

Rezania, A., Riedel, M. J., Wideman, R. D., Karanu, F., Ao, Z., Warnock, G. L., et al. (2011). Production of functional glucagon-secreting alpha-cells from human embryonic stem cells. *Diabetes*, *60*, 239–247.

Riedel, M. J., Asadi, A., Wang, R., Ao, Z., Warnock, G. L., & Kieffer, T. J. (2012). Immunohistochemical characterisation of cells co-producing insulin and glucagon in the developing human pancreas. *Diabetologia*, *55*, 372–381.

Rother, K. I., & Harlan, D. M. (2004). Challenges facing islet transplantation for the treatment of type 1 diabetes mellitus. *The Journal of Clinical Investigation*, *114*, 877–883.

Rovira, M., Huang, W., Yusuff, S., Shim, J. S., Ferrante, A. A., Liu, J. O., et al. (2011). Chemical screen identifies FDA-approved drugs and target pathways that induce precocious pancreatic endocrine differentiation. *Proceedings of the National Academy of Sciences of the United States of America*, *108*, 19264–19269.

Rukstalis, J. M., & Habener, J. F. (2007). Snail2, a mediator of epithelial-mesenchymal transitions, expressed in progenitor cells of the developing endocrine pancreas. *Gene Expression Patterns*, *7*, 471–479.

Russ, H. A., Parent, A. V., Ringler, J. J., Hennings, T. G., Nair, G. G., Shveygert, M., et al. (2015). Controlled induction of human pancreatic progenitors produces functional beta-like cells in vitro. *The EMBO Journal*, *34*, 1759–1772.

Sakaguchi, T., Kikuchi, Y., Kuroiwa, A., Takeda, H., & Stainier, D. Y. (2006). The yolk syncytial layer regulates myocardial migration by influencing extracellular matrix assembly in zebrafish. *Development*, *133*, 4063–4072.

Sakaue-Sawano, A., Kurokawa, H., Morimura, T., Hanyu, A., Hama, H., Osawa, H., et al. (2008). Visualizing spatiotemporal dynamics of multicellular cell-cycle progression. *Cell*, *132*, 487–498.

Sanvito, F., Herrera, P. L., Huarte, J., Nichols, A., Montesano, R., Orci, L., et al. (1994). TGF-beta 1 influences the relative development of the exocrine and endocrine pancreas in vitro. *Development*, *120*, 3451–3462.

Sarras, M. P., Jr., Leontovich, A. A., Olsen, A. S., & Intine, R. V. (2013). Impaired tissue regeneration corresponds with altered expression of developmental genes that persists in the metabolic memory state of diabetic zebrafish. *Wound Repair and Regeneration*, *21*, 320–328.

Schier, A. F., Neuhauss, S. C., Helde, K. A., Talbot, W. S., & Driever, W. (1997). The one-eyed pinhead gene functions in mesoderm and endoderm formation in zebrafish and interacts with no tail. *Development*, *124*, 327–342.

Schlegel, R. D., Tiscornia, O., de Vedia y Mitre, E., Lembeye, A. R., Coqui, R., Macagno, D., et al. (2000). [Is there pancreatic regeneration? Morphological and functional certification after a corporocaudal splenopancreatectomy/]. *Acta Gastroenterologica Latinoamericana*, *30*, 107–113.

Schonhoff, S. E., Giel-Moloney, M., & Leiter, A. B. (2004). Neurogenin 3-expressing progenitor cells in the gastrointestinal tract differentiate into both endocrine and non-endocrine cell types. *Developmental Biology*, *270*, 443–454.

Seguin, C. A., Draper, J. S., Nagy, A., & Rossant, J. (2008). Establishment of endoderm progenitors by SOX transcription factor expression in human embryonic stem cells. *Cell Stem Cell*, *3*, 182–195.

Seth, A., Stemple, D. L., & Barroso, I. (2013). The emerging use of zebrafish to model metabolic disease. *Disease Models & Mechanisms*, *6*, 1080–1088.

Shapiro, A. M., Lakey, J. R., Ryan, E. A., Korbutt, G. S., Toth, E., Warnock, G. L., et al. (2000). Islet transplantation in seven patients with type 1 diabetes mellitus using a glucocorticoid-free immunosuppressive regimen. *The New England Journal of Medicine*, *343*, 230–238.

Shin, C. H., Chung, W. S., Hong, S. K., Ober, E. A., Verkade, H., Field, H. A., et al. (2008). Multiple roles for Med12 in vertebrate endoderm development. *Developmental Biology*, *317*, 467–479.

Solar, M., Cardalda, C., Houbracken, I., Martin, M., Maestro, M. A., De Medts, N., et al. (2009). Pancreatic exocrine duct cells give rise to insulin-producing beta cells during embryogenesis but not after birth. *Developmental Cell*, *17*, 849–860.

Stafford, D., Hornbruch, A., Mueller, P. R., & Prince, V. E. (2004). A conserved role for retinoid signaling in vertebrate pancreas development. *Development Genes and Evolution*, *214*, 432–441.

Stafford, D., & Prince, V. E. (2002). Retinoic acid signaling is required for a critical early step in zebrafish pancreatic development. *Current Biology*, *12*, 1215–1220.

Strahle, U., Blader, P., Henrique, D., & Ingham, P. W. (1993). Axial, a zebrafish gene expressed along the developing body axis, shows altered expression in cyclops mutant embryos. *Genes & Development*, *7*, 1436–1446.

Sun, Z., & Hopkins, N. (2001). vhnf1, the MODY5 and familial GCKD-associated gene, regulates regional specification of the zebrafish gut, pronephros, and hindbrain. *Genes & Development*, *15*, 3217–3229.

Sun, L., Mathews, L. A., Cabarcas, S. M., Zhang, X., Yang, A., Zhang, Y., et al. (2013). Epigenetic regulation of SOX9 by the NF-kappaB signaling pathway in pancreatic cancer stem cells. *Stem Cells*, *31*, 1454–1466.

Takahashi, K., & Yamanaka, S. (2006). Induction of pluripotent stem cells from mouse embryonic and adult fibroblast cultures by defined factors. *Cell*, *126*, 663–676.

Teo, A. K., Ali, Y., Wong, K. Y., Chipperfield, H., Sadasivam, A., Poobalan, Y., et al. (2012). Activin and BMP4 synergistically promote formation of definitive endoderm in human embryonic stem cells. *Stem Cells*, *30*, 631–642.

Thisse, B., Wright, C. V., & Thisse, C. (2000). Activin- and Nodal-related factors control antero-posterior patterning of the zebrafish embryo. *Nature*, *403*, 425–428.

Thorel, F., Nepote, V., Avril, I., Kohno, K., Desgraz, R., Chera, S., et al. (2010). Conversion of adult pancreatic alpha-cells to beta-cells after extreme beta-cell loss. *Nature*, *464*, 1149–1154.

Tiso, N., Filippi, A., Pauls, S., Bortolussi, M., & Argenton, F. (2002). BMP signalling regulates anteroposterior endoderm patterning in zebrafish. *Mechanisms of Development*, *118*, 29–37.

Tiso, N., Moro, E., & Argenton, F. (2009). Zebrafish pancreas development. *Molecular and Cellular Endocrinology*, *312*, 24–30.

Toyoda, T., Mae, S., Tanaka, H., Kondo, Y., Funato, M., Hosokawa, Y., et al. (2015). Cell aggregation optimizes the differentiation of human ESCs and iPSCs into pancreatic bud-like progenitor cells. *Stem Cell Research*, *14*, 185–197.

Tsiotos, G. G., Barry, M. K., Johnson, C. D., & Sarr, M. G. (1999). Pancreas regeneration after resection: Does it occur in humans? *Pancreas*, *19*, 310–313.

Tsuji, N., Ninov, N., Delawary, M., Osman, S., Roh, A. S., Gut, P., et al. (2014). Whole organism high content screening identifies stimulators of pancreatic beta-cell proliferation. *PLoS One*, *9*, e104112.

Varshney, G. K., Sood, R., & Burgess, S. M. (2015). Understanding and editing the zebrafish genome. *Advances in Genetics*, *92*, 1–52.

Verbruggen, V., Ek, O., Georlette, D., Delporte, F., Von Berg, V., Detry, N., et al. (2010). The Pax6b homeodomain is dispensable for pancreatic endocrine cell differentiation in zebrafish. *The Journal of Biological Chemistry*, *285*, 13863–13873.

Villas-Boas, S. G., Rasmussen, S., & Lane, G. A. (2005). Metabolomics or metabolite profiles? *Trends in Biotechnology*, *23*, 385–386.

Wan, H., Korzh, S., Li, Z., Mudumana, S. P., Korzh, V., Jiang, Y. J., et al. (2006). Analyses of pancreas development by generation of gfp transgenic zebrafish using an exocrine pancreas-specific elastaseA gene promoter. *Experimental Cell Research*, *312*, 1526–1539.

Wang, Y. J., Park, J. T., Parsons, M. J., & Leach, S. D. (2015). Fate mapping of ptf1a-expressing cells during pancreatic organogenesis and regeneration in zebrafish. *Developmental Dynamics*, *244*, 724–735.

Wang, G., Rajpurohit, S. K., Delaspre, F., Walker, S. L., White, D. T., Ceasrine, A., et al. (2015). First quantitative high-throughput screen in zebrafish identifies novel pathways for increasing pancreatic beta-cell mass. *eLife*, *4*, e08261.

Wang, Y., Rovira, M., Yusuff, S., & Parsons, M. J. (2011). Genetic inducible fate mapping in larval zebrafish reveals origins of adult insulin-producing beta-cells. *Development*, *138*, 609–617.

Warga, R. M., & Nusslein-Volhard, C. (1999). Origin and development of the zebrafish endoderm. *Development*, *126*, 827–838.

Wendik, B., Maier, E., & Meyer, D. (2004). Zebrafish mnx genes in endocrine and exocrine pancreas formation. *Developmental Biology*, *268*, 372–383.

Wright, J. R., Jr., Abraham, C., Dickson, B. C., Yang, H., & Morrison, C. M. (1999). Streptozotocin dose–response curve in tilapia, a glucose-responsive teleost fish. *General and Comparative Endocrinology*, *114*, 431–440.

Wu, J., Yang, X., Chen, B., & Xu, X. (2015). Pancreas beta cell regeneration and type 1 diabetes (Review). *Experimental and Therapeutic Medicine*, *9*, 653–657.

Xiao, X., Guo, P., Shiota, C., Prasadan, K., El-Gohary, Y., Wiersch, J., et al. (2013). Neurogenin3 activation is not sufficient to direct duct-to-beta cell transdifferentiation in the adult pancreas. *The Journal of Biological Chemistry*, *288*, 25297–25308.

Xu, J., Cui, J., Del Campo, A., & Shin, C. H. (2016). Four and a Half LIM Domains 1b (Fhl1b) Is Essential for Regulating the Liver versus Pancreas Fate Decision and for beta-Cell Regeneration. *PLoS Genetics*, *12*, e1005831.

Xu, X., D'Hoker, J., Stange, G., Bonne, S., De Leu, N., Xiao, X., et al. (2008). Beta cells can be generated from endogenous progenitors in injured adult mouse pancreas. *Cell*, *132*, 197–207.

Xu, P. F., Zhu, K. Y., Jin, Y., Chen, Y., Sun, X. J., Deng, M., et al. (2010). Setdb2 restricts dorsal organizer territory and regulates left-right asymmetry through suppressing fgf8 activity. *Proceedings of the National Academy of Sciences of the United States of America*, *107*, 2521–2526.

Yabe, S. G., Fukuda, S., Takeda, F., Nashiro, K., Shimoda, M., & Okochi, H. (2016). Efficient generation of functional pancreatic beta cells from human iPS cells. *Journal of Diabetes*. http://dx.doi.org/10.1111/1753-0407.12400.

Yang, H., & Wright, J. R., Jr. (2002). Human beta cells are exceedingly resistant to streptozotocin in vivo. *Endocrinology*, *143*, 2491–2495.

Ye, L., Robertson, M. A., Hesselson, D., Stainier, D. Y., & Anderson, R. M. (2015). Glucagon is essential for alpha cell transdifferentiation and beta cell neogenesis. *Development*, *142*, 1407–1417.

Yee, N. S., Yusuff, S., & Pack, M. (2001). Zebrafish pdx1 morphant displays defects in pancreas development and digestive organ chirality, and potentially identifies a multipotent pancreas progenitor cell. *Genesis*, *30*, 137–140.

Zang, L., Shimada, Y., Nishimura, Y., Tanaka, T., & Nishimura, N. (2013). A novel, reliable method for repeated blood collection from aquarium fish. *Zebrafish*, *10*, 425–432.

Zecchin, E., Filippi, A., Biemar, F., Tiso, N., Pauls, S., Ellertsdottir, E., et al. (2007). Distinct delta and jagged genes control sequential segregation of pancreatic cell types from precursor pools in zebrafish. *Developmental Biology*, *301*, 192–204.

Zecchin, E., Mavropoulos, A., Devos, N., Filippi, A., Tiso, N., Meyer, D., et al. (2004). Evolutionary conserved role of ptf1a in the specification of exocrine pancreatic fates. *Developmental Biology, 268*, 174–184.
Zhou, X., Sasaki, H., Lowe, L., Hogan, B. L., & Kuehn, M. R. (1993). Nodal is a novel TGF-beta-like gene expressed in the mouse node during gastrulation. *Nature, 361*, 543–547.
Zhu, Z., Li, Q. V., Lee, K., Rosen, B. P., Gonzalez, F., Soh, C. L., et al. (2016). Genome editing of lineage determinants in human pluripotent stem cells reveals mechanisms of pancreatic development and diabetes. *Cell Stem Cell, 18*, 755–768.

CHAPTER EIGHT

Modeling Infectious Diseases in the Context of a Developing Immune System

Samrah Masud, Vincenzo Torraca, Annemarie H. Meijer[1]
Institute of Biology, Leiden University, Leiden, The Netherlands
[1]Corresponding author: e-mail address: a.h.meijer@biology.leidenuniv.nl

Contents

Current Topics in Developmental Biology, Volume 124
ISSN 0070-2153
http://dx.doi.org/10.1016/bs.ctdb.2016.10.006

Abstract

Zebrafish has been used for over a decade to study the mechanisms of a wide variety of inflammatory disorders and infections, with models ranging from bacterial, viral, to fungal pathogens. Zebrafish has been especially relevant to study the differentiation, specialization, and polarization of the two main innate immune cell types, the macrophages and the neutrophils. The optical accessibility and the early appearance of myeloid cells that can be tracked with fluorescent labels in zebrafish embryos and the ability to use genetics to selectively ablate or expand immune cell populations have permitted studying the interaction between infection, development, and metabolism. Additionally, zebrafish embryos are readily colonized by a commensal flora, which facilitated studies that emphasize the requirement for immune training by the natural microbiota to properly respond to pathogens. The remarkable conservation of core mechanisms required for the recognition of microbial and danger signals and for the activation of the immune defenses illustrates the high potential of the zebrafish model for biomedical research. This review will highlight recent insight that the developing zebrafish has contributed to our understanding of host responses to invading microbes and the involvement of the microbiome in several physiological processes. These studies are providing a mechanistic basis for developing novel therapeutic approaches to control infectious diseases.

1. INTRODUCTION

Infectious diseases remain a major global health problem, with tuberculosis (TB) and HIV/AIDS as the biggest killers, each responsible for over a million deaths annually according to the reports of the World Health Organization (www.who.int). The increasing occurrence of multidrug-resistant strains of *Mycobacterium tuberculosis*, the bacterial pathogen causing TB, indicates that current antibiotic treatment regimens are ineffective. Antibiotic resistances represent a serious problem also in hospital settings, with methicillin-resistant *Staphylococcus aureus* as a notable example of a pathogen causing opportunistic infections in immunocompromised patients. Despite intense research efforts, there are no effective vaccines against some of the major human bacterial pathogens, including *M. tuberculosis* and *S. aureus*. Furthermore, vaccines are not yet available for newly emerging viral diseases, which can spread rapidly due to transmission by insect vectors, as exemplified by the recent Zika virus outbreak. Development of novel therapeutic approaches for the treatment of infectious diseases requires detailed understanding of the mechanisms by which pathogens subvert the immune system of the infected host. As we discuss in this review, the zebrafish is a valuable addition to the range of animal models used for preclinical research into infectious disease biology.

The immune system of vertebrates functions by cooperative mechanisms of innate and adaptive immunity. During infection, innate immunity is activated by the recognition of microbial molecules and danger signals released by damaged host cells. Across species, innate immunity is mediated primarily by phagocytic cells, including macrophages, neutrophils, and dendritic cells. Activated innate immune cells represent an important line of defense against a large spectrum of pathogens as they provide an immediate response to invading microbes. Additionally, cells of the innate immune system, by functioning as antigen presenting cells and by providing stimulatory signals, are essential to alert the adaptive immune system to mount a more specific immune response mediated by antibody-producing B-lymphocytes and cytotoxic T-lymphocytes. These cells collaborate to target, isolate, or kill infected cells to prevent infection spreading throughout the organism.

Developing organisms rely more heavily on innate immunity, because the adaptive immune system takes longer to mature. For instance, it is well known that human neonates depend on maternal antibodies for adequate protection against infectious diseases. In zebrafish larvae, the first immature T-cell precursors are the first signs of an adaptive immunity, detected by 3 days postfertilization (dpf) (Langenau et al., 2004); however, functional phagocytes are present earlier, at 1 dpf (Fig. 1) (Herbomel, Thisse, & Thisse, 1999). B cells emerge from the pronephros of juvenile zebrafish only at 19 dpf and (Langenau et al., 2004) and antibody production does not occur until at least 21 dpf (Page et al., 2013). As a result, the zebrafish embryo and early larval stages have become widely used as an in vivo model to study innate immunity in separation from adaptive immunity (Harvie & Huttenlocher, 2015; Levraud, Palha, Langevin, & Boudinot, 2014; Meijer & Spaink, 2011; Ramakrishnan, 2013; Renshaw & Trede, 2012).

The different cell types of the immune system are generated by hematopoiesis, defined as the differentiation of multipotent, self-renewing stem cells into all cellular components of the blood (Davidson & Zon, 2004; Jagannathan-Bogdan & Zon, 2013). In all vertebrates, hematopoiesis is a highly conserved process that involves successive waves of primitive, intermediate, and definitive generation of hematopoietic progenitor cells

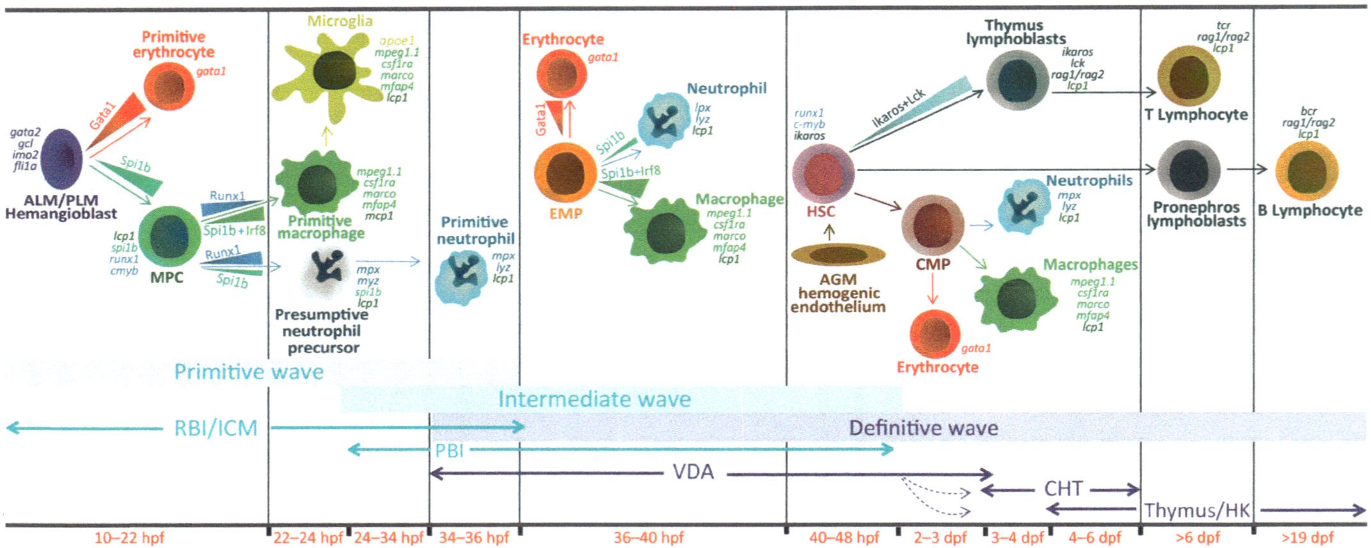

Fig. 1 Development of zebrafish immune system: In zebrafish, immune cells are generated via a primitive, intermediate, and definitive wave of hematopoiesis, which are active in the indicated tissues in the developmental windows reported on the timeline. The figure also indicates the key transcriptional regulators controlling the differentiation fate and the distinctive markers expressed by each cell type (described in more detail in the main text). Abbreviations: *AGM*, aorta-gonad-mesonephros; *ALM*, anterior lateral mesoderm; *CHT*, caudal hematopoietic tissue; *CMP*, common myeloid progenitor; *EMP*, erythromyeloid progenitor; *HK*, head kidney; *HSC*, hematopoietic stem cells; *ICM*, intermediate cell mass; *MPC*, myeloid progenitor cell; *PBI*, posterior blood island; *PLM*, posterior lateral mesoderm; *RBI*, rostral blood island; *VDA*, ventral wall of dorsal aorta.

during ontogeny (Fig. 1) (Bertrand et al., 2007; Galloway & Zon, 2003). Hematopoiesis can be further differentiated into erythropoiesis (the development of red blood cells), myelopoiesis (the development of leukocytes mediating innate immunity), and lymphopoiesis (the generation of the leukocytes (lymphocytes) of the adaptive immune system). Myeloid cells consist of two main categories based on cellular contents: (i) granulocytes and (ii) agranulated cells. Granulocytes (including neutrophils, eosinophils, basophils, and mast cells) display characteristic secretory granules in the cytoplasm containing antimicrobial molecules and inflammatory mediators. Furthermore, granulocytes can be recognized by a polymorphic nucleus, while agranulated cells, including monocytes and macrophages, are mononuclear.

In zebrafish embryos and early larval stages, all mononuclear cells are commonly referred to as (primitive) macrophages, irrespective of whether these cells are circulating in the blood or have invaded tissues (Herbomel et al., 1999; Herbomel, Thisse, & Thisse, 2001). The specialized macrophages resident in the brain (microglia) are also already present in the early life stages of zebrafish and their progenitors can be distinguished as early as 1 dpf (Fig. 1). Neutrophils are the main granulocyte cell type in embryos and larvae (Lieschke et al., 2002). Mast cells can also be distinguished, but eosinophils are only described in adult zebrafish and basophils have not been identified (Balla et al., 2010; Dobson et al., 2008).

In this review, we describe how innate immune cell types arise during the normal course of zebrafish embryo and larval development, and how the production, differentiation, and function of these cells can be affected by infection, inflammation, and the presence of the gut microbiota. We discuss recent studies that show how innate immune responses are intricately linked with the regulation of energy metabolism and homeostasis, in which autophagy plays a major role. Furthermore, we review work that contributed to develop zebrafish infection models (Table 1), which has been particularly helpful to dissect the specific implications of different innate immune cell types in infectious disease pathologies. To illustrate this, we highlight recent studies of bacterial infections, including causative agents of human infectious diseases or opportunistic infections, such as *Mycobacteria*, *Listeria*, *Shigella*, Staphylococci, and a range of viral and fungal pathogens. These studies are providing new insight into host–pathogen interaction mechanisms that hold promise for translation into novel therapeutic strategies for human infectious diseases.

Table 1 Human Infection Diseases Modeled in Zebrafish

Infectious Agents	Human Disease	Zebrafish Infection Model	First Description
Bacteria	Tuberculosis	*Mycobacterium marinum* surrogate model for *Mycobacterium tuberculosis*	Davis et al. (2002)
	Salmonellosis	*Salmonella enterica* serovar Typhimurium	van der Sar et al. (2003)
	Shigellosis	*Shigella flexneri*	Mostowy et al. (2013)
	Listeriosis	*Listeria monocytogenes*	Levraud et al. (2009)
	Opportunistic infections	*Burkholderia cenocepacia*	Vergunst, Meijer, Renshaw, and O'Callaghan (2010)
		Pseudomonas aeruginosa	Clatworthy et al. (2009)
		Staphylococcus aureus	Prajsnar, Cunliffe, Foster, and Renshaw (2008)
Viruses	Influenza	Influenza A virus	Gabor et al. (2014)
	Herpes simplex	Herpes simplex virus type 1	Burgos, Ripoll-Gomez, Alfaro, Sastre, and Valdivieso (2008)
	Chikungunya fever	Chikungunya virus	Palha et al. (2013)
Fungi	Candidiasis	*Candida albicans*	Chao et al. (2010)
	Aspergillosis	*Aspergillus fumigatus*	Knox et al. (2014)
	Mucormycosis	*Mucor circinelloides*	Voelz, Gratacap, and Wheeler (2015)
	Cryptococcosis	*Cryptococcus neoformans*	Tenor, Oehlers, Yang, Tobin, and Perfect (2015)

2. DEVELOPMENT OF THE CELL TYPES OF THE INNATE IMMUNE SYSTEM

To understand how the immune system works, we must first understand how the cells in the innate immune system form, and zebrafish have provided an outstanding system for such studies. This is covered in depth in the chapter by Kwan and North on "Studying Hematopoiesis in Zebrafish Provides Insight to Hematologic Malignancies" in this volume. Here, we review the developmental aspects of innate immunity that are relevant to understanding the response to infection.

2.1 Generation of Primitive Myeloid Cells

The development of the zebrafish immune system mirrors processes observed in other vertebrates, including mammals, but at an accelerated scale (Fig. 1). The first innate immune cells of the zebrafish embryo are generated during primitive hematopoiesis, which occurs in two locations of the zebrafish embryo: the anterior lateral mesoderm (ALM) and posterior lateral mesoderm (PLM). As the development proceeds, the ALM and PLM differentiate into the rostral blood island (RBI) and intermediate cell mass (ICM), respectively (Bertrand et al., 2007). The primitive myeloid cells develop from the RBI, while primitive erythrocytes originate from the ICM. By the 6-somite stage, expression of *spi1b* (*pu.1*) is detected, which encodes Pu.1, a master transcriptional regulator of myelopoiesis (Lieschke et al., 2002; Rhodes et al., 2005). By 16 h postfertilization (hpf), Pu.1-positive myeloid progenitors originating from the RBI start to migrate over the yolk sac (Fig. 1) (Bennett et al., 2001; Lieschke et al., 2002). This process requires granulocyte colony-stimulating factor receptor (Gcsfr) signaling (Liongue, Hall, O'Connell, Crosier, & Ward, 2009). During migration, these myeloid progenitors turn on the pan-leukocyte marker L-plastin (*lcp1*) (Bennett et al., 2001; Herbomel et al., 1999, 2001; Liu & Wen, 2002). Morphologically distinguishable macrophages are observed as early as 22 hpf on the yolk sac and enter the blood circulation by 26 hpf. Some macrophages migrate into the cephalic mesenchyme from 22 hpf onward in a *csf1ra*-dependent manner and can eventually develop into microglia (Herbomel et al., 2001; Peri & Nusslein-Volhard, 2008). These macrophages are functional and are capable of phagocytosing apoptotic debris, senescent red blood cells, and experimentally injected bacteria (Herbomel et al., 1999). Thus, as early as 1 dpf, zebrafish embryos can be used to study the response to infection.

The genes *csf1ra*, *mpeg1.1*, *marco*, and *mfap4* are marker genes that are predominantly expressed in macrophages in comparison with other leukocytes (Benard, Roobol, Spaink, & Meijer, 2014; Ellett, Pase, Hayman, Andrianopoulos, & Lieschke, 2011; Mathias et al., 2009; Walton, Cronan, Beerman, & Tobin, 2015; Zakrzewska et al., 2010). Several of these markers have been used to generate transgenic reporter lines that are frequently used in infectious disease research (Table 2) (Ellett et al., 2011; Gray et al., 2011; Walton et al., 2015).

Morphologically distinguishable neutrophils appear later than macrophages (Le Guyader et al., 2008). Using an in vivo photoactivatable cell tracer, it has been demonstrated that primitive neutrophils originate from the RBI-derived hemangioblasts, the same lineage as the primitive macrophages, after the dispersal of the progenitors into the tissues (Fig. 1) (Le Guyader et al., 2008). At 34 hpf, differentiated neutrophils are detectable by electron microscopy (Willett, Cortes, Zuasti, & Zapata, 1999). In agreement, granules are observed under video-enhanced differential interference contrast microscopy around 35 hpf, a time when neutrophils can also be detected by staining with Sudan Black, a lipid marker for granules (Le Guyader et al., 2008). Sudan Black-positive neutrophils also stain positive for myeloperoxidase (Mpx) enzyme activity (Le Guyader et al., 2008; Lieschke, Oates, Crowhurst, Ward, & Layton, 2001) as early as 24 hpf, along with the expression of the other neutrophil marker lysosome C (*lyz*) (Le Guyader et al., 2008; Meijer et al., 2008). Transgenic reporter lines for the *mpx* and *lyz* marker genes are widely used to study neutrophil behavior (Table 2) (Hall, Flores, Storm, Crosier, & Crosier, 2007; Mathias et al., 2006; Renshaw et al., 2006). The *mpx*/*lyz*-positive phagocytes first appear as migrating cells on the yolk sac, and these are most likely progenitors of the neutrophils that can be detected in tissues of older embryos using Sudan Black staining (Harvie & Huttenlocher, 2015; Le Guyader et al., 2008).

An important study in zebrafish has revealed previously underappreciated differences in phagocytic behavior between macrophages and neutrophils that are very relevant for the design of infection models (Colucci-Guyon, Tinevez, Renshaw, & Herbomel, 2011). This study showed that, in contrast to macrophages, neutrophils possess limited ability to phagocytose fluid-borne bacteria, but can quickly migrate to wounded or infected tissues and efficiently remove surface-associated bacteria (Colucci-Guyon et al., 2011). A previous study describes a similar "surface phagocytosis" behavior for mammalian neutrophils (Wood, 1960). This property is likely to be relevant for human infectious disease, since the first encounter of microbes

Table 2 Markers for Cell Types of the Zebrafish Innate Immune System

Cell Type	Transgenic Marker[a]	Gene Marker	Antibody/Cell Staining	Functional Assay
Pan-leukocytic	—	*lcp1*	Anti-L-plastin	Morphological and functional characterization of macrophages and neutrophils
	Tg(coro1a:EGFP)	*coro1a*	—	
Myeloid cell precursors	*Tg(-5.3spi1b:EGFP)* *Tg(-9.0spi1b:EGFP)* *Tg(-4spi1b:Gal4)* *Tg(-4spi1:LY-EGFP)*	*spi1b/ pu.1*	—	Marker of macrophage and neutrophil precursors
Macrophages	*Tg(mpeg1:EGFP)* *Tg(mpeg1:Gal4-VP16)* *Tg(mpeg1:mCherry-F)* *Tg(mpeg1:Dendra2)*	*mpeg1.1*	—	Specific marker of macrophages, but downregulated by several infections; also labels microglia
	TgBAC(csf1ra:Gal4-VP16)	*csf1ra/fms*	—	Specific marker of macrophages; also labels nonmotile pigment cells (xanthophores)
	Tg(mfap4:dLanYFP-CAAX) *Tg(mfap4:mTurquoise)*	*mfap4*	—	Specific marker of macrophages; less sensitive to infection downregulation than *mpeg1.1*
Neutrophils	*TgBAC(mpx:EGFP)* *Tg(mpx:GFP)* *Tg(mpx:mCherry)* *Tg(mpx:EGFP-F)* *Tg(mpx:DsRed-F)* *Tg(mpx:Dendra2)*	*mpx*	Anti-Mpx/Mpx enzyme activity staining	Specific marker of neutrophils

Continued

Table 2 Markers for Cell Types of the Zebrafish Innate Immune System—cont'd

Cell Type	Transgenic Marker	Gene Marker	Antibody/Cell Staining	Functional Assay
	Tg(lyz:EGFP) *Tg(lyz:DsRed2)* *Tg(lyz:Gal4-VP16)*	*lyz/lysc*	—	Specific marker of neutrophils; some overlap with macrophages at early developmental stages
	—	—	Sudan black	Staining of neutrophil granules
Activated macrophages/ neutrophils	*Tg(il1b:GFP-F)*	*il1b*	Anti-Il1b	Reporter to *distinguish inflammatory phenotypes of macrophages (M1) and neutrophils*
	Tg(tnfa:eGFP-F)	*tnfa*	—	Marker for activated macrophages (M1)
	Tg(irg1:EGFP)	*irg1*	—	Marker for activated macrophages (M1)
	Tg(CMV:EGFP-map1lc3b)	*map1lc3b*	—	Marker for autophagy activation
	Tg(Myd88:EGFP) *Tg(Myd88:Dsred2)*	*myd88*	—	Marker for TLR signaling potential
	Tg(NFκB:EGFP)	*nfκB*	—	Marker for transcriptional induction of innate immune response
Microglia	*Tg(apoeb:lynEGFP)*	*apoeb*	—	Specifically marker of microglia
	—	—	Neutral red	Efficient staining of microglia; partially effective staining of macrophages
Mast cells	—	*cpa5*	—	Marks a subpopulation of L-plastin positive myeloid cells by in situ hybridization

[a]Only the most frequently used transgenic lines are indicated; for additional lines and references, we refer to the Zebrafish Model Organism Database (http://zfin.org/).

with phagocytes is critical for the outcome of infection (Colucci-Guyon et al., 2011). In zebrafish embryos and larvae, phagocytosis by macrophages is favored when microbes are injected into the blood or into a body cavity such as the hindbrain ventricle, whereas subcutaneous, muscle or tail fin injections will provide the conditions for efficient engagement of neutrophils (Colucci-Guyon et al., 2011). The technical options allowed by using zebrafish, where the initial infection site can be varied to investigate how macrophages and neutrophils respond differently, is a strength of zebrafish infection models.

In addition to neutrophil and macrophage lineages, also mast cells are thought to be generated from the RBI (Dobson et al., 2008). The activation of mast cells at sites of infection can have direct effector functions or contribute to the regulation of innate and adaptive immune responses (Prykhozhij & Berman, 2014). As the gene encoding carboxypeptidase A5 (*cpa5*), a marker for mast cells, is expressed as early as 24 hpf (Dobson et al., 2008), zebrafish embryos could become a valuable model to study the function of mast cells in the context of infection. However, to date, studies in zebrafish infection models have concentrated on macrophage and neutrophil functions, where work has uncovered novel insights into how these cells respond to infection, and into the genes required for mounting an immune response, as further discussed later.

2.2 Generation of Myeloid Cells by the Intermediate and Definitive Waves of Hematopoiesis

As in all vertebrates, hematopoiesis in zebrafish occurs in waves (Jagannathan-Bogdan & Zon, 2013). The second wave of hematopoiesis is identified as an intermediate wave (Fig. 1), occurring at the posterior blood island (PBI) at the most posterior part of the ICM. The PBI is a temporary location of hematopoiesis in zebrafish (24–48 hpf), analogous to the mammalian fetal liver. The intermediate wave of hematopoiesis generates the first committed erythromyeloid progenitors (EMPs) that are capable of giving rise to both erythroid and myeloid lineage cells (Bertrand et al., 2007), including macrophages, neutrophils, and mast cells (Fig. 1) (Bertrand et al., 2007). The primitive and intermediate waves cannot sustain hematopoiesis for a long time. Only the final wave that occurs during embryogenesis, namely definitive hematopoiesis, is able to produce hematopoietic stem cells (HSCs) that can generate all types of hematopoietic cells for the whole life span. The development of HSCs is dependent on transcription factor Runx1 (Lam et al., 2009). In zebrafish, HSCs are generated from about 1 to 2.5 dpf in the ventral wall of the dorsal aorta (VDA) (Fig. 1). This hematopoietic site derives from the aorta-gonad-mesonephros (AGM), which is also the origin of HSC in mammals. HSCs emerging from the VDA migrate to and colonize the three sites of definitive hematopoiesis: the caudal hematopoietic tissue (CHT), the thymus, and the anterior part of the kidney (pronephros). From 3 to 6 dpf, the CHT is the main hematopoietic tissue of the larvae. However, the CHT does not produce lymphoid progenitors and is readily exhausted. From approximately 4 dpf, the thymus and the pronephros (which will later develop into the adult head kidney) start to contribute to hematopoiesis and only these organs will maintain erythroid, myeloid, and lymphoid hematopoiesis throughout the life span of the fish (Jin, Xu, & Wen, 2007; Kissa et al., 2008; Murayama et al., 2006; Willett et al., 1999).

In the VDA, HSCs are shown to originate from hemogenic endothelial cells via a developmental process termed endothelial hematopoietic transition (Bertrand et al., 2010; Kissa & Herbomel, 2010). The hemogenic cells are bipotential precursors that can differentiate into both hematopoietic and endothelial cells (Vogeli, Jin, Martin, & Stainier, 2006). These HSCs undergo limited divisions to either maintain the stem cell pool throughout the life of the host or give rise to multipotent and lineage-committed hematopoietic progenitor cells (HSCs) that generate all mature blood cell lineages (Takizawa, Boettcher, & Manz, 2012). Macrophages originating from the primitive and the intermediate wave play a decisive role in the expansion and specification of definitive HSCs. They colonize the AGM during the HSCs emergence stage, start patrolling between the dorsal aorta and the posterior caudal vein, and intimately interact with the HSCs. Genetic or chemical depletion of macrophages derived from the nondefinitive waves impairs the accumulation of the definitive HSCs in the AGM and their colonization of the CHT (Travnickova et al., 2015). Furthermore, it has been shown that the mobilization of HSCs and the intravasation and colonization of tissues are dependent on the function of matrix metalloproteinases (MMPs), in particular Mmp9, which can be produced by myeloid and surrounding tissue cells (Travnickova et al., 2015). Mmp9 is known as a strongly inducible component of the proinflammatory response to infections, facilitating leukocyte migration and cytokine processing (Stockhammer, Zakrzewska, Hegedus, Spaink, & Meijer, 2009; Van Lint & Libert, 2007; Volkman et al., 2010).

Therefore, the role of Mmp9 in HSC mobilization is likely to be significant also under conditions of infection, which demands enhanced hematopoiesis.

2.3 Functional Diversification of Myeloid Subtypes

It is not precisely known to what extent the zebrafish macrophages or neutrophils generated by primitive, intermediate, or definitive hematopoiesis have different functional competencies when dealing with infections. It is clear, however, that zebrafish embryos are less competent to combat infections at 1 dpf than at later stages, which likely can be attributed for a major part to the fact that neutrophils are still undergoing differentiation between 1 and 2 dpf (Fig. 1) (Clatworthy et al., 2009). Indeed, these early neutrophils have been shown to phagocytose less well than neutrophils at later developmental stages (Le Guyader et al., 2008). Nevertheless, zebrafish embryos infected at 1 dpf are already capable of inducing a robust innate immune response with expression of genes for cytokines, complement factors, proteases, and other mediators of pathogen defense (Stockhammer et al., 2009; van der Vaart, Spaink, & Meijer, 2012).

A pioneering study using zebrafish showed, for the first time in a living vertebrate, that macrophages undergo polarization to develop into functional M1- (classically activated) and M2-like (alternatively activated) subtypes (Nguyen-Chi et al., 2015). M1 macrophages promote inflammation, while M2 macrophages are involved in the resolution of inflammation and wound healing. Therefore, in many diseases, the persistence of M1 macrophages signifies an inflammatory state that can promote a range of negative outcomes, including inflammatory disorders (Mills, 2012). On the other hand, tumor-associated macrophages often display an M2 phenotype linked with properties that stimulate tumor growth, angiogenesis, tissue invasion, and metastasis (Noy & Pollard, 2014). Nguyen-Chi et al. used live imaging of a zebrafish fluorescent reporter line for tumor necrosis factor alpha (Tnfα), a distinctive proinflammatory marker for M1 macrophages. They showed that a subset of macrophages start to express the *tnfa* reporter in response to wounding, or in response to a tissue infection with *Escherichia coli*. Moreover, these *tnfa*-positive macrophages revert back to an M2-like phenotype when the inflammation is resolving (Nguyen-Chi et al., 2015). By separating *tnfa*-expressing and *tnfa*-negative macrophages using fluorescent cell sorting, it was found that *tnfa*-positive cells express other typical M1 markers, such as interleukin 1β and 6 (*il1b* and *il6*), while negative cells express M2 markers, such as tumor growth factor β (*tgfb*), CC-motif chemokine receptor 2 (*ccr2*), and CXC-motif chemokine receptor 4b (*cxcr4b*).

Macrophage activation has also been demonstrated using a fluorescent reporter fish line (Table 2) for immunoresponsive gene 1 (*irg1*), which is strongly induced by injection of bacterial lipopolysaccharide (LPS) (Sanderson et al., 2015). Arginase-2 (*arg2*) is considered to be a reliable M2 marker for teleost fish and a reporter line for this gene would thus be a valuable addition to further study M1/M2 polarization in zebrafish (Wiegertjes, Wentzel, Spaink, Elks, & Fink, 2016).

There is increasing interest also in neutrophil subtypes, which by analogy with macrophage subtypes are referred to as N1 and N2 (Mantovani, 2009). With new transgenic lines being generated by several labs (Table 2), zebrafish embryos and larvae provide a unique opportunity to carry out live imaging of such possible neutrophil polarization and of neutrophil-specific defense mechanisms, like the formation of neutrophil extracellular traps (NETs) (Palic, Andreasen, Ostojic, Tell, & Roth, 2007). The release of NETs coincides with a specific type of neutrophil cell death, named NETosis, resulting in an extracellular network of chromatin and granular proteins that can entrap and kill microbes. Besides this direct antimicrobial function, NETosis is thought to deliver danger signals that alert the innate immune system, and, if not properly controlled, NETosis may contribute to inflammatory and autoimmune diseases (Brinkmann & Zychlinsky, 2012). A newly established zebrafish notochord infection model is very useful to address neutrophil-specific defenses (Nguyen-Chi et al., 2014). The notochord is the developmental precursor of the vertebral column and this structure is inaccessible to phagocytes. However, injection of *E. coli* bacteria into this tissue induces massive macrophage and neutrophil accumulation in the surrounding area. The accumulating neutrophils are polarized to express high levels of *il1b* and a significant proportion of them show release of the Mpx-containing granules. This response results in rapid elimination of the bacterial infection, but the inflammatory reaction is persistent and has long-term consequences leading to notochord damage and vertebral column malformations (Nguyen-Chi et al., 2014). This study provided the first in vivo evidence that neutrophils can degranulate without making direct contact with a pathogen. Furthermore, the zebrafish notochord model developed in this study provides a new tool to study human inflammatory and infectious diseases of cartilage and bone, such as osteomyelitis and septic arthritis.

3. GENETIC CONTROL AND EXPERIMENTAL MANIPULATION OF THE ZEBRAFISH INNATE IMMUNE SYSTEM

3.1 Development and Differentiation of Innate Immune Cells

Primitive myelopoiesis in zebrafish is genetically controlled by two parallel pathways, the *cloche–estrp–scl* pathway and the *bmp/alk8* pathway (Hogan et al., 2006; Liao et al., 1998). Cloche is required very early for development of normal hemangioblasts as *cloche* mutants have defects in both endothelial and hematopoietic (erythroid and myeloid) lineages (Stainier, Weinstein, Detrich, Zon, & Fishman, 1995). The *estrp* and *scl* genes act downstream of *cloche* to regulate hematopoietic and endothelial development (Liao et al., 1998; Liu & Patient, 2008; Sumanas et al., 2008; Sumanas & Lin, 2006). The Bmp receptor Alk8 specifically regulates primitive myelopoiesis in the RBI but is not required for erythropoiesis. In agreement with an instructive role of the *bmp/alk8* pathway in myelopoiesis, the expression of *pu.1* is lost in the absence of *alk8*, while constitutively expressed *alk8* can increase *pu.1* expression (Hogan et al., 2006). The differentiation of EMPs is controlled by the orchestrated expression of transcription factors, where Pu.1 is the master regulator of the myelopoiesis and Gata1 is the key regulator of the erythroid cell lineage. Pu.1 and Gata1 negatively regulate each other and an interplay between these two transcription factors determines myeloid vs erythroid cell fate (Fig. 1) (Galloway, Wingert, Thisse, Thisse, & Zon, 2005; Rhodes et al., 2005).

Myeloid progenitors need additional factors to differentiate into any of the innate immune cell-type populations. Some of these factors are required for pan-myeloid development, while some are required for a specific lineage development. The *spi1l* gene encodes an ETS transcription factor, closely related to Pu.1. It functions downstream of Pu.1 and promotes myeloid development (Bukrinsky, Griffin, Zhao, Lin, & Banerjee, 2009). Extrinsic factors like granulocyte-colony-stimulating factor (Gcsf) also play a critical role in myeloid cell development (Liongue et al., 2009). Pu.1, Runx1, and Irf8 are important for the cell fate determination between macrophages and neutrophils. High levels of Pu.1 promote macrophage fate, whereas low levels promote neutrophil fate during primitive myelopoiesis (Jin et al., 2012; Su et al., 2007). Increased levels of Runx1 promote the expansion of the neutrophil population, whereas low levels of Runx1 result in more macrophages at the expense of the neutrophil progeny (Jin et al., 2012). In contrast to Runx1, Irf8 is necessary for macrophage fate determination. Suppressing *irf8* leads to reduced macrophage and increased neutrophil numbers, while increased *irf8* expression has the opposite effect (Li, Jin, Xu, Shi, & Wen, 2011). The regulation of mast cell fate is less well understood, but it has recently been shown to be influenced by Gata2, which functions downstream of the Notch pathway. Pu.1 is also required for mast cell development, independent from Gata2 and the Notch pathway (Da'as et al., 2012). As discussed later, the knowledge of the genetic pathways that control myeloid development can be exploited in infection studies to determine the specific roles of macrophages and neutrophils in host defense and pathology.

3.2 Genetic and Chemical Approaches to Manipulating the Zebrafish Innate Immune System

The transcription factor, Pu.1, is essential for development of both macrophages and neutrophils. A low dose of a *pu.1* morpholino can block macrophage development up to 3 dpf and can also block neutrophil development when injected at a higher dose (Su et al., 2007). *pu.1* morphants are more susceptible to various pathogens such as *Mycobacterium marinum*, *Salmonella enterica* Typhimurium, *S. aureus*, and Chikungunya virus (CHIKV), indicating that macrophages are essential for defense against these. Additionally, similar experiments demonstrated that macrophages are critical vectors for dissemination of *M. marinum* (Clay et al., 2007; Palha et al., 2013; Prajsnar et al., 2012; van der Vaart et al., 2012).

Not only macrophages but also neutrophils are critical for the defense against *M. marinum*, which has been shown using a transgenic zebrafish line, which mimics the WHIM (Warts, Hypogammaglobulinemia, Immunodeficiency, and Myelokathexis) syndrome. In the WHIM zebrafish line, the neutrophil-specific *mpx* promoter is used to overexpress a constitutively active form of *cxcr4b*, which is an important retention factor for myeloid progenitors that permits their maintenance in the hematopoietic tissues. As a result, mature neutrophils are retained in the hematopoietic tissues that express Cxcl12a, the chemotactic ligand of Cxcr4b. Thus, neutrophils are unable to reach the tissue infection sites, resulting in increased growth of *M. marinum* (Yang et al., 2012). However, neutrophils cannot control *M. marinum* infection in the absence of macrophages, as shown by using *irf8* morpholino to expand neutrophils at the expense of macrophages (Elks, Renshaw, Meijer, Walmsley, & van Eeden, 2015; Pagan et al., 2015).

In contrast, the essential role for neutrophils in controlling viral infection was shown by knockdown of *csf3r* (*gcsfr*), which mostly depletes the neutrophil population and renders embryos more susceptible to CHIKV infection (Palha et al., 2013). The selective depletion of neutrophils can also be achieved with *cebp1* morpholino, an approach used in a study demonstrating the importance of neutrophils as a source for inflammatory cytokines promoting hematopoiesis (He et al., 2015).

Alternative to examples of genetic manipulation of macrophage/neutrophil ratios, transgenic drug-inducible cell ablation systems have been applied in zebrafish infection studies. For example, selective ablation of macrophages demonstrated that these cells are less important than neutrophils in defense against CHIKV (Palha et al., 2013). The same approach showed that both macrophages and neutrophils are required for defense against *S. aureus*, but that neutrophils also function as a potential reservoir where the pathogen finds a protected niche that enables it to subsequently cause a disseminated and fatal infection (Prajsnar et al., 2012). Finally, macrophages have been selectively depleted using clodronate-containing liposomes, showing their essential role in control of *Mycobacterium abscessus* and *Cryptococcus neoformans* infections (Bernut et al., 2014; Bojarczuk et al., 2016). Together, these examples demonstrate the advantage of zebrafish infection models for in vivo dissection of innate immune cell functions, due to the ease of genetic and chemical manipulation of macrophage vs neutrophil ratios in this model.

4. PATHWAYS REQUIRED FOR PATHOGEN RECOGNITION AND ACTIVATION OF THE INNATE IMMUNE RESPONSE

Cells composing the innate immune system can recognize invading microbes by expressing a series of pattern recognition receptors (PRRs). PRRs were evolved to sense and respond to recurrent molecular patterns that are found in microbes (e.g., LPS, peptidoglycan, lipoprotein, flagellin, exogenous nucleic acids) or that are derived from the host as a consequence of the infection (e.g., heat-shock proteins and aberrantly processed, exposed, or localized cell components). These signals are collectively referred to as pathogen-/damage-associated molecular patterns (P/DAMPs) (Akira, Uematsu, & Takeuchi, 2006). PRRs belong to different families, which comprise membrane proteins on the cell surface or endosomal compartments, cytosolic proteins as well as secreted proteins. PPRs are essential not only for innate immune responses but also for the activation of adaptive immunity, and defects or polymorphisms in these receptors have been linked to numerous immune-related diseases in human (Caruso, Warner, Inohara, & Nunez, 2014; Netea, Wijmenga, & O'Neill, 2012). The major families of PRRs are well conserved between mammals and zebrafish. However, as reviewed later, the current knowledge of PRRs and downstream signaling in zebrafish is still relatively limited.

4.1 Families of PRRs

4.1.1 Scavenger Receptors

Scavenger receptors represent a heterogeneous group of surface PRR receptors, able to recognize a broad spectrum of molecules from bacterial/fungal wall, viral capsid parasite glycocalyx, as well as host-derived ligands. The interaction of these receptors with their ligands can directly mediate phagocytosis of the pathogen or can contribute as costimulatory signal for the activation of downstream signaling pathways, such as cytokine responses mediated by NFκB signaling (Bowdish et al., 2009). The zebrafish homologs of human macrophage receptor with collagen structure (Marco) and Cd36 were recently characterized (Benard et al., 2014; Fink et al., 2015). Marco expression by macrophages is important for rapid phagocytosis of *M. marinum* and mediates an initial transient proinflammatory response to this pathogen (Benard et al., 2014). Consequently, knockdown of this receptor impairs bacterial growth control. Although not highly expressed by macrophage and neutrophils, also the knockdown of Cd36 in zebrafish larvae led to higher bacterial burden upon *M. marinum* infection (Fink et al., 2015).

4.1.2 C-Type Lectin Receptors

The mammalian C-type lectin receptors (CLRs) include cell surface as well as secreted proteins (collectins) that are able to bind to different surface carbohydrate moieties from viruses, bacteria, fungi, or eukaryotic parasites, and similar to scavenger receptors, they can guide phagocytosis of nonopsonized bacteria, and their destruction in acidified phagolysosomes. Several homologs of CLRs have been detected in zebrafish, but a real functional characterization of this class of receptors in zebrafish is still missing. Only recently the zebrafish mannose receptor was cloned and found to be highly induced upon infection with *Aeromonas sobria* (Zheng et al., 2015). In addition to this cell surface receptor for mannose-rich glycans, mannose recognition is also mediated extracellularly by the mannose-binding lectin (MBL).

Zebrafish embryos express a homolog of mammalian MBL and this molecule can opsonize both Gram-negative and Gram-positive bacteria, promoting their phagocytosis by macrophages, like its mammalian counterpart (Yang, Bu, Sun, Hu, & Zhang, 2014). Neutralization of this molecule could also increase mortality of embryos infected with *Aeromonas hydrophila*, while the injection of the recombinant protein promotes resistance to this pathogen. This study also suggests that the lectin pathway may be already functional in the early embryos in zebrafish before their cell-mediated innate immunity is fully matured, and largely contributes to the protection of the developing embryos.

4.1.3 Toll-Like Receptors

Toll-like receptors (TLRs) are a family of PRRs located on the plasma membrane or on the endosome/phagosome membranes that can sense a wide variety of PAMPs and DAMPs. Their extracellular ligand-binding domain contains conserved leucine-rich repeat motifs and their cytoplasmic signaling domain consists of a TIR (toll-interleukin-1 receptor) homology domain. TLRs are known to essentially signal as hetero- or homodimers, via coupling with downstream adaptor molecules (Akira et al., 2006). In mammals, five adaptors have been identified, namely MYD88 (myeloid differentiation factor 88), TIRAP, TRIF, TRAM, and SARM1 (Akira et al., 2006). Among these, MYD88 represents the most central mediator, since most of the TLRs rely heavily on MYD88 to activate their downstream signaling pathway. This consists mostly of modulation of gene expression via activation and translocation of transcription factors such as NFκB, ATFs, IRFs, AP-1, and STATs (Akira et al., 2006). Stimulation of these factors triggers profound modification of gene expression, especially upregulation of an array of proinflammatory effector molecules, including cytokines, chemokines, antimicrobials, and activators of adaptive immunity (Kanwal, Wiegertjes, Veneman, Meijer, & Spaink, 2014).

Orthologs of TLR1–2–3–4–5–7–8–9 and of their adaptor intermediates (Myd88, Tirap, Trif, and Sarm1) and other downstream signaling intermediates (e.g., Traf6) have been identified and studied in zebrafish too (Kanwal et al., 2014). However, for some of them it is still unclear what ligands they respond to. The zebrafish Tlr2–3–5–9 maintain ligand specificity consistent with their mammalian counterparts, yet the closest orthologs to mammalian TLR4 in zebrafish are unable to respond to LPS, its ligand in mammals (Kanwal et al., 2014). Several functional and fish-specific Tlrs also exist, such as Tlr21 and Tlr22, which can respond to dsRNA and CpG-oligodeoxynucleotides, respectively (Kanwal et al., 2014). Another fish-specific Tlr cluster is represented by Tlr20, which phylogenetically seems related to mammalian Tlr11–12 (Kanwal et al., 2014). In agreement with studies in mammalian models, transcriptional analysis of the responses to bacterial infections has demonstrated that activation of downstream transcription factors and proinflammatory immune response genes is largely dependent on the function of the Myd88, which serves as an adaptor in both Tlr and interleukin 1 receptor signaling (Gay, Gangloff, & O'Neill, 2011; van der Vaart, van Soest, Spaink, & Meijer, 2013).

A reporter zebrafish line (Table 2) containing promoter elements of the zebrafish *myd88* gene (Hall et al., 2009) has helped to define that the innate immune cells have the highest potential for MyD88-dependent/TLR-mediated signaling. *Myd88*:GFP-labeled cells include a set of myeloid leukocytes, which not only are highly responsive to wounding and infections but also express a full battery of Tlrs and other Tlr-downstream adaptors together with *myd88*.

Application of the zebrafish model has recently also contributed to define common and specific downstream signaling targets controlled by several Tlrs. While a large part of well-defined inflammatory markers such as *il1b*, *tnfa*, *mmp9*, and *Cxcl18b/Cxcl-c1c* were inducible by either Tlr2 or Tlr5 stimulation at a similar extents, other infection-responsive genes, especially transcription factors (e.g., *fosb*, *egr3*, *cebpb*, *hnf4a*) but also some effector molecules, including *il6* and *il10* were found to rely more heavily on one or the other signaling system. Comparative studies of Tlr signaling in zebrafish with other teleost and mammalian species have been more comprehensively reviewed in Kanwal et al. (2014), and these studies, in summary, demonstrate how zebrafish genetics can be used to dissect the specific molecules that contribute to a robust immune response.

4.1.4 Nod-Like Receptors

Differently from scavenger receptors and TLRs, nucleotide-binding-oligomerization-domain (NOD)-like receptors (NLRs) are soluble receptors and can detect PAMPs and DAMPs in the cytosol, such as those deriving from pathogens escaping from phagosomes (Akira et al., 2006). NOD1 and NOD2 have been implicated in the recognition of bacterial cell wall, although several studies suggest a broader range of ligands for these NLRs, since they seemed implicated also into recognition of intracellular eukaryotic parasites (Silva et al., 2010). Other NLR include IPAF, NALP1, and NALP3, which can assemble in the inflammasome, a cytosolic multicomponent complex, which is involved in the activation of procaspase 1 to caspase 1 (Martinon, Burns, & Tschopp, 2002). The active form of caspase 1, in turn, can process pro-IL1β and pro-IL18 into IL1β and IL18 (Martinon et al., 2002). Most of NLRs are conserved in zebrafish in addition to another large teleost-specific subfamily of NLRs (Stein, Caccamo, Laird, & Leptin, 2007). The functional conservation of NOD1–2 was demonstrated by depletion of these genes during *S. enterica* Typhimurium infection, which resulted in increased burden and decreased host survival (Oehlers et al., 2011). Investigation of the NLR-dependent inflammasome activation and Il1β processing still requires a more detailed characterization in this species (Ogryzko, Renshaw, & Wilson, 2014; Varela et al., 2014).

4.1.5 RIG-I-Like Receptors

RIG-I-like receptors (RLRs) are another family of cytosolic PRRs that activate the inflammasome (Kell & Gale, 2015). RLRs can detect the presence of RNA from a broad range of viruses. The downstream signaling cascade is cooperative with Tlr signaling and induces activation of transcription factors like IRF3, IRF7, and NFκB, leading to high production of interferons (IFN) and interferon-stimulated genes (ISGs) (Kell & Gale, 2015). Both type I and type II IFNs exist in zebrafish, and like in humans, these molecules are key for the antiviral response. However, direct homologies with the mammalian systems cannot be definitively traced. Zebrafish Ifnγ1 and Ifnγ2 are the type II homologs, while Ifnφ1 and Ifnφ2, members of a large Ifnφ family in zebrafish, represent a fish-specific type of IFNs that more closely resemble the mammalian type I IFN molecules (Aggad et al., 2009; Langevin et al., 2013). The zebrafish homologs for RIG-I and other members of RLRs are predicted in the zebrafish genome, but functional characterization in zebrafish is still incomplete. The RLRs were shown to be involved in IFN gene induction in zebrafish by overexpression of the key RLR-adaptor IPS-1/MAVS. This led to massive induction of ISGs, similar to what was found in mammalian models (Biacchesi et al., 2009). This role in IFN induction places RLRs as a central factor in containing viral infections. Studies in zebrafish suggest that they might also have a significant function in defense against bacterial infections (Zou, Wen, Yang, & Wei, 2013).

4.2 Inflammatory Signaling Initiated by PRRs

The downstream mediators activated by most PRR signaling include pro- and antiinflammatory protein and lipid molecules secreted at the infection site. Cytokines are small secreted proteins exerting central modulatory activities in both adaptive and innate immunity. This heterogeneous group of peptides includes TNF, interleukins, and chemokines (CCLs, CXCLs, CX3CLs, and XCLs). All these classes exist in zebrafish and other teleosts. However, expansions and diversifications have occurred (Nomiyama et al., 2008).

Similar to mammalian models, a large number of these mediators are transcriptionally modulated by infection with different pathogens (Stockhammer et al., 2009; Veneman et al., 2013) or cleaved to their mature/active form. In zebrafish, functional similarities are proven for the Tnf, Il1β, Il8/Cxcl8, Cxcl11, Il6, and Il10 (Roca & Ramakrishnan, 2013). Knockdowns or full knockouts of several of these molecules or their cognate receptors led to significant aberrancies in the containment of infections (Roca & Ramakrishnan, 2013). For example, knockdown of the Tnfa receptor *tnfrsf1a* in mycobacterial infection revealed a key function of this axis to control the host inflammatory status (Roca & Ramakrishnan, 2013). The chemokines Il8/Cxcl8 and Cxcl11, like in mammalian species, were found to recruit neutrophils (via Cxcr2) and macrophages (via Cxcr3.2), respectively, and impacted on the mobilization and response of phagocytes to infection.

The mechanisms for lipid inflammatory/antiinflammatory mediators, including prostaglandins, leukotrienes, and lypoxins, are highly conserved from zebrafish to human. The importance and functional conservation of these molecules are exemplified by the results of a zebrafish genetic screen for genes causing hypersusceptibility to *M. marinum,* which uncovered the gene encoding Lta4h (leukotriene A4 hydrolase) (Tobin et al., 2010). Lta4h catalyzes the final step of synthesis of the lipid mediator leukotriene B4 (LTB4) and its deficiency in zebrafish impairs the balance between antiinflammatory and proinflammatory lipid mediators (Tobin et al., 2010). Similarly, polymorphisms in the human *LTA4H* locus have been reported to associate with susceptibility to *M. tuberculosis* (Tobin et al., 2010). LTB4 synergizes with Tnfα in order to maintain a balanced level of inflammation. Via its cognate receptor (Tnfr), Tnfα mediates activation of Rip1/2 kinases and release of reactive oxygen species (ROS) by increasing mitochondrion permeability (Roca & Ramakrishnan, 2013). ROS act as a double-edged sword, by both exerting a microbicidal function and mediating activation of necroptosis of the host cell. Therefore, impaired (too high or too low) inflammatory statuses lead to increased susceptibility to mycobacterial infection in zebrafish (Roca & Ramakrishnan, 2013). A tight control of the inflammatory status is critically important also in human TB and other infectious diseases (Dorhoi & Kaufmann, 2014).

4.3 Complement System

In addition to the PRR-mediated cellular responses of the innate immune system, zebrafish embryos highly upregulate components of the complement system upon challenge with a variety of pathogens, indicating that soluble complement factors and complement receptors may be critical for opsonization, recognition, and lysis of pathogens in this developmental window. In early zebrafish embryos, extracellular *S. enterica* Typhimurium LPS mutant and heat-killed bacteria are rapidly lysed, a phenomenon that was suggested to be complement mediated, since LPS mutants were found to be highly susceptible to complement killing in other models (van der Sar et al., 2003). Bacteriolytic mechanisms ascribed to complement are also proposed to contribute to the antibacterial activity in zebrafish egg cytosol (Wang & Zhang, 2010). Mostly complement components are known to derive from the liver. However, complement components are infection inducible in the early embryos long before hepatic development (Wang, Zhang, & Wang, 2008). In line with these observations, we have found by transcriptional profiling of sorted phagocytes during infections that these cells can be a relevant source of extrahepatic production of complement components (unpublished results). Additionally, many of the complement factors in zebrafish can be transferred from mothers to eggs at either protein or mRNA level (Hu, Pan, Xiang, & Shao, 2010). Maternal immunization with *A. hydrophila* also resulted in increased protein transfer of complement factors to their offspring (Wang, Zhang, Tong, Li, & Wang, 2009) and contributed to immunoprotection of the early embryo against this pathogen (Wang et al., 2008).

5. EFFECTS OF COMMENSAL MICROBES ON DEVELOPMENT OF THE IMMUNE SYSTEM

The impact of the gut microbiota on development of the mammalian immune system is well known (Kaplan, Shi, & Walker, 2011). Following a large body of work in rodents, methods for growing zebrafish in a germ-free environment or in the presence of defined microbial communities (gnotobiotic) are now well established (Pham, Kanther, Semova, & Rawls, 2008). Comparison of studies in germ-free and gnotobiotic zebrafish and rodent models has revealed strong similarities among vertebrates in how microbes shape the development of the gut epithelium and the mucosal immune system and influence the expression of genes involved in processes such as cell proliferation, metabolism, and inflammation (Cheesman & Guillemin, 2007; Rawls, Samuel, & Gordon, 2004).

Inside the chorion, the zebrafish embryo develops in an axenic environment, but the intestine of larvae hatching around 3 dpf is rapidly colonized by microbes (Kanther & Rawls, 2010). Zebrafish larvae reared in germ-free water were shown to express lower levels of the proinflammatory cytokine gene *il1b* compared to larvae reared under conventional conditions (Galindo-Villegas, Garcia-Moreno, de Oliveira, Meseguer, & Mulero, 2012). This microbiota-induced *il1b* expression is mediated by the TLR/MyD88 signaling pathway described in Section 4 (Galindo-Villegas et al., 2012). This microbial recognition pathway can also be activated before hatching under conditions of experimental infection with bacterial pathogens (van der Vaart et al., 2013). Microbial colonization leads to activation of a reporter for NFκB (Table 2), a master transcriptional regulator of the immune response downstream of Tlr/Myd88 signaling (Kanther et al., 2011). Furthermore, the presence of a microbiota has been shown to result in increased numbers of neutrophils and systemic alterations in neutrophil localization and migratory behavior, which were found to be dependent on the microbiota-induced acute-phase protein serum amyloid A (Kanther et al., 2014). In another study, commensal microbes were not found to promote a higher rate of myelopoiesis, but did affect neutrophil activity in response to injury (Galindo-Villegas et al., 2012). In addition, this study showed that the presence of commensal microbes primes the innate immune system of zebrafish larvae resulting in an increased resistance to experimental infections.

Independent from the effect of commensal microbes, the expression of proinflammatory genes appears to be controlled by epigenetic mechanisms that likely serve to protect of zebrafish larvae against infectious agents before adaptive immunity has developed and prevent pathologies associated with excessive inflammation during development (Galindo-Villegas et al., 2012). This is corroborated by a recent study, showing that mutation in the epigenetic regulator *uhrf1* leads to a strong induction of the proinflammatory cytokine gene *tnfa* in zebrafish larvae (Marjoram et al., 2015). The *tnfa* induction in these *uhrf1* mutants is associated with severe damage of the intestinal epithelium and infiltration by neutrophils, mimicking the chronic inflammation seen in inflammatory bowel diseases (IBDs). The development of zebrafish models for IBDs provides new avenues to study the factors that contribute to the onset of these complex multifactorial diseases where inappropriate responses of the immune system to the intestinal microbiota are thought to play a major role (Marjoram & Bagnat, 2015).

6. ADAPTATION TO INFECTION AND INFLAMMATION

In response to infection or inflammation, the hematopoietic system can mount an adaptive response that is known as demand-driven hematopoiesis or emergency hematopoiesis (Takizawa et al., 2012). This response serves in the first place to replenish neutrophils, which due to their short life span are rapidly consumed during infections. Both the expansion of HSCs and the skewing of myeloid cell specification into the direction of granulopoiesis play a role in demand-driven adjustments of hematopoiesis in zebrafish larvae (Hall, Crosier, & Crosier, 2016; Hall et al., 2012; Herbomel, 2012).

That zebrafish embryos can mount an emergency granulopoietic response was first recognized in the study, showing that intravenous administration of LPS at 2 dpf led to a Gcsf/Gcsfr-dependent increase in the numbers of neutrophils within 8 h (Liongue et al., 2009). A recent report shows that phagocyte numbers can be modulated by immune stimulation even at an earlier stage. In this case, a host defense peptide, chicken cathelicidin-2, was injected into the yolk of embryos shortly after fertilization, resulting in a 30% increase of *lcp1*-positive cells at 2 dpf and an increased resistance of embryos to bacterial infection (Schneider et al., 2016). We review recent work in zebrafish that has brought new insights into the molecular pathway underlying emergency hematopoiesis and has revealed roles for several proinflammatory mediators as well as Tlr signaling in hematopoiesis.

6.1 Molecular Mediators of Emergency Granulopoiesis

Embryos infected with *S. enterica* Typhimurium into the hindbrain at 2 dpf develop neutropenia within 1 day and counter this within 2 days by emergency granulopoiesis throughout the VDA/AGM and CHT regions (Hall et al., 2012). While this Gcsf/Gcsfr-dependent response is at the expense of lymphoid progenitors, it is not only due to an increased commitment of HSCs to myeloid rather than lymphoid fate but also due to an increase in the number of Gcsfr-expressing HSCs (Hall et al., 2012). The zebrafish ortholog of CCAAT-enhancer binding protein (Cebpb), a well-known transcriptional regulator of emergency granulopoiesis in mammals, is required for the expansion of the HSC compartment (Hall et al., 2012). Importantly, the study in zebrafish revealed that inducible nitric oxide synthase (iNOS, Nos2a) functions downstream of Cebpb in the emergency granulopoiesis pathway (Hall et al., 2012). Knockdown of *nos2a* to block the infection-induced expansion of neutrophils was subsequently shown to be associated with increased viral replication and mortality of embryos during CHIKV infection (Palha et al., 2013). It is currently unknown if the role of NO in emergency hematopoiesis is conserved across species, but this is to be expected in view of the fact that NO is involved steady state hematopoiesis in both zebrafish and mouse (Hall et al., 2012). The newly discovered roles of Cebpb and NO therefore prompt further investigations into the possibilities of therapeutic targeting of these factors in human disease (Hall et al., 2012).

Through work in zebrafish, a highly conserved myeloid-specific microRNA, *miR-142a-3p*, has recently been linked with Gcsf/Gcsfr and NO-dependent signaling (Lu et al., 2013). Depletion of *miR-142a-3p* was found to reduce the numbers of HSCs in the VDA/AGM and CHT, associated with reduced expression of *gcsfr* as well as decreased production of NO (Lu et al., 2013). The inflammatory transcription factor interferon regulatory factor 7 (Irf7) is a potential target of this microRNA, suggesting that this pathway might also be relevant not only to steady state but also to infection-induced hematopoiesis. When *irf7* and *miR-142a-3p* were depleted simultaneously *gcsfr* expression and NO production could be restored, suggesting that Irf7 acts as a repressor of Gcsfr/NO signaling and that in turn *miR-142a-3p* can repress Irf7 function to promote HSC development (Lu et al., 2013). This mechanism is conserved in mouse and therefore also of potential interest for therapeutic targeting (Lu et al., 2013).

6.2 Implication of Cytokines and IFNs in Hematopoiesis

Macrophages are thought to be the source of Gcsf that promotes emergency granulopoiesis (Hall et al., 2012). It has recently been shown that also several proinflammatory cytokines that are produced by macrophages and neutrophils can influence the production of HSCs in the embryo. These cytokines include Tnfα, Ifng1–1, Ifng1–2, and Il1β (Espin-Palazon et al., 2014; He et al., 2015; Li et al., 2014). Tnfα in zebrafish is encoded by two genes, *tnfa* and *tnfb*, and the expression of both genes is inducible by infections (van der Vaart et al., 2013). Tnfα is expressed as a transmembrane protein functional on the cell surface and signals through two receptors, Tnfr1 (Tnfrsf1a) and Tnfr2 (Tnfrsf1b). Signaling through Tnfr1 is important for resistance to mycobacterial infection as it prevents necrosis of infected macrophages (Clay, Volkman, & Ramakrishnan, 2008), whereas Tnfr2 is the receptor that has been implicated in hematopoiesis (Espin-Palazon et al., 2014). Primitive neutrophils were found to be the primary source of Tnfα, which was found to promote the specification and emergence of HSCs through Tnfr2 and the Notch and NFκB signaling pathways (Espin-Palazon et al., 2014).

Similar to Tnfα, interferon gamma (IFNγ) is another important activator of macrophages that has been implicated in hematopoiesis. Overexpression of a zebrafish homolog of IFNγ, Ifngl–2, increases HSC counts in embryos with an intact Notch signaling pathway (Sawamiphak, Kontarakis, & Stainier, 2014). Ifngl–2 specifically controls the endothelial-to-HSC transition by activating signal transducer and activator of transcription 3 (Stat3) (Sawamiphak et al., 2014). In agreement, knockdown of interferon regulatory factor 2 (Irf2) increases HSC production in zebrafish (Li et al., 2014). The other zebrafish homolog of IFNγ, Ifngl–1, and fish-specific type I interferons (*ifnphi1–2–3* and *4*) also contributes to HSC development and expansion (Li et al., 2014). Thus, Ifns are key regulators of HSC behavior and this suggests that HSCs are a prime response to an infection that stimulates Ifns.

6.3 Role of Tlr Signaling in Hematopoiesis

The primary pathway of pathogen recognition, namely Tlr4–MyD88–NFκB signaling, has recently been linked to HSC development (He et al., 2015). Expression of *runx1* in the VDA/AGM at 1 dpf and *cmyb* in the CHT at 2 dpf is significantly reduced in *tlr4bb*- or *myd88*-deficient embryos when compared to controls (He et al., 2015). However, *myd88* mutant larvae at 3 dpf show no significant alterations in macrophage or neutrophil numbers (van der Vaart et al., 2013), suggesting that the defect in HSC development is compensated for by Myd88-independent mechanisms. Embryos deficient in *tlr4bb* or *myd88* show a reduction in the expression of Notch target genes, and overexpression of the intracellular domain of Notch in endothelial cells can rescue *runx1* expression in *tlr4bb* and *myd88* morphants (He et al., 2015). As discussed earlier, Notch signaling can regulate NFκB, and therefore, it is likely that the Tlr4–MyD88–NFκB and Notch–NFκB signaling routes function cooperatively in HSC development (Espin-Palazon et al., 2014; He et al., 2015). The HSC defect in *tlr4bb* and *myd88* morphants can also be rescued by overexpression of the gene for Il1β, adding also this cytokine to the list of proinflammatory mediators that modulate hematopoiesis and the production of innate immune cells (He et al., 2015). Studies in *tlr4*$^{-/-}$ knockout mice confirmed that TLR-mediated inflammatory signaling plays an evolutionary conserved role in HSC development (He et al., 2015). In conclusion, a number of recent studies in zebrafish and mouse models support a previously unrecognized link between inflammatory signaling and hematopoiesis that might be translated into new approaches for treatment of immune-related diseases or to improve the success of HSC transplantations (Espin-Palazon et al., 2014; He et al., 2015; Li et al., 2014; Sawamiphak et al., 2014).

7. THE INTERFACE OF IMMUNITY AND METABOLISM

During the first 5 days of development the zebrafish embryo/larva derives all its nutrients from the yolk and it has to adapt its metabolism to switch to external feeding when yolk proteins become limiting. How this metabolic adaptation might affect the immune system is currently unknown and worthy of exploration, especially considering new links between immunity and metabolism that have recently been revealed in zebrafish (Hall et al., 2013; Marin-Juez, Jong-Raadsen, Yang, & Spaink, 2014; van der Vaart et al., 2013). The relevance of immunometabolism for human disease is emerging strongly from recent studies that have revealed extensive metabolic reprogramming of human macrophages and dendritic cells in response to environmental conditions and during activation of innate and adaptive immune responses (O'Neill & Pearce, 2016).

7.1 Lipid and Glucose Metabolism as Fuels for Fighting Infection

Fatty acid metabolism has been shown to fuel the production of mitochondrial ROS in zebrafish macrophages following infection of embryos with *S. enterica* Typhimurium (Hall et al., 2013). Immunoresponsive gene 1 (*irg1*), an infection-inducible and macrophage-specific gene encoding a homolog of bacterial 2-methylcitrate dehydratase, was found to be required for the utilization of fatty acids during this response, and knockdown of this gene increased the susceptibility to infection (Hall et al., 2013). This study showed that also murine macrophages require fatty acid β-oxidation for infection-induced mitochondrial ROS production and bactericidal activity. ROS production is also dependent on glucose metabolism and overproduction of ROS, which can have tissue-damaging effects, has been associated with diabetes (Coughlan & Sharma, 2016). Studies in a zebrafish model for hyperinsulinemia suggest that the metabolic switch between insulin-sensitive and insulin-resistant states is mediated by protein tyrosine phosphatase nonreceptor type 6 (Ptpn6), which is well known as a negative regulator of the innate immune response (Kanwal et al., 2013; Marin-Juez et al., 2014). The dual role of this phosphatase in the regulation of glucose metabolism and immunity is particularly interesting in the light of the emerging coepidemic of TB and diabetes (Pizzol et al., 2016). There are many ongoing efforts to develop zebrafish models for metabolic diseases, including diabetic complications, providing new opportunities to study the relation with infectious diseases (Schlegel & Gut, 2015).

7.2 Autophagy

The process of autophagy might be considered as the most important link between metabolism and immune function. Autophagy is a cellular process of self-degradation that functions to regulate energy metabolism and it can be activated by nutrient stress, such as the depletion of the yolk during zebrafish larval development (Varga, Fodor, & Vellai, 2015). During autophagy (or strictly macroautophagy), the cytosolic material is entrapped in double membrane structures (autophagosomes) and delivered to lysosomes for degradation. Autophagy has an important housekeeping function in removing and recycling aggregates of misfolded proteins and damaged organelles (Levine, Mizushima, & Virgin, 2011). The same machinery can also target intracellular microbes to lysosomal degradation and therefore several pathogens are thought to have evolved mechanisms to counteract the autophagic defenses (Huang & Brumell, 2014). Autophagy also controls inflammation, cytokine secretion, antigen presentation, and the regulation of innate and adaptive immune responses (Deretic, Saitoh, & Akira, 2013).

Some recent work showing that autophagy is important in infection has used knockdown of the autophagy receptor p62 (Sqstm1), which mediates selective autophagy of ubiquitinated cargo. Sqstm1 morphants have an impaired defense against *Shigella* and *Mycobacterium* infections (Mostowy et al., 2013; van der Vaart et al., 2014). These bacterial pathogens have the ability to escape from phagosomes into the cytosol, where they can be tagged by ubiquitin (Ub) ligation and subsequently targeted to autophagy by p62 (Huang & Brumell, 2014). The susceptibility of p62-deficient zebrafish larvae to these pathogens clearly shows that autophagy is an essential cellular process for effective immunity against some deadly bacteria. Similarly, many studies in human cells have shown increased replication of *M. tuberculosis* under conditions of autophagy inhibition. In contrast, loss of p62 and other essential autophagy genes did not correlate with susceptibility to *M. tuberculosis* in mice (Kimmey et al., 2015). This suggests that, in the context of full adaptive immunity, *M. tuberculosis* might be less subject to phagosomal escape and autophagic targeting and that this pathogen is capable of effectively inhibiting the antibacterial function of the autophagy process.

The microtubule-associated light chain 3 protein (Lc3) is widely used as a marker of autophagosomes and the generation of a zebrafish reporter line (Table 2) expressing a GFP-Lc3 fusion protein *Tg(CMV:GFP-Lc3)* allows to monitor the process of autophagy in vivo (He, Bartholomew, Zhou, & Klionsky, 2009). The zebrafish GFP-Lc3 reporter is activated by autophagy-inducing drugs (such as rapamycin), in different tissues of the developing embryo (e.g., the heart), and in response to infections with *Shigella* and *Mycobacterium* (He et al., 2009; Hosseini et al., 2014; Lee et al., 2014; Mostowy et al., 2013; van der Vaart et al., 2014). The autophagic morphology of *M. marinum*-containing GFP-Lc3-positive vesicles in zebrafish has been confirmed by correlative light and electron microscopy (Hosseini et al., 2014). Furthermore, small GFP-Lc3 vesicles are frequently seen to accumulate around mycobacterial aggregates in infected zebrafish hosts (Hosseini et al., 2014; van der Vaart et al., 2014). These autophagosomes might serve to deliver neo-antimicrobial peptides and enhance the bactericidal properties of the autolysosomal compartment (Ponpuak et al., 2010).

From studies in human and mammalian cells, autophagy is known to be induced downstream of pathogen recognition by TLR signaling (Deretic et al., 2013). The DNA damage-regulated autophagy modulator 1 (Dram1) was discovered in zebrafish as a novel mechanistic link between autophagy induction and the TLR/Il1R–MydD88–NFκB innate immune sensing pathway (van der Vaart et al., 2014). Dram1 overexpression in the zebrafish host promotes the formation of autophagosomes and the p62-dependent selective autophagy targeting of *M. marinum*. Although the molecular mechanism remains to be elucidated, this host protective role of Dram1 might be exploited as a therapeutic strategy for treatment of mycobacterial disease in humans (van der Vaart et al., 2014). In further support of autophagy modulation as a therapeutic approach, a clinically approved anticonvulsant drug, carbamazepine, was recently shown to trigger autophagy in zebrafish embryos and protect against *M. marinum* infection (Schiebler et al., 2015). This drug was also shown to be effective against *M. tuberculosis* within primary human macrophages and in a mouse model of TB. Therefore, despite recent findings that deficiency in essential autophagy genes did not correlate with *M. tuberculosis* deficiency in mice, pharmacological activation of autophagy still remains a promising therapeutic strategy to be further explored (Kimmey et al., 2015).

8. RECENT INSIGHTS FROM MODELING INFECTIOUS DISEASES IN DEVELOPING EMBRYOS AND LARVAE

8.1 Bacterial Infections

Zebrafish infection models have been established for a wide variety of bacterial pathogens that are the causative agents of human infectious diseases or opportunistic infections, including species of the Mycobacteria, *Listeria*, *Shigella*, *Salmonella*, Streptococci, *Burkholderia*, and other genera (Table 1). Since most of these models have been reviewed elsewhere (Cronan & Tobin, 2014; Meijer, 2016; Ramakrishnan, 2013; Saralahti & Ramet, 2015; Torraca, Masud, Spaink, & Meijer, 2014; Vergunst et al., 2010), we focus here on some examples of recent work showing how these models are contributing to a better understanding of macrophage and neutrophil functions in the containment or the promotion of specific disease features.

8.1.1 Listeria *and* Shigella *Infections*

Listeria monocytogenes and *Shigella flexneri* are two human pathogens that can cause serious gastrointestinal infections (food poisoning), especially in infants, the elderly, and immunocompromised patients. These bacteria share the capability to extensively manipulate the host cytoskeleton. Despite not being natural fish pathogens, these species were seen to escape into the cytosol after phagocytosis and to induce in the heterologous host the same cytoskeleton rearrangements, including actin tails and septin cages (Levraud et al., 2009; Mostowy et al., 2013). Mechanistically, *Shigella* and *Listeria* models in zebrafish mimic the main disease-causing feature of human shigellosis and listeriosis. *Shigella* bacteria are phagocytized by both neutrophils and macrophages, but while well contained by the first cell type, they rapidly induce cell death in the second. Both *Shigella* and *Listeria*, in human and in zebrafish tissue, can largely exploit host actin polymerization to be propelled from the infected cell and invade new cells. These findings emphasize how these mechanisms of pathogenicity are shared across distant bacterial species and across vertebrates.

Similar to *Shigella*, *Salmonella* (*S. enterica*) is an enterobacterial species that does not generally infect ectothermic animals. However, injection of *S. enterica* Typhimurium establishes severe infection in zebrafish, which cannot be contained in most of the infected embryos and goes together with profound transcriptional induction of inflammatory genes (cytokine storm), a response that is largely dependent on Myd88-dependent signaling and negatively regulated by Ptpn6 phosphatase (Kanwal et al., 2013; van der Vaart et al., 2013). Deficiency in either of these signaling factors is detrimental to the resistance of zebrafish embryos to *Salmonella* infection (Kanwal et al., 2013; van der Vaart et al., 2013), indicating that the inflammatory response is protective when properly controlled but leads to lethality when inhibitory mechanisms are lost.

8.1.2 Staphylococcal Infections

S. aureus causes a range of serious infections in human and mammalian models, including skin ulceration, osteomyelitis, pneumonia, and septicemia. Injections of large inoculums of this bacterium in zebrafish embryos also provoke septicemic death (Prajsnar et al., 2008, 2012). Histologically, in zebrafish like in mammals, *S. aureus* determines formation of necrotizing lesions, the abscesses. Interestingly, the zebrafish model revealed that the bacteria forming the individual abscesses derive from clonal expansion of persistent bacteria. While the vast majority of the injected Staphylococci are cleared by macrophages and neutrophils, some bacteria gain the capability to avoid intracellular killing and will secondarily expand, forming the localized lesion (Prajsnar et al., 2012). This study identified the neutrophils as the main niche necessary to establish this immunological bottleneck and to determine the emergence of clonal infection foci.

Experimental *S. aureus* infection in zebrafish has also been instrumental in a recent study that revealed an unexpected role of the Spaetzle-like nerve growth factor β (NGFβ) in pathogen-specific host immunity to Staphylococcal infection (Hepburn et al., 2014). Spaetzle is a key mediator of the immune response to Gram-positive bacteria in *Drosophila* and is required for the activation of the Toll signaling pathway (Lemaitre & Hoffmann, 2007) . While Spaetzle had always been thought to lack a vertebrate equivalent, chordate NGFβ in fact shares remarkable structure similarities to this mediator. It was found that Staphylococcal infection triggers release of NGFβ in human macrophages, a mechanism that depends on recognition of pathogen exoproducts and on activation of the immune response via NOD-like receptor signaling. Subsequently, knockdown in zebrafish of tropomyosin-related kinase receptor A, the corresponding receptor of NGFβ, was found to impair neutrophil recruitment and to increase susceptibility to *S. aureus* infection (Hepburn et al., 2014). This study supports an evolutionary conserved role for NGFβ acting as an alarm signal in the inflammatory response to *S. aureus* infection. Moreover, this work suggests that variation between individuals in secretion of NGFβ by macrophages might determine susceptibility to Staphylococcal disease.

8.1.3 Tuberculosis

One-third of the world population carries *M. tuberculosis* and more than 10 million people fell ill with TB in 2015 (www.who.int). The zebrafish model for TB is by far the best-studied zebrafish infection model and has made great contributions to our understanding of TB pathology (Fig. 2) (Cronan & Tobin, 2014; Meijer, 2016; Ramakrishnan, 2013). *M. marinum is* a natural pathogen of zebrafish and is phylogenetically very close to *M. tuberculosis,* the causative agent of human TB. The establishment of *M. marinum* pathogenesis in the zebrafish host is strikingly similar to human TB (Fig. 2). The disease hallmark in both host–pathogen systems consists of granulomas, essentially consisting of organized collections of immune cells that engulf and confine the bacteria.

Granulomas were previously considered relatively static structures generated by the host as a protective mechanism to restrict bacterial spread. Use of the zebrafish *M. marinum* infection model demonstrated that these structures are instead highly dynamic (Fig. 2) (Ramakrishnan, 2012). Noninvasive imaging in live zebrafish has shown that granulomas are characterized by a continuous trafficking of innate immune cells and that the pathogen takes advantage of infected macrophages to disseminate secondary lesions (Clay et al., 2007; Oehlers et al., 2015; Torraca et al., 2015). This model also helped to reconsider the contribution of the innate and the adaptive branches of the immune system in initiating the formation of granulomas. Imaging the earliest stages of granuloma formation in zebrafish embryos has shown that this process is initially driven by macrophages and occurs before lymphocyte differentiation, demonstrating that cells of the adaptive immune system are not required for granuloma formation (Davis et al., 2002).

Establishing TB infection in the form of granulomas depends on both pathogen and host factors, including mycobacterial virulence components, macrophage chemoattractants, and inflammatory mediators. The Region of Difference 1 (*RD1*) is a virulence-associated locus covering the ESX-1 bacterial secretion system and is notably shared between pathogenic mycobacteria, including *M. marinum*. ESX-1 is crucial for the establishment of granulomas and the zebrafish model helped to understand that the ESAT-6 virulence released via ESX-1 mediates macrophage aggregation in the early granulomas by stimulating production of Mmp9 in the epithelium surrounding the infection focus. By digesting the extracellular matrix,

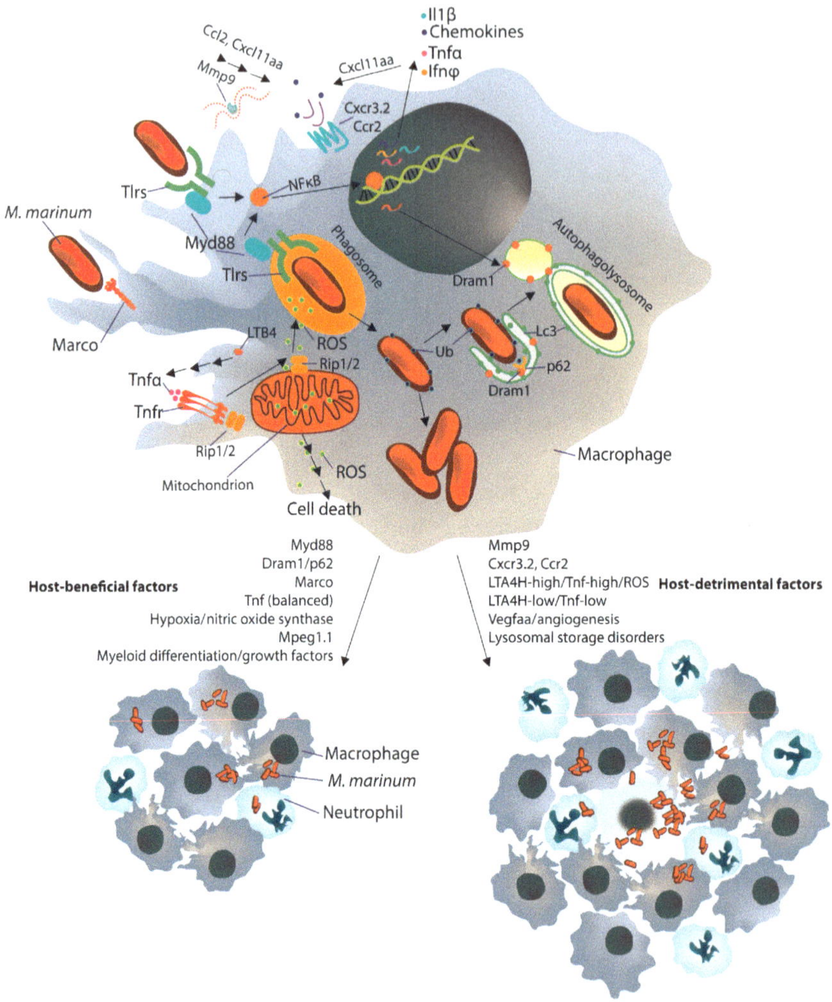

Fig. 2 Mechanistic insight into mycobacterial pathogenesis provided by the zebrafish *M. marinum* infection model. The host factors implicated in *M. marinum* infection of macrophages in the zebrafish host are summarized in this figure. The factors limiting (host-beneficial) or promoting (host-detrimental) the early expansion of granulomas are indicated below the schematic drawing of the macrophage. Macrophage-recruitment and tissue-inflammation mediators (such as Ccl2 and Mmp9) are also produced by neighboring cells as indicated by the *arrows* above the macrophage. Genes, pathways, and molecular functions depicted in the figure: Marco, scavenger receptor important for efficient phagocytosis and immune recognition; Tlr/Myd88/NFκB signaling pathway, leading to induction of inflammatory cytokines (e.g., Tnfα, Il1β), interferons (e.g., Ifnφ), chemokines (e.g., Cxcl11aa), and autophagy modulators (e.g., Dram 1); Mmp9, matrix metalloproteinase 9 facilitating macrophage migration; Ccl2/Ccr2 and Cxcl11aa/Cxcr3.2, chemokine ligand/receptor signaling axes implicated in macrophage migration; LTB4, lipid mediator of inflammation; Tnfr–Rip1/2 pathway, mediating release of reactive oxygen species (ROS) from mitochondria; Dram1, lysosomal/autophagosomal membrane protein stimulating autophagic flux; p62, pattern recognition receptor targeting ubiquitin-tagged (Ub) mycobacteria (escaped from the phagosomal compartment into the cytosol) to autophagy; Lc3, marker for autophagic activity. Vegfaa, angiogenesis-promoting factor. See text for further details.

Mmp9 in turn facilitates infiltration of macrophages and establishment of chronic intracellular parasitosis (Fig. 2) (Ramakrishnan, 2013). MMP9 is highly expressed in human TB and other inflammatory conditions; therefore, the observation that Mmp9 depletion confers resistance to mycobacterial infection in zebrafish highlights MMP9 as a potential therapeutic target (Volkman et al., 2010).

In addition to bacteria-driven mechanisms of granuloma expansion, chemokine signaling affecting macrophage recruitment is important to establish mycobacterial infection and to sustain granuloma expansion and secondary dissemination. Using the zebrafish model it was shown that deficiency in Ccr2/Ccl2 signaling reduces the chances of successful establishment of infection and that abrogation of Cxcr3–Cxcl11 signaling delays granuloma formation and attenuates seeding of the pathogen throughout the host (Fig. 2) (Cambier et al., 2014; Torraca et al., 2015). However, the equilibrium controlling macrophage supply to the granuloma is very delicate, and while slight perturbations lead to host-beneficial effects, more drastic alterations can promote bacterial growth.

Macrophages that are engorged with undigested contents, such as in lysosomal storage disorder (LSD) patients and in smokers, display severe migratory aberrations, which can be mimicked in the zebrafish model by knockdown of LSD-associated genes (*gba, arsa, hexa*), by filling macrophages with indigestible particles or by compromising the levels of lysosomal cathepsins. These paralyzed macrophages cannot sufficiently contain the infection and will permit extracellular growth of the pathogen (Berg et al., 2016). Similarly, blockade of key macrophage differentiation regulators, such as *spi1, csf1ra,* or *irf8,* leads to severe depletion of macrophages, with the consequent massive noncellular bacterial growth (Clay et al., 2007; Elks et al., 2015; Pagan et al., 2015). Conversely, drastic increase in macrophage supply, evoked by overexpression of myeloid growth factors, can encourage resistance to mycobacterial infection, by preventing granuloma necroptosis (Pagan et al., 2015; Ramakrishnan, 2012). Taken together, recent findings from the zebrafish model are helping to critically dissect the highly debated dual role of macrophages in TB pathogenesis (Clay et al., 2007).

Human granulomas are amply vascularized, which suggested that, similar to affecting tumor growth, curtailing vascularization might help to restrict granuloma formation. By injecting bacteria in the poorly vascularized zebrafish trunk tissue, the granuloma-driven promotion of angiogenesis could be mimicked in this model (Oehlers et al., 2015). Establishment of the intramacrophage parasitosis, the production of RD1-encoded virulence factors, and the induction of local hypoxia are critical to mediate this response, which coincides with local induction of the angiogenic mediator *vegfaa*. In turn, depletion of Vegf signaling, which suppresses pathological angiogenesis, leads to contained granuloma expansion (Oehlers et al., 2015). Using the zebrafish-*Mm* model and genetic tools to control the function of Hif-1α/Hif-2α (the two main variants of hypoxia inducible factor alpha), it was found that hypoxia signaling controls not only angiogenesis but also the production of NO by neutrophils, an important signaling mediator, and antimicrobial factors (Elks et al., 2013). Interestingly, stabilization of Hif-1α stimulated activity of the nitric oxide synthase (Nos2a), while stabilization of the Hif-2α variant could antagonize NO production, with consequent opposing effects in inhibiting or promoting bacterial growth (Elks et al., 2013, 2015). These studies suggest angiogenic and hypoxia signaling pathways as possible targets for TB treatment. Several other host-directed therapeutic strategies have been proposed based on work in the zebrafish model and these are extensively covered in previous reviews (Cronan & Tobin, 2014; Ramakrishnan, 2012; Torraca et al., 2014). In conclusion, the zebrafish *M. marinum* model has provided mechanistic insight into host factors that have been implicated either in protection against human TB or in the pathology of the disease, and provides a valuable antitubercular drug testing platform to develop novel therapeutic approaches.

8.2 Viral Infections

Viral epidemics, with influenza and HIV/AIDS as prominent examples, have had devastating effects throughout human history and emerging viral diseases such as Dengue, Chikungunya, and, most recently Zika are a growing concern (Tilak, Ray, Tilak, & Mukherji, 2016). While bacterial infections have been modeled in zebrafish for about two decades, the concept that the heterologous zebrafish model could be useful also to address viral infection with natural human pathogens emerged relatively recently in the field (Goody, Sullivan, & Kim, 2014; Levraud et al., 2014; Meijer & Spaink, 2011). In fact, while the zebrafish model proved immediately very useful to address economically relevant fish-specific viral infections, three main aspects represented a limitation into the use of zebrafish to model human viral disease. These include the tight and evolutionary rapid adaptation of viruses to their natural hosts, the large implication of a mature adaptive immunity during virus pathogenesis, and the fact that the IFN-mediated signaling (the main pathway used by innate immune cells to counteract viral infections) remains poorly characterized in (zebra)fish and diverges in some aspects from mammalian systems (Briolat et al., 2014; Langevin et al., 2013; Levraud et al., 2014).

Despite these considerations, zebrafish models for several important human viral disease have now been established, including Chickungunya, Influenza, and Herpes simplex (Antoine, Jones, Dale, Shukla, & Tiwari, 2014; Burgos et al., 2008; Gabor et al., 2014; Goody et al., 2014; Levraud et al., 2014; Palha et al., 2013).

CHIKV is a mosquito-transmitted virus, causing serious and sometimes deadly illness in humans with acute fever, persistent rash, and debilitating muscle and joint pain. Infection of 3-day-old zebrafish larvae with CHIKV showed that the pathogen can invade multiple host tissues such as muscles, liver, jaws, and spinal cord cartilages, gills, fins, vascular endothelium, and even eyes and brain (Palha et al., 2013). Thus, in some tissues, CHIKV infection in zebrafish mimics the pattern in humans. Interestingly, in zebrafish CHIKV infection persists persistent in the brain, while other tissues mostly clear the infection (Palha et al., 2013). Use of an *ifnφ1* fluorescent reporter line demonstrated that neutrophils are important to mediate an antiviral response to CHIKV infection via Ifn signaling (Palha et al., 2013). The fact that CHIKV displays a remarkable brain tropism and persistence suggests that in humans too this pathogen might persist in this organ. The hypothesis of a brain reservoir in humans is in line with the fact that, in adults, some CHIKV symptoms can persist for years, even after the apparent eradication of the pathogen. Furthermore, CHIKV is known to cause encephalitis in newborns (Gerardin et al., 2016; Rajapakse, Rodrigo, & Rajapakse, 2010). Therefore, further use of the zebrafish model could elucidate how CHIKV crosses the blood–brain barrier and persists in the central nervous system (CNS).

Influenza A virus (IAV) is the causative agent of annual epidemics of influenza. Similar to CHIKV infection, IAV infection could be followed over time in zebrafish, using fluorescently labeled viruses (Gabor et al., 2014). Strikingly, the viral kinetics and tissue tropisms in zebrafish recapitulate those observed in other models. Heart and skeletal muscles, blood endothelium, and the mucosa-associated epithelium of the swim bladder accumulate the GFP-labeled virus, which is consistent with the fact that IAV preferentially infects human muscle, epithelial, and endothelial cells in vitro. The pathology evoked in zebrafish shows relevant parallels also at the molecular level, since the viremia coincides with upregulation of the antiviral transcripts of *ifnφ1* and Myxovirus influenza resistance a (*mxa*), the latter being a close fish ortholog of human MX1. The study also successfully proved that the zebrafish disease can be reverted by treatment with the known human antiinfluenza drug Zanamivir, which indicates that zebrafish has a potential use as a screening platform for the discovery of novel antiviral compounds (Gabor et al., 2014).

Adult zebrafish have been used to study Herpes simplex virus type 1 (HSV-1) infection, a common cause of mucocutaneous orolabial, ocular, and genital infections in humans (Antoine et al., 2014; Burgos et al., 2008). HSV-1 can also invade and damage the CNS, persist in nervous ganglia, and lead to severe complications such as blindness and encephalitis. Following injection into the zebrafish abdominal cavity, the viral infection could spread to the midbody and ultimately reach the head, where it replicated abundantly in the CNS (Burgos et al., 2008). The current model of HSV-1 entry is that surface heparan sulfate derivatives mediate the initial viral adhesion, which in turn permits the fusion of the viral envelope with the host cell. These heparan sulfate moieties that act as viral receptors are remarkably conserved in zebrafish and are widely expressed in the CNS, like in mammals (Baldwin, Antoine, Shukla, & Tiwari, 2013). The entry in the CNS causes the most severe HSV-1 complications and the penetration in nervous ganglia is a well-known mechanism by which this pathogen can establish latent infections. Therefore, the zebrafish model can be used to address the mechanisms responsible for HSV-1 CNS invasion and provide new insight into how HSV-1 establishes latency and provokes repetitive episodes of disease reactivation.

Together, these studies have demonstrated that the possibility to longitudinally follow the infection course with fluorescently labeled viruses in developing zebrafish embryos or adult fish is very attractive to model important aspects of human viral infections, such as the cellular and molecular bases of tissue and organ-specific viral tropisms. These successes indicate that it will be also be worth to explore the possibility of developing a zebrafish model for other problematic human viral infections, including Zika virus. Studying Zika infection in developing zebrafish embryos and larvae could be a valuable addition to mouse models that have only recently been established and could provide new opportunities for studying the mechanistic basis of the association of this virus with microencephaly in newborns (Cugola et al., 2016; Li et al., 2016).

8.3 Fungal Infections

A variety of fungi are present in the commensal flora of human mucosae and skin. Most of them represent opportunistic pathogens and can cause mycotic disease in immunocompromised individuals or when subjects are exposed to large doses. The increasing number of people following immunosuppressive regimens or that are HIV-positive have made fungal disease an important cause of illness, especially in hospitalized settings.

The zebrafish model has been used to study several fungal pathogens of global health interest, which include *Candida albicans*, *Aspergillus fumigatus*, *Mucor circinelloides*, and *C. neoformans*. All these studies have shown that an appropriate competency of the innate immunity is important to curtail fungal infections (Chao et al., 2010; Knox et al., 2014; Tenor et al., 2015; Voelz et al., 2015). However, the involvement and relevance of macrophage and neutrophils in the response to each of these pathogens (or at least to the particular strains used in these studies) shows interesting specificity. During *Mucor* and *Candida* infection, both macrophages and neutrophils are highly recruited to the infection site and both phagocytose the fungal spores (Chao et al., 2010; Voelz et al., 2015). In sharp contrast, it is observed that *Aspergillus* conidia (asexual fungal spores) and *Cryptococcus* cells are essentially engulfed by macrophages, with neutrophils playing only a marginal function in counteracting these pathogens (Knox et al., 2014; Tenor et al., 2015). *A. fumigatus* is a dimorphic fungus that grows in yeast and hyphal forms. Infected zebrafish showed that neutrophils did not engulf the fungal spores (conidia), but can tightly associate with the hyphal form of the fungus (Knox et al., 2014). This suggests differential specificity of macrophage and neutrophil responses to the vegetative and reproductive fungal forms (Knox et al., 2014).

Similar to *A. fumigatus*, *C. albicans* is an opportunistic dimorphic fungus and most of humans are healthy carries of this commensal. The most frequent *Candida* infections are those that remain localized to the mucosal tissues, but life-threatening conditions can derive from systemic dissemination, especially in immunocompromised individuals (Brothers et al., 2013). Interestingly, when *C. albicans* is injected locally in the zebrafish hindbrain, it readily causes disseminated infection and high mortality, which is associated to its germination from yeast to hypha. Both zebrafish macrophages and neutrophils can phagocytose *Candida* (Brothers et al., 2013). Uptake of the yeast form is important to contain the transition to the hyphal and more invasive form, indicating that immune cells are also crucial to counteract the yeast-to-hyphal transition of dimorphic fungi. While this model mimics human systemic candidiasis, injection of *C. albicans* into the swimbladder of zebrafish larvae can be used to model mucosal *Candida* colonization and to study the distinctive immune mechanisms at play on the mucosal surfaces (Gratacap & Wheeler, 2014).

Recent use of the zebrafish model has been critical to better characterize the mechanism of virulence of *C. neoformans*, which represents an emerging and often fatal human pathogen (Bojarczuk et al., 2016; Tenor et al., 2015). Cryptococcal infection in humans generally initiates in the lung. However, the pathogen displays a remarkable tropism for the CNS, which is the main life-threatening complication of this fungal disease. Live imaging in zebrafish demonstrated that the predisposition of this pathogen to colonize the brain is maintained in this host and that the capability of the pathogen to cross the blood–brain barrier depends on the virulence gene FNX1 and on a known cryptococcal invasion-promoting pathway previously identified in a murine model (Tenor et al., 2015). Additionally, longitudinal studies in zebrafish showed that macrophages are important to counteract the acute infection with this pathogen (Bojarczuk et al., 2016). However, it was observed that cryptococci can still largely proliferate intracellularly in macrophages, and, within 24 h, they can counteract macrophage phagocytosis by progressively increasing their capsule size until this reaches an extent that severely limits further phagocytosis. This study suggests that the early proinflammatory activation of macrophages can control cryptococcal infection in healthy individuals, while intracellular survival and modification of the cryptococcal capsule will lead to uncontrolled progression of infection in immunocompromised patients (Bojarczuk et al., 2016).

9. CONCLUDING REMARKS

Modeling of infectious diseases using the early life stages of zebrafish is continuing to demonstrate striking similarities in the mechanism of action of the innate immune system across vertebrates, which not only is evolutionary relevant but also adds a high biomedical value to the use of the zebrafish model. Notably, in many cases the zebrafish platform has served as a valuable springboard to more extended studies in mammals. In other cases, the zebrafish has worked well as a surrogate system to model certain disease features that have otherwise been difficult to reproduce or study in mammalian models. Considering the expanding genetic toolbox for zebrafish research and the advanced use for noninvasive intravital imaging, it is to be expected that the zebrafish model will attract an increasingly larger scientific audience and continue to enforce its position in translational research. With state-of-art genome editing techniques now being successfully applied in zebrafish, it will be possible to generate a collection of key immune gene knockouts that will help to better understand the core mechanisms of immune recognition and pathogen virulence and to generate knowledge that can be exploited for developing novel therapeutic strategies to combat infectious and inflammatory diseases.

ACKNOWLEDGMENTS

S.M. is supported by a fellowship from the Higher Education Commission of Pakistan and the Bahauddin Zakariya University, Multan. V.T. is a Marie Curie fellow in the Initial Training Network FishForPharma (PITN-GA-2011-289209) funded by the 7th Framework Programme of the European Commission. Work that led to this review was additionally supported by the European 7th framework project ZF-Health (FP7-Health-2009-242048), by the Smart Mix Program (SSM06010) of The Netherlands Ministry of Economic Affairs and The Netherlands Ministry of Education, Culture and Science, and by the Technology Foundation STW (project 13259).

REFERENCES

Aggad, D., Mazel, M., Boudinot, P., Mogensen, K. E., Hamming, O. J., Hartmann, R., … Levraud, J. P. (2009). The two groups of zebrafish virus-induced interferons signal via distinct receptors with specific and shared chains. *Journal of Immunology*, *183*(6), 3924–3931. http://dx.doi.org/10.4049/jimmunol.0901495.

Akira, S., Uematsu, S., & Takeuchi, O. (2006). Pathogen recognition and innate immunity. *Cell*, *124*(4), 783–801. http://dx.doi.org/10.1016/j.cell.2006.02.015.

Antoine, T. E., Jones, K. S., Dale, R. M., Shukla, D., & Tiwari, V. (2014). Zebrafish: Modeling for herpes simplex virus infections. *Zebrafish*, *11*(1), 17–25. http://dx.doi.org/10.1089/zeb.2013.0920.

Baldwin, J., Antoine, T. E., Shukla, D., & Tiwari, V. (2013). Zebrafish encoded 3-O-sulfotransferase-2 generated heparan sulfate serves as a receptor during HSV-1 entry and spread. *Biochemical and Biophysical Research Communications*, *432*(4), 672–676. http://dx.doi.org/10.1016/j.bbrc.2013.02.020.

Balla, K. M., Lugo-Villarino, G., Spitsbergen, J. M., Stachura, D. L., Hu, Y., Banuelos, K., … Traver, D. (2010). Eosinophils in the zebrafish: Prospective isolation, characterization, and eosinophilia induction by helminth determinants. *Blood*, *116*(19), 3944–3954. http://dx.doi.org/10.1182/blood-2010-03-267419.

Benard, E. L., Roobol, S. J., Spaink, H. P., & Meijer, A. H. (2014). Phagocytosis of mycobacteria by zebrafish macrophages is dependent on the scavenger receptor Marco, a key control factor of pro-inflammatory signalling. *Developmental and Comparative Immunology*, *47*(2), 223–233. http://dx.doi.org/10.1016/j.dci.2014.07.022.

Bennett, C. M., Kanki, J. P., Rhodes, J., Liu, T. X., Paw, B. H., Kieran, M. W., … Look, A. T. (2001). Myelopoiesis in the zebrafish, Danio rerio. *Blood*, *98*(3), 643–651.

Berg, R. D., Levitte, S., O'Sullivan, M. P., O'Leary, S. M., Cambier, C. J., Cameron, J., … Ramakrishnan, L. (2016). Lysosomal disorders drive susceptibility to tuberculosis by compromising macrophage migration. *Cell*, *165*(1), 139–152. http://dx.doi.org/10.1016/j.cell.2016.02.034.

Bernut, A., Herrmann, J. L., Kissa, K., Dubremetz, J. F., Gaillard, J. L., Lutfalla, G., & Kremer, L. (2014). Mycobacterium abscessus cording prevents phagocytosis and promotes abscess formation. *Proceedings of the National Academy of Sciences of the United States of America*, *111*(10), E943–E952. http://dx.doi.org/10.1073/pnas.1321390111.

Bertrand, J. Y., Chi, N. C., Santoso, B., Teng, S., Stainier, D. Y., & Traver, D. (2010). Haematopoietic stem cells derive directly from aortic endothelium during development. *Nature*, *464*(7285), 108–111. http://dx.doi.org/10.1038/nature08738.

Bertrand, J. Y., Kim, A. D., Violette, E. P., Stachura, D. L., Cisson, J. L., & Traver, D. (2007). Definitive hematopoiesis initiates through a committed erythromyeloid progenitor in the zebrafish embryo. *Development*, *134*(23), 4147–4156. http://dx.doi.org/10.1242/dev.012385.

Biacchesi, S., LeBerre, M., Lamoureux, A., Louise, Y., Lauret, E., Boudinot, P., & Bremont, M. (2009). Mitochondrial antiviral signaling protein plays a major role in induction of the fish innate immune response against RNA and DNA viruses. *Journal of Virology*, *83*(16), 7815–7827. http://dx.doi.org/10.1128/jvi.00404-09.

Bojarczuk, A., Miller, K. A., Hotham, R., Lewis, A., Ogryzko, N. V., Kamuyango, A. A., ... Johnston, S. A. (2016). Cryptococcus neoformans intracellular proliferation and capsule size determines early macrophage control of infection. *Scientific Reports*, *6*, 21489. http://dx.doi.org/10.1038/srep21489.

Bowdish, D. M., Sakamoto, K., Kim, M. J., Kroos, M., Mukhopadhyay, S., Leifer, C. A., ... Russell, D. G. (2009). MARCO, TLR2, and CD14 are required for macrophage cytokine responses to mycobacterial trehalose dimycolate and Mycobacterium tuberculosis. *PLoS Pathogens*, *5*(6). e1000474http://dx.doi.org/10.1371/journal.ppat.1000474.

Brinkmann, V., & Zychlinsky, A. (2012). Neutrophil extracellular traps: Is immunity the second function of chromatin? *The Journal of Cell Biology*, *198*(5), 773–783. http://dx.doi.org/10.1083/jcb.201203170.

Briolat, V., Jouneau, L., Carvalho, R., Palha, N., Langevin, C., Herbomel, P., ... Boudinot, P. (2014). Contrasted innate responses to two viruses in zebrafish: Insights into the ancestral repertoire of vertebrate IFN-stimulated genes. *Journal of Immunology*, *192*(9), 4328–4341. http://dx.doi.org/10.4049/jimmunol.1302611.

Brothers, K. M., Gratacap, R. L., Barker, S. E., Newman, Z. R., Norum, A., & Wheeler, R. T. (2013). NADPH oxidase-driven phagocyte recruitment controls Candida albicans filamentous growth and prevents mortality. *PLoS Pathogens*, *9*(10). e1003634http://dx.doi.org/10.1371/journal.ppat.1003634.

Bukrinsky, A., Griffin, K. J., Zhao, Y., Lin, S., & Banerjee, U. (2009). Essential role of spi-1-like (spi-1 l) in zebrafish myeloid cell differentiation. *Blood*, *113*(9), 2038–2046. http://dx.doi.org/10.1182/blood-2008-06-162495.

Burgos, J. S., Ripoll-Gomez, J., Alfaro, J. M., Sastre, I., & Valdivieso, F. (2008). Zebrafish as a new model for herpes simplex virus type 1 infection. *Zebrafish*, *5*(4), 323–333. http://dx.doi.org/10.1089/zeb.2008.0552.

Cambier, C. J., Takaki, K. K., Larson, R. P., Hernandez, R. E., Tobin, D. M., Urdahl, K. B., ... Ramakrishnan, L. (2014). Mycobacteria manipulate macrophage recruitment through coordinated use of membrane lipids. *Nature*, *505*(7482), 218–222. http://dx.doi.org/10.1038/nature12799.

Caruso, R., Warner, N., Inohara, N., & Nunez, G. (2014). NOD1 and NOD2: Signaling, host defense, and inflammatory disease. *Immunity*, *41*(6), 898–908. http://dx.doi.org/10.1016/j.immuni.2014.12.010.

Chao, C. C., Hsu, P. C., Jen, C. F., Chen, I. H., Wang, C. H., Chan, H. C., ... Chuang, Y. J. (2010). Zebrafish as a model host for Candida albicans infection. *Infection and Immunity*, *78*(6), 2512–2521. http://dx.doi.org/10.1128/iai.01293-09.

Cheesman, S. E., & Guillemin, K. (2007). We know you are in there: Conversing with the indigenous gut microbiota. *Research in Microbiology*, *158*(1), 2–9. http://dx.doi.org/10.1016/j.resmic.2006.10.005.

Clatworthy, A. E., Lee, J. S., Leibman, M., Kostun, Z., Davidson, A. J., & Hung, D. T. (2009). Pseudomonas aeruginosa infection of zebrafish involves both host and pathogen determinants. *Infection and Immunity*, *77*(4), 1293–1303. http://dx.doi.org/10.1128/iai.01181-08.

Clay, H., Davis, J. M., Beery, D., Huttenlocher, A., Lyons, S. E., & Ramakrishnan, L. (2007). Dichotomous role of the macrophage in early Mycobacterium marinum infection of the zebrafish. *Cell Host & Microbe*, *2*(1), 29–39. http://dx.doi.org/10.1016/j.chom.2007.06.004.

Clay, H., Volkman, H. E., & Ramakrishnan, L. (2008). Tumor necrosis factor signaling mediates resistance to mycobacteria by inhibiting bacterial growth and macrophage death. *Immunity*, *29*(2), 283–294. http://dx.doi.org/10.1016/j.immuni.2008.06.011.

Colucci-Guyon, E., Tinevez, J. Y., Renshaw, S. A., & Herbomel, P. (2011). Strategies of professional phagocytes in vivo: Unlike macrophages, neutrophils engulf only surface-associated microbes. *Journal of Cell Science*, *124*(Pt 18), 3053–3059. http://dx.doi.org/10.1242/jcs.082792.

Coughlan, M. T., & Sharma, K. (2016). Challenging the dogma of mitochondrial reactive oxygen species overproduction in diabetic kidney disease. *Kidney International*, *90*(2), 272–279. http://dx.doi.org/10.1016/j.kint.2016.02.043.

Cronan, M. R., & Tobin, D. M. (2014). Fit for consumption: Zebrafish as a model for tuberculosis. *Disease Models & Mechanisms*, 7(7), 777–784. http://dx.doi.org/10.1242/dmm.016089.

Cugola, F. R., Fernandes, I. R., Russo, F. B., Freitas, B. C., Dias, J. L., Guimaraes, K. P., ... Beltrao-Braga, P. C. (2016). The Brazilian Zika virus strain causes birth defects in experimental models. *Nature*, *534*(7606), 267–271. http://dx.doi.org/10.1038/nature18296.

Da'as, S. I., Coombs, A. J., Balci, T. B., Grondin, C. A., Ferrando, A. A., & Berman, J. N. (2012). The zebrafish reveals dependence of the mast cell lineage on Notch signaling in vivo. *Blood*, *119*(15), 3585–3594. http://dx.doi.org/10.1182/blood-2011-10-385989.

Davidson, A. J., & Zon, L. I. (2004). The 'definitive' (and 'primitive') guide to zebrafish hematopoiesis. *Oncogene*, *23*(43), 7233–7246. http://dx.doi.org/10.1038/sj.onc.1207943.

Davis, J. M., Clay, H., Lewis, J. L., Ghori, N., Herbomel, P., & Ramakrishnan, L. (2002). Real-time visualization of mycobacterium-macrophage interactions leading to initiation of granuloma formation in zebrafish embryos. *Immunity*, *17*(6), 693–702.

Deretic, V., Saitoh, T., & Akira, S. (2013). Autophagy in infection, inflammation and immunity. *Nature Reviews. Immunology*, *13*(10), 722–737. http://dx.doi.org/10.1038/nri3532.

Dobson, J. T., Seibert, J., Teh, E. M., Da'as, S., Fraser, R. B., Paw, B. H., ... Berman, J. N. (2008). Carboxypeptidase A5 identifies a novel mast cell lineage in the zebrafish providing new insight into mast cell fate determination. *Blood*, *112*(7), 2969–2972. http://dx.doi.org/10.1182/blood-2008-03-145011.

Dorhoi, A., & Kaufmann, S. H. (2014). Tumor necrosis factor alpha in mycobacterial infection. *Seminars in Immunology*, *26*(3), 203–209. http://dx.doi.org/10.1016/j.smim.2014.04.003.

Elks, P. M., Brizee, S., van der Vaart, M., Walmsley, S. R., van Eeden, F. J., Renshaw, S. A., & Meijer, A. H. (2013). Hypoxia inducible factor signaling modulates susceptibility to mycobacterial infection via a nitric oxide dependent mechanism. *PLoS Pathogens*, *9*(12). e1003789. http://dx.doi.org/10.1371/journal.ppat.1003789.

Elks, P. M., Renshaw, S. A., Meijer, A. H., Walmsley, S. R., & van Eeden, F. J. (2015). Exploring the HIFs, buts and maybes of hypoxia signalling in disease: Lessons from zebrafish models. *Disease Models & Mechanisms*, *8*(11), 1349–1360. http://dx.doi.org/10.1242/dmm.021865.

Ellett, F., Pase, L., Hayman, J. W., Andrianopoulos, A., & Lieschke, G. J. (2011). mpeg1 promoter transgenes direct macrophage-lineage expression in zebrafish. *Blood*, *117*(4), e49–e56. http://dx.doi.org/10.1182/blood-2010-10-314120.

Espin-Palazon, R., Stachura, D. L., Campbell, C. A., Garcia-Moreno, D., Del Cid, N., Kim, A. D., ... Traver, D. (2014). Proinflammatory signaling regulates hematopoietic stem cell emergence. *Cell*, *159*(5), 1070–1085. http://dx.doi.org/10.1016/j.cell.2014.10.031.

Fink, I. R., Benard, E. L., Hermsen, T., Meijer, A. H., Forlenza, M., & Wiegertjes, G. F. (2015). Molecular and functional characterization of the scavenger receptor CD36 in zebrafish and common carp. *Molecular Immunology*, *63*(2), 381–393. http://dx.doi.org/10.1016/j.molimm.2014.09.010.

Gabor, K. A., Goody, M. F., Mowel, W. K., Breitbach, M. E., Gratacap, R. L., Witten, P. E., & Kim, C. H. (2014). Influenza A virus infection in zebrafish recapitulates mammalian infection and sensitivity to anti-influenza drug treatment. *Disease Models & Mechanisms*, 7(11), 1227–1237. http://dx.doi.org/10.1242/dmm.014746.

Galindo-Villegas, J., Garcia-Moreno, D., de Oliveira, S., Meseguer, J., & Mulero, V. (2012). Regulation of immunity and disease resistance by commensal microbes and chromatin modifications during zebrafish development. *Proceedings of the National Academy of Sciences of the United States of America*, *109*(39), E2605–E2614. http://dx.doi.org/10.1073/pnas.1209920109.

Galloway, J. L., Wingert, R. A., Thisse, C., Thisse, B., & Zon, L. I. (2005). Loss of gata1 but not gata2 converts erythropoiesis to myelopoiesis in zebrafish embryos. *Developmental Cell*, *8*(1), 109–116. http://dx.doi.org/10.1016/j.devcel.2004.12.001.

Galloway, J. L., & Zon, L. I. (2003). Ontogeny of hematopoiesis: Examining the emergence of hematopoietic cells in the vertebrate embryo. *Current Topics in Developmental Biology*, *53*, 139–158.

Gay, N. J., Gangloff, M., & O'Neill, L. A. (2011). What the Myddosome structure tells us about the initiation of innate immunity. *Trends in Immunology*, *32*(3), 104–109. http://dx.doi.org/10.1016/j.it.2010.12.005.

Gerardin, P., Couderc, T., Bintner, M., Tournebize, P., Renouil, M., Lemant, J., … Michault, A. (2016). Chikungunya virus-associated encephalitis: A cohort study on La Reunion Island, 2005–2009. *Neurology*, *86*(1), 94–102. http://dx.doi.org/10.1212/wnl.0000000000002234.

Goody, M. F., Sullivan, C., & Kim, C. H. (2014). Studying the immune response to human viral infections using zebrafish. *Developmental and Comparative Immunology*, *46*(1), 84–95. http://dx.doi.org/10.1016/j.dci.2014.03.025.

Gratacap, R. L., & Wheeler, R. T. (2014). Utilization of zebrafish for intravital study of eukaryotic pathogen-host interactions. *Developmental and Comparative Immunology*, *46*(1), 108–115. http://dx.doi.org/10.1016/j.dci.2014.01.020.

Gray, C., Loynes, C. A., Whyte, M. K., Crossman, D. C., Renshaw, S. A., & Chico, T. J. (2011). Simultaneous intravital imaging of macrophage and neutrophil behaviour during inflammation using a novel transgenic zebrafish. *Thrombosis and Haemostasis*, *105*(5), 811–819. http://dx.doi.org/10.1160/th10-08-0525.

Hall, C. J., Boyle, R. H., Astin, J. W., Flores, M. V., Oehlers, S. H., Sanderson, L. E., … Crosier, P. S. (2013). Immunoresponsive gene 1 augments bactericidal activity of macrophage-lineage cells by regulating beta-oxidation-dependent mitochondrial ROS production. *Cell Metabolism*, *18*(2), 265–278. http://dx.doi.org/10.1016/j.cmet.2013.06.018.

Hall, C., Crosier, P., & Crosier, K. (2016). Inflammatory cytokines provide both infection-responsive and developmental signals for blood development: Lessons from the zebrafish. *Molecular Immunology*, *69*, 113–122. http://dx.doi.org/10.1016/j.molimm.2015.10.020.

Hall, C., Flores, M. V., Chien, A., Davidson, A., Crosier, K., & Crosier, P. (2009). Transgenic zebrafish reporter lines reveal conserved Toll-like receptor signaling potential in embryonic myeloid leukocytes and adult immune cell lineages. *Journal of Leukocyte Biology*, *85*(5), 751–765. http://dx.doi.org/10.1189/jlb.0708405.

Hall, C. J., Flores, M. V., Oehlers, S. H., Sanderson, L. E., Lam, E. Y., Crosier, K. E., & Crosier, P. S. (2012). Infection-responsive expansion of the hematopoietic stem and progenitor cell compartment in zebrafish is dependent upon inducible nitric oxide. *Cell Stem Cell*, *10*(2), 198–209. http://dx.doi.org/10.1016/j.stem.2012.01.007.

Hall, C., Flores, M. V., Storm, T., Crosier, K., & Crosier, P. (2007). The zebrafish lysozyme C promoter drives myeloid-specific expression in transgenic fish. *BMC Developmental Biology*, 7, 42. http://dx.doi.org/10.1186/1471-213x-7-42.

Harvie, E. A., & Huttenlocher, A. (2015). Neutrophils in host defense: New insights from zebrafish. *Journal of Leukocyte Biology*, *98*(4), 523–537. http://dx.doi.org/10.1189/jlb.4MR1114-524R.

He, C., Bartholomew, C. R., Zhou, W., & Klionsky, D. J. (2009). Assaying autophagic activity in transgenic GFP-Lc3 and GFP-Gabarap zebrafish embryos. *Autophagy*, *5*(4), 520–526.

He, Q., Zhang, C., Wang, L., Zhang, P., Ma, D., Lv, J., & Liu, F. (2015). Inflammatory signaling regulates hematopoietic stem and progenitor cell emergence in vertebrates. *Blood*, *125*(7), 1098–1106. http://dx.doi.org/10.1182/blood-2014-09-601542.

Hepburn, L., Prajsnar, T. K., Klapholz, C., Moreno, P., Loynes, C. A., Ogryzko, N. V., … Floto, R. A. (2014). Innate immunity. A Spaetzle-like role for nerve growth factor beta in vertebrate immunity to Staphylococcus aureus. *Science*, *346*(6209), 641–646. http://dx.doi.org/10.1126/science.1258705.

Herbomel, P. (2012). Infection-induced hematopoiesis: A zebrafish perspective. *Cell Stem Cell*, *10*(2), 105–106. http://dx.doi.org/10.1016/j.stem.2012.01.015.

Herbomel, P., Thisse, B., & Thisse, C. (1999). Ontogeny and behaviour of early macrophages in the zebrafish embryo. *Development*, *126*(17), 3735–3745.

Herbomel, P., Thisse, B., & Thisse, C. (2001). Zebrafish early macrophages colonize cephalic mesenchyme and developing brain, retina, and epidermis through a M-CSF receptor-dependent invasive process. *Developmental Biology*, *238*(2), 274–288. http://dx.doi.org/10.1006/dbio.2001.0393.

Hogan, B. M., Layton, J. E., Pyati, U. J., Nutt, S. L., Hayman, J. W., Varma, S., … Lieschke, G. J. (2006). Specification of the primitive myeloid precursor pool requires signaling through Alk8 in zebrafish. *Current Biology*, *16*(5), 506–511. http://dx.doi.org/10.1016/j.cub.2006.01.047.

Hosseini, R., Lamers, G. E., Hodzic, Z., Meijer, A. H., Schaaf, M. J., & Spaink, H. P. (2014). Correlative light and electron microscopy imaging of autophagy in a zebrafish infection model. *Autophagy*, *10*(10), 1844–1857. http://dx.doi.org/10.4161/auto.29992.

Hu, Y. L., Pan, X. M., Xiang, L. X., & Shao, J. Z. (2010). Characterization of C1q in teleosts: Insight into the molecular and functional evolution of C1q family and classical pathway. *The Journal of Biological Chemistry*, *285*(37), 28777–28786. http://dx.doi.org/10.1074/jbc.M110.131318.

Huang, J., & Brumell, J. H. (2014). Bacteria-autophagy interplay: A battle for survival. *Nature Reviews. Microbiology*, *12*(2), 101–114. http://dx.doi.org/10.1038/nrmicro3160.

Jagannathan-Bogdan, M., & Zon, L. I. (2013). Hematopoiesis. *Development*, *140*(12), 2463–2467. http://dx.doi.org/10.1242/dev.083147.

Jin, H., Li, L., Xu, J., Zhen, F., Zhu, L., Liu, P. P., … Wen, Z. (2012). Runx1 regulates embryonic myeloid fate choice in zebrafish through a negative feedback loop inhibiting Pu.1 expression. *Blood*, *119*(22), 5239–5249. http://dx.doi.org/10.1182/blood-2011-12-398362.

Jin, H., Xu, J., & Wen, Z. (2007). Migratory path of definitive hematopoietic stem/progenitor cells during zebrafish development. *Blood*, *109*(12), 5208–5214. http://dx.doi.org/10.1182/blood-2007-01-069005.

Kanther, M., & Rawls, J. F. (2010). Host-microbe interactions in the developing zebrafish. *Current Opinion in Immunology*, *22*(1), 10–19. http://dx.doi.org/10.1016/j.coi.2010.01.006.

Kanther, M., Sun, X., Muhlbauer, M., Mackey, L. C., Flynn, E. J., 3rd, Bagnat, M., … Rawls, J. F. (2011). Microbial colonization induces dynamic temporal and spatial patterns of NF-kappaB activation in the zebrafish digestive tract. *Gastroenterology*, *141*(1), 197–207. http://dx.doi.org/10.1053/j.gastro.2011.03.042.

Kanther, M., Tomkovich, S., Xiaolun, S., Grosser, M. R., Koo, J., Flynn, E. J., 3rd, … Rawls, J. F. (2014). Commensal microbiota stimulate systemic neutrophil migration through induction of serum amyloid A. *Cellular Microbiology*, *16*(7), 1053–1067. http://dx.doi.org/10.1111/cmi.12257.

Kanwal, Z., Wiegertjes, G. F., Veneman, W. J., Meijer, A. H., & Spaink, H. P. (2014). Comparative studies of toll-like receptor signalling using zebrafish. *Developmental and Comparative Immunology*, *46*(1), 35–52. http://dx.doi.org/10.1016/j.dci.2014.02.003.

Kanwal, Z., Zakrzewska, A., den Hertog, J., Spaink, H. P., Schaaf, M. J., & Meijer, A. H. (2013). Deficiency in hematopoietic phosphatase ptpn6/Shp1 hyperactivates the innate immune system and impairs control of bacterial infections in zebrafish embryos. *Journal of Immunology*, *190*(4), 1631–1645. http://dx.doi.org/10.4049/jimmunol.1200551.

Kaplan, J. L., Shi, H. N., & Walker, W. A. (2011). The role of microbes in developmental immunologic programming. *Pediatric Research*, *69*(6), 465–472. http://dx.doi.org/10.1203/PDR.0b013e318217638a.

Kell, A. M., & Gale, M., Jr. (2015). RIG-I in RNA virus recognition. *Virology*, *479–480*, 110–121. http://dx.doi.org/10.1016/j.virol.2015.02.017.

Kimmey, J. M., Huynh, J. P., Weiss, L. A., Park, S., Kambal, A., Debnath, J., … Stallings, C. L. (2015). Unique role for ATG5 in neutrophil-mediated immunopathology during M. tuberculosis infection. *Nature*, *528*(7583), 565–569. http://dx.doi.org/10.1038/nature16451.

Kissa, K., & Herbomel, P. (2010). Blood stem cells emerge from aortic endothelium by a novel type of cell transition. *Nature*, *464*(7285), 112–115. http://dx.doi.org/10.1038/nature08761.

Kissa, K., Murayama, E., Zapata, A., Cortes, A., Perret, E., Machu, C., & Herbomel, P. (2008). Live imaging of emerging hematopoietic stem cells and early thymus colonization. *Blood*, *111*(3), 1147–1156. http://dx.doi.org/10.1182/blood-2007-07-099499.

Knox, B. P., Deng, Q., Rood, M., Eickhoff, J. C., Keller, N. P., & Huttenlocher, A. (2014). Distinct innate immune phagocyte responses to Aspergillus fumigatus conidia and hyphae in zebrafish larvae. *Eukaryotic Cell*, *13*(10), 1266–1277. http://dx.doi.org/10.1128/ec.00080-14.

Lam, E. Y., Chau, J. Y., Kalev-Zylinska, M. L., Fountaine, T. M., Mead, R. S., Hall, C. J., … Flores, M. V. (2009). Zebrafish runx1 promoter-EGFP transgenics mark discrete sites of definitive blood progenitors. *Blood*, *113*(6), 1241–1249. http://dx.doi.org/10.1182/blood-2008-04-149898.

Langenau, D. M., Ferrando, A. A., Traver, D., Kutok, J. L., Hezel, J. P., Kanki, J. P., … Trede, N. S. (2004). In vivo tracking of T cell development, ablation, and engraftment in transgenic zebrafish. *Proceedings of the National Academy of Sciences of the United States of America*, *101*(19), 7369–7374. http://dx.doi.org/10.1073/pnas.0402248101.

Langevin, C., Aleksejeva, E., Passoni, G., Palha, N., Levraud, J. P., & Boudinot, P. (2013). The antiviral innate immune response in fish: Evolution and conservation of the IFN system. *Journal of Molecular Biology*, *425*(24), 4904–4920. http://dx.doi.org/10.1016/j.jmb.2013.09.033.

Lee, E., Koo, Y., Ng, A., Wei, Y., Luby-Phelps, K., Juraszek, A., … Amatruda, J. F. (2014). Autophagy is essential for cardiac morphogenesis during vertebrate development. *Autophagy*, *10*(4), 572–587. http://dx.doi.org/10.4161/auto.27649.

Le Guyader, D., Redd, M. J., Colucci-Guyon, E., Murayama, E., Kissa, K., Briolat, V., … Herbomel, P. (2008). Origins and unconventional behavior of neutrophils in developing zebrafish. *Blood*, *111*(1), 132–141. http://dx.doi.org/10.1182/blood-2007-06-095398.

Lemaitre, B., & Hoffmann, J. (2007). The host defense of Drosophila melanogaster. *Annual Review of Immunology*, *25*, 697–743. http://dx.doi.org/10.1146/annurev.immunol.25.022106.141615.

Levine, B., Mizushima, N., & Virgin, H. W. (2011). Autophagy in immunity and inflammation. *Nature*, *469*(7330), 323–335. http://dx.doi.org/10.1038/nature09782.

Levraud, J. P., Disson, O., Kissa, K., Bonne, I., Cossart, P., Herbomel, P., & Lecuit, M. (2009). Real-time observation of listeria monocytogenes-phagocyte interactions in living zebrafish larvae. *Infection and Immunity*, 77(9), 3651–3660. http://dx.doi.org/10.1128/iai.00408-09.

Levraud, J. P., Palha, N., Langevin, C., & Boudinot, P. (2014). Through the looking glass: Witnessing host-virus interplay in zebrafish. *Trends in Microbiology*, *22*(9), 490–497. http://dx.doi.org/10.1016/j.tim.2014.04.014.

Li, Y., Esain, V., Teng, L., Xu, J., Kwan, W., Frost, I. M., … Speck, N. A. (2014). Inflammatory signaling regulates embryonic hematopoietic stem and progenitor cell production. *Genes & Development*, *28*(23), 2597–2612. http://dx.doi.org/10.1101/gad.253302.114.

Li, L., Jin, H., Xu, J., Shi, Y., & Wen, Z. (2011). Irf8 regulates macrophage versus neutrophil fate during zebrafish primitive myelopoiesis. *Blood*, *117*(4), 1359–1369. http://dx.doi.org/10.1182/blood-2010-06-290700.

Li, C., Xu, D., Ye, Q., Hong, S., Jiang, Y., Liu, X., … Xu, Z. (2016). Zika virus disrupts neural progenitor development and leads to microcephaly in mice. *Cell Stem Cell*, *19*(1), 120–126. http://dx.doi.org/10.1016/j.stem.2016.04.017.

Liao, E. C., Paw, B. H., Oates, A. C., Pratt, S. J., Postlethwait, J. H., & Zon, L. I. (1998). SCL/Tal-1 transcription factor acts downstream of cloche to specify hematopoietic and vascular progenitors in zebrafish. *Genes & Development*, *12*(5), 621–626.

Lieschke, G. J., Oates, A. C., Crowhurst, M. O., Ward, A. C., & Layton, J. E. (2001). Morphologic and functional characterization of granulocytes and macrophages in embryonic and adult zebrafish. *Blood*, *98*(10), 3087–3096.

Lieschke, G. J., Oates, A. C., Paw, B. H., Thompson, M. A., Hall, N. E., Ward, A. C., … Layton, J. E. (2002). Zebrafish SPI-1 (PU.1) marks a site of myeloid development independent of primitive erythropoiesis: Implications for axial patterning. *Developmental Biology*, *246*(2), 274–295. http://dx.doi.org/10.1006/dbio.2002.0657.

Liongue, C., Hall, C. J., O'Connell, B. A., Crosier, P., & Ward, A. C. (2009). Zebrafish granulocyte colony-stimulating factor receptor signaling promotes myelopoiesis and myeloid cell migration. *Blood*, *113*(11), 2535–2546. http://dx.doi.org/10.1182/blood-2008-07-171967.

Liu, F., & Patient, R. (2008). Genome-wide analysis of the zebrafish ETS family identifies three genes required for hemangioblast differentiation or angiogenesis. *Circulation Research*, *103*(10), 1147–1154. http://dx.doi.org/10.1161/circresaha.108.179713.

Liu, F., & Wen, Z. (2002). Cloning and expression pattern of the lysozyme C gene in zebrafish. *Mechanisms of Development*, *113*(1), 69–72.

Lu, X., Li, X., He, Q., Gao, J., Gao, Y., Liu, B., & Liu, F. (2013). miR-142-3p regulates the formation and differentiation of hematopoietic stem cells in vertebrates. *Cell Research*, *23*(12), 1356–1368. http://dx.doi.org/10.1038/cr.2013.145.

Mantovani, A. (2009). The yin-yang of tumor-associated neutrophils. *Cancer Cell*, *16*(3), 173–174. http://dx.doi.org/10.1016/j.ccr.2009.08.014.

Marin-Juez, R., Jong-Raadsen, S., Yang, S., & Spaink, H. P. (2014). Hyperinsulinemia induces insulin resistance and immune suppression via Ptpn6/Shp1 in zebrafish. *The Journal of Endocrinology*, *222*(2), 229–241. http://dx.doi.org/10.1530/joe-14-0178.

Marjoram, L., Alvers, A., Deerhake, M. E., Bagwell, J., Mankiewicz, J., Cocchiaro, J. L., … Bagnat, M. (2015). Epigenetic control of intestinal barrier function and inflammation in zebrafish. *Proceedings of the National Academy of Sciences of the United States of America*, *112*(9), 2770–2775. http://dx.doi.org/10.1073/pnas.1424089112.

Marjoram, L., & Bagnat, M. (2015). Infection, inflammation and healing in zebrafish: Intestinal inflammation. *Current Pathobiology Reports*, *3*(2), 147–153. http://dx.doi.org/10.1007/s40139-015-0079-x.

Martinon, F., Burns, K., & Tschopp, J. (2002). The inflammasome: A molecular platform triggering activation of inflammatory caspases and processing of proIL-beta. *Molecular Cell*, *10*(2), 417–426.

Mathias, J. R., Dodd, M. E., Walters, K. B., Yoo, S. K., Ranheim, E. A., & Huttenlocher, A. (2009). Characterization of zebrafish larval inflammatory macrophages. *Developmental and Comparative Immunology*, *33*(11), 1212–1217. http://dx.doi.org/10.1016/j.dci.2009.07.003.

Mathias, J. R., Perrin, B. J., Liu, T. X., Kanki, J., Look, A. T., & Huttenlocher, A. (2006). Resolution of inflammation by retrograde chemotaxis of neutrophils in transgenic zebrafish. *Journal of Leukocyte Biology*, *80*(6), 1281–1288. http://dx.doi.org/10.1189/jlb.0506346.

Meijer, A. H. (2016). Protection and pathology in TB: Learning from the zebrafish model. *Seminars in Immunopathology*, *38*(2), 261–273. http://dx.doi.org/10.1007/s00281-015-0522-4.

Meijer, A. H., & Spaink, H. P. (2011). Host-pathogen interactions made transparent with the zebrafish model. *Current Drug Targets*, *12*(7), 1000–1017.

Meijer, A. H., van der Sar, A. M., Cunha, C., Lamers, G. E., Laplante, M. A., Kikuta, H., … Spaink, H. P. (2008). Identification and real-time imaging of a myc-expressing neutrophil population involved in inflammation and mycobacterial granuloma formation in zebrafish. *Developmental and Comparative Immunology*, *32*(1), 36–49. http://dx.doi.org/10.1016/j.dci.2007.04.003.

Mills, C. D. (2012). M1 and M2 macrophages: Oracles of health and disease. *Critical Reviews in Immunology*, *32*(6), 463–488.

Mostowy, S., Boucontet, L., Mazon Moya, M. J., Sirianni, A., Boudinot, P., Hollinshead, M., … Colucci-Guyon, E. (2013). The zebrafish as a new model for the in vivo study of Shigella flexneri interaction with phagocytes and bacterial autophagy. *PLoS Pathogens*, *9*(9). e1003588http://dx.doi.org/10.1371/journal.ppat.1003588.

Murayama, E., Kissa, K., Zapata, A., Mordelet, E., Briolat, V., Lin, H. F., … Herbomel, P. (2006). Tracing hematopoietic precursor migration to successive hematopoietic organs during zebrafish development. *Immunity*, *25*(6), 963–975. http://dx.doi.org/10.1016/j.immuni.2006.10.015.

Netea, M. G., Wijmenga, C., & O'Neill, L. A. (2012). Genetic variation in Toll-like receptors and disease susceptibility. *Nature Immunology*, *13*(6), 535–542. http://dx.doi.org/10.1038/ni.2284.

Nguyen-Chi, M., Laplace-Builhe, B., Travnickova, J., Luz-Crawford, P., Tejedor, G., Phan, Q. T., … Djouad, F. (2015). Identification of polarized macrophage subsets in zebrafish. *eLife*, *4*. e07288. http://dx.doi.org/10.7554/eLife.07288.

Nguyen-Chi, M., Phan, Q. T., Gonzalez, C., Dubremetz, J. F., Levraud, J. P., & Lutfalla, G. (2014). Transient infection of the zebrafish notochord with E. coli induces chronic inflammation. *Disease Models & Mechanisms*, 7(7), 871–882. http://dx.doi.org/10.1242/dmm.014498.

Nomiyama, H., Hieshima, K., Osada, N., Kato-Unoki, Y., Otsuka-Ono, K., Takegawa, S., … Yoshie, O. (2008). Extensive expansion and diversification of the chemokine gene family in zebrafish: Identification of a novel chemokine subfamily CX. *BMC Genomics*, *9*, 222. http://dx.doi.org/10.1186/1471-2164-9-222.

Noy, R., & Pollard, J. W. (2014). Tumor-associated macrophages: From mechanisms to therapy. *Immunity*, *41*(1), 49–61. http://dx.doi.org/10.1016/j.immuni.2014.06.010.

Oehlers, S. H., Cronan, M. R., Scott, N. R., Thomas, M. I., Okuda, K. S., Walton, E. M., … Tobin, D. M. (2015). Interception of host angiogenic signalling limits mycobacterial growth. *Nature*, *517*(7536), 612–615. http://dx.doi.org/10.1038/nature13967.

Oehlers, S. H., Flores, M. V., Hall, C. J., Swift, S., Crosier, K. E., & Crosier, P. S. (2011). The inflammatory bowel disease (IBD) susceptibility genes NOD1 and NOD2 have conserved anti-bacterial roles in zebrafish. *Disease Models & Mechanisms*, *4*(6), 832–841. http://dx.doi.org/10.1242/dmm.006122.

Ogryzko, N. V., Renshaw, S. A., & Wilson, H. L. (2014). The IL-1 family in fish: Swimming through the muddy waters of inflammasome evolution. *Developmental and Comparative Immunology*, *46*(1), 53–62. http://dx.doi.org/10.1016/j.dci.2014.03.008.

O'Neill, L. A., & Pearce, E. J. (2016). Immunometabolism governs dendritic cell and macrophage function. *The Journal of Experimental Medicine*, *213*(1), 15–23. http://dx.doi.org/10.1084/jem.20151570.

Pagan, A. J., Yang, C. T., Cameron, J., Swaim, L. E., Ellett, F., Lieschke, G. J., & Ramakrishnan, L. (2015). Myeloid growth factors promote resistance to mycobacterial infection by curtailing granuloma necrosis through macrophage replenishment. *Cell Host & Microbe*, *18*(1), 15–26. http://dx.doi.org/10.1016/j.chom.2015.06.008.

Page, D. M., Wittamer, V., Bertrand, J. Y., Lewis, K. L., Pratt, D. N., Delgado, N., … Traver, D. (2013). An evolutionarily conserved program of B-cell development and activation in zebrafish. *Blood*, *122*(8), e1–e11. http://dx.doi.org/10.1182/blood-2012-12-471029.

Palha, N., Guivel-Benhassine, F., Briolat, V., Lutfalla, G., Sourisseau, M., Ellett, F., … Levraud, J. P. (2013). Real-time whole-body visualization of Chikungunya virus infection and host interferon response in zebrafish. *PLoS Pathogens*, *9*(9). e1003619. http://dx.doi.org/10.1371/journal.ppat.1003619.

Palic, D., Andreasen, C. B., Ostojic, J., Tell, R. M., & Roth, J. A. (2007). Zebrafish (Danio rerio) whole kidney assays to measure neutrophil extracellular trap release and degranulation of primary granules. *Journal of Immunological Methods*, *319*(1–2), 87–97. http://dx.doi.org/10.1016/j.jim.2006.11.003.

Peri, F., & Nusslein-Volhard, C. (2008). Live imaging of neuronal degradation by microglia reveals a role for v0-ATPase a1 in phagosomal fusion in vivo. *Cell*, *133*(5), 916–927. http://dx.doi.org/10.1016/j.cell.2008.04.037.

Pham, L. N., Kanther, M., Semova, I., & Rawls, J. F. (2008). Methods for generating and colonizing gnotobiotic zebrafish. *Nature Protocols*, *3*(12), 1862–1875. http://dx.doi.org/10.1038/nprot.2008.186.

Pizzol, D., Di Gennaro, F., Chhaganlal, K. D., Fabrizio, C., Monno, L., Putoto, G., & Saracino, A. (2016). Tuberculosis and diabetes: Current state and future perspectives. *Tropical Medicine & International Health*, *21*, 694–702. http://dx.doi.org/10.1111/tmi.12704.

Ponpuak, M., Davis, A. S., Roberts, E. A., Delgado, M. A., Dinkins, C., Zhao, Z., … Deretic, V. (2010). Delivery of cytosolic components by autophagic adaptor protein p62 endows autophagosomes with unique antimicrobial properties. *Immunity*, *32*(3), 329–341. http://dx.doi.org/10.1016/j.immuni.2010.02.009.

Prajsnar, T. K., Cunliffe, V. T., Foster, S. J., & Renshaw, S. A. (2008). A novel vertebrate model of Staphylococcus aureus infection reveals phagocyte-dependent resistance of zebrafish to non-host specialized pathogens. *Cellular Microbiology*, *10*(11), 2312–2325. http://dx.doi.org/10.1111/j.1462-5822.2008.01213.x.

Prajsnar, T. K., Hamilton, R., Garcia-Lara, J., McVicker, G., Williams, A., Boots, M., … Renshaw, S. A. (2012). A privileged intraphagocyte niche is responsible for disseminated infection of Staphylococcus aureus in a zebrafish model. *Cellular Microbiology*, *14*(10), 1600–1619. http://dx.doi.org/10.1111/j.1462-5822.2012.01826.x.

Prykhozhij, S. V., & Berman, J. N. (2014). The progress and promise of zebrafish as a model to study mast cells. *Developmental and Comparative Immunology*, *46*(1), 74–83. http://dx.doi.org/10.1016/j.dci.2014.01.023.

Rajapakse, S., Rodrigo, C., & Rajapakse, A. (2010). Atypical manifestations of chikungunya infection. *Transactions of the Royal Society of Tropical Medicine and Hygiene*, *104*(2), 89–96. http://dx.doi.org/10.1016/j.trstmh.2009.07.031.

Ramakrishnan, L. (2012). Revisiting the role of the granuloma in tuberculosis. *Nature Reviews. Immunology*, *12*(5), 352–366. http://dx.doi.org/10.1038/nri3211.

Ramakrishnan, L. (2013). The zebrafish guide to tuberculosis immunity and treatment. *Cold Spring Harbor Symposia on Quantitative Biology*, *78*, 179–192. http://dx.doi.org/10.1101/sqb.2013.78.023283.

Rawls, J. F., Samuel, B. S., & Gordon, J. I. (2004). Gnotobiotic zebrafish reveal evolutionarily conserved responses to the gut microbiota. *Proceedings of the National Academy of Sciences of the United States of America*, *101*(13), 4596–4601. http://dx.doi.org/10.1073/pnas.0400706101.

Renshaw, S. A., Loynes, C. A., Trushell, D. M., Elworthy, S., Ingham, P. W., & Whyte, M. K. (2006). A transgenic zebrafish model of neutrophilic inflammation. *Blood*, *108*(13), 3976–3978. http://dx.doi.org/10.1182/blood-2006-05-024075.

Renshaw, S. A., & Trede, N. S. (2012). A model 450 million years in the making: Zebrafish and vertebrate immunity. *Disease Models & Mechanisms*, *5*(1), 38–47. http://dx.doi.org/10.1242/dmm.007138.

Rhodes, J., Hagen, A., Hsu, K., Deng, M., Liu, T. X., Look, A. T., & Kanki, J. P. (2005). Interplay of pu.1 and gata1 determines myelo-erythroid progenitor cell fate in zebrafish. *Developmental Cell*, *8*(1), 97–108. http://dx.doi.org/10.1016/j.devcel.2004.11.014.

Roca, F. J., & Ramakrishnan, L. (2013). TNF dually mediates resistance and susceptibility to mycobacteria via mitochondrial reactive oxygen species. *Cell*, *153*(3), 521–534. http://dx.doi.org/10.1016/j.cell.2013.03.022.

Sanderson, L. E., Chien, A. T., Astin, J. W., Crosier, K. E., Crosier, P. S., & Hall, C. J. (2015). An inducible transgene reports activation of macrophages in live zebrafish larvae. *Developmental and Comparative Immunology*, *53*(1), 63–69. http://dx.doi.org/10.1016/j.dci.2015.06.013.

Saralahti, A., & Ramet, M. (2015). Zebrafish and streptococcal infections. *Scandinavian Journal of Immunology*, *82*(3), 174–183. http://dx.doi.org/10.1111/sji.12320.

Sawamiphak, S., Kontarakis, Z., & Stainier, D. Y. (2014). Interferon gamma signaling positively regulates hematopoietic stem cell emergence. *Developmental Cell*, *31*(5), 640–653. http://dx.doi.org/10.1016/j.devcel.2014.11.007.

Schiebler, M., Brown, K., Hegyi, K., Newton, S. M., Renna, M., Hepburn, L., … Floto, R. A. (2015). Functional drug screening reveals anticonvulsants as enhancers of mTOR-independent autophagic killing of Mycobacterium tuberculosis through inositol depletion. *EMBO Molecular Medicine*, 7(2), 127–139. http://dx.doi.org/10.15252/emmm.201404137.

Schlegel, A., & Gut, P. (2015). Metabolic insights from zebrafish genetics, physiology, and chemical biology. *Cellular and Molecular Life Sciences*, 72(12), 2249–2260. http://dx.doi.org/10.1007/s00018-014-1816-8.

Schneider, V. A., van Dijk, A., van der Sar, A. M., Kraaij, M. D., Veldhuizen, E. J., & Haagsman, H. P. (2016). Prophylactic administration of chicken cathelicidin-2 boosts zebrafish embryonic innate immunity. *Developmental and Comparative Immunology*, *60*, 108–114. http://dx.doi.org/10.1016/j.dci.2016.02.023.

Silva, G. K., Gutierrez, F. R., Guedes, P. M., Horta, C. V., Cunha, L. D., Mineo, T. W., … Zamboni, D. S. (2010). Cutting edge: Nucleotide-binding oligomerization domain 1-dependent responses account for murine resistance against Trypanosoma cruzi infection. *Journal of Immunology*, *184*(3), 1148–1152. http://dx.doi.org/10.4049/jimmunol.0902254.

Stainier, D. Y., Weinstein, B. M., Detrich, H. W., 3rd, Zon, L. I., & Fishman, M. C. (1995). Cloche, an early acting zebrafish gene, is required by both the endothelial and hematopoietic lineages. *Development*, *121*(10), 3141–3150.

Stein, C., Caccamo, M., Laird, G., & Leptin, M. (2007). Conservation and divergence of gene families encoding components of innate immune response systems in zebrafish. *Genome Biology*, *8*(11), R251. http://dx.doi.org/10.1186/gb-2007-8-11-r251.

Stockhammer, O. W., Zakrzewska, A., Hegedus, Z., Spaink, H. P., & Meijer, A. H. (2009). Transcriptome profiling and functional analyses of the zebrafish embryonic innate immune response to Salmonella infection. *Journal of Immunology*, *182*(9), 5641–5653. http://dx.doi.org/10.4049/jimmunol.0900082.

Su, F., Juarez, M. A., Cooke, C. L., Lapointe, L., Shavit, J. A., Yamaoka, J. S., & Lyons, S. E. (2007). Differential regulation of primitive myelopoiesis in the zebrafish by Spi-1/Pu.1 and C/ebp1. *Zebrafish*, *4*(3), 187–199. http://dx.doi.org/10.1089/zeb.2007.0505.

Sumanas, S., Gomez, G., Zhao, Y., Park, C., Choi, K., & Lin, S. (2008). Interplay among Etsrp/ER71, Scl, and Alk8 signaling controls endothelial and myeloid cell formation. *Blood*, *111*(9), 4500–4510. http://dx.doi.org/10.1182/blood-2007-09-110569.

Sumanas, S., & Lin, S. (2006). Ets1-related protein is a key regulator of vasculogenesis in zebrafish. *PLoS Biology*, *4*(1). e10http://dx.doi.org/10.1371/journal.pbio.0040010.

Takizawa, H., Boettcher, S., & Manz, M. G. (2012). Demand-adapted regulation of early hematopoiesis in infection and inflammation. *Blood*, *119*(13), 2991–3002. http://dx.doi.org/10.1182/blood-2011-12-380113.

Tenor, J. L., Oehlers, S. H., Yang, J. L., Tobin, D. M., & Perfect, J. R. (2015). Live imaging of host-parasite interactions in a zebrafish infection model reveals cryptococcal determinants of virulence and central nervous system invasion. *MBio*, *6*(5). http://dx.doi.org/10.1128/mBio.01425-15. e01425-15.

Tilak, R., Ray, S., Tilak, V. W., & Mukherji, S. (2016). Dengue, chikungunya … the missing entity—Zika fever: A new emerging threat. *Medical Journal, Armed Forces India*, *72*(2), 157–163. http://dx.doi.org/10.1016/j.mjafi.2016.02.017.

Tobin, D. M., Vary, J. C., Jr., Ray, J. P., Walsh, G. S., Dunstan, S. J., Bang, N. D., … Ramakrishnan, L. (2010). The lta4h locus modulates susceptibility to mycobacterial infection in zebrafish and humans. *Cell*, *140*(5), 717–730. http://dx.doi.org/10.1016/j.cell.2010.02.013.

Torraca, V., Cui, C., Boland, R., Bebelman, J. P., van der Sar, A. M., Smit, M. J., … Meijer, A. H. (2015). The CXCR3-CXCL11 signaling axis mediates macrophage recruitment and dissemination of mycobacterial infection. *Disease Models & Mechanisms*, *8*(3), 253–269. http://dx.doi.org/10.1242/dmm.017756.

Torraca, V., Masud, S., Spaink, H. P., & Meijer, A. H. (2014). Macrophage-pathogen interactions in infectious diseases: New therapeutic insights from the zebrafish host model. *Disease Models & Mechanisms*, 7(7), 785–797. http://dx.doi.org/10.1242/dmm.015594.

Travnickova, J., Tran Chau, V., Julien, E., Mateos-Langerak, J., Gonzalez, C., Lelievre, E., … Kissa, K. (2015). Primitive macrophages control HSPC mobilization and definitive haematopoiesis. *Nature Communications*, *6*, 6227. http://dx.doi.org/10.1038/ncomms7227.

van der Sar, A. M., Musters, R. J., van Eeden, F. J., Appelmelk, B. J., Vandenbroucke-Grauls, C. M., & Bitter, W. (2003). Zebrafish embryos as a model host for the real time analysis of Salmonella typhimurium infections. *Cellular Microbiology*, *5*(9), 601–611.

van der Vaart, M., Korbee, C. J., Lamers, G. E., Tengeler, A. C., Hosseini, R., Haks, M. C., … Meijer, A. H. (2014). The DNA damage-regulated autophagy modulator DRAM1 links mycobacterial recognition via TLR-MYD88 to autophagic defense [corrected]. *Cell Host & Microbe*, *15*(6), 753–767. http://dx.doi.org/10.1016/j.chom.2014.05.005.

van der Vaart, M., Spaink, H. P., & Meijer, A. H. (2012). Pathogen recognition and activation of the innate immune response in zebrafish. *Advances in Hematology*, *2012*, 159807. http://dx.doi.org/10.1155/2012/159807.

van der Vaart, M., van Soest, J. J., Spaink, H. P., & Meijer, A. H. (2013). Functional analysis of a zebrafish myd88 mutant identifies key transcriptional components of the innate immune system. *Disease Models & Mechanisms*, *6*(3), 841–854. http://dx.doi.org/10.1242/dmm.010843.

Van Lint, P., & Libert, C. (2007). Chemokine and cytokine processing by matrix metalloproteinases and its effect on leukocyte migration and inflammation. *Journal of Leukocyte Biology*, *82*(6), 1375–1381. http://dx.doi.org/10.1189/jlb.0607338.

Varela, M., Romero, A., Dios, S., van der Vaart, M., Figueras, A., Meijer, A. H., & Novoa, B. (2014). Cellular visualization of macrophage pyroptosis and interleukin-1beta release in a viral hemorrhagic infection in zebrafish larvae. *Journal of Virology*, *88*(20), 12026–12040. http://dx.doi.org/10.1128/jvi.02056-14.

Varga, M., Fodor, E., & Vellai, T. (2015). Autophagy in zebrafish. *Methods*, *75*, 172–180. http://dx.doi.org/10.1016/j.ymeth.2014.12.004.

Veneman, W. J., Stockhammer, O. W., de Boer, L., Zaat, S. A., Meijer, A. H., & Spaink, H. P. (2013). A zebrafish high throughput screening system used for Staphylococcus epidermidis infection marker discovery. *BMC Genomics*, *14*, 255. http://dx.doi.org/10.1186/1471-2164-14-255.

Vergunst, A. C., Meijer, A. H., Renshaw, S. A., & O'Callaghan, D. (2010). Burkholderia cenocepacia creates an intramacrophage replication niche in zebrafish embryos, followed by bacterial dissemination and establishment of systemic infection. *Infection and Immunity*, *78*(4), 1495–1508. http://dx.doi.org/10.1128/iai.00743-09.

Voelz, K., Gratacap, R. L., & Wheeler, R. T. (2015). A zebrafish larval model reveals early tissue-specific innate immune responses to Mucor circinelloides. *Disease Models & Mechanisms*, *8*(11), 1375–1388. http://dx.doi.org/10.1242/dmm.019992.

Vogeli, K. M., Jin, S. W., Martin, G. R., & Stainier, D. Y. (2006). A common progenitor for haematopoietic and endothelial lineages in the zebrafish gastrula. *Nature*, *443*(7109), 337–339. http://dx.doi.org/10.1038/nature05045.

Volkman, H. E., Pozos, T. C., Zheng, J., Davis, J. M., Rawls, J. F., & Ramakrishnan, L. (2010). Tuberculous granuloma induction via interaction of a bacterial secreted protein with host epithelium. *Science*, *327*(5964), 466–469. http://dx.doi.org/10.1126/science.1179663.

Walton, E. M., Cronan, M. R., Beerman, R. W., & Tobin, D. M. (2015). The macrophage-specific promoter mfap4 allows live, long-term analysis of macrophage behavior during mycobacterial infection in zebrafish. *PLoS One*, *10*(10). e0138949. http://dx.doi.org/10.1371/journal.pone.0138949.

Wang, Z., & Zhang, S. (2010). The role of lysozyme and complement in the antibacterial activity of zebrafish (Danio rerio) egg cytosol. *Fish & Shellfish Immunology*, *29*(5), 773–777. http://dx.doi.org/10.1016/j.fsi.2010.07.002.

Wang, Z., Zhang, S., Tong, Z., Li, L., & Wang, G. (2009). Maternal transfer and protective role of the alternative complement components in zebrafish Danio rerio. *PLoS One*, *4*(2). e4498. http://dx.doi.org/10.1371/journal.pone.0004498.

Wang, Z., Zhang, S., & Wang, G. (2008). Response of complement expression to challenge with lipopolysaccharide in embryos/larvae of zebrafish Danio rerio: Acquisition of immunocompetent complement. *Fish & Shellfish Immunology*, *25*(3), 264–270. http://dx.doi.org/10.1016/j.fsi.2008.05.010.

Wiegertjes, G. F., Wentzel, A. S., Spaink, H. P., Elks, P. M., & Fink, I. R. (2016). Polarization of immune responses in fish: The 'macrophages first' point of view. *Molecular Immunology*, *69*, 146–156. http://dx.doi.org/10.1016/j.molimm.2015.09.026.

Willett, C. E., Cortes, A., Zuasti, A., & Zapata, A. G. (1999). Early hematopoiesis and developing lymphoid organs in the zebrafish. *Developmental Dynamics*, *214*(4), 323–336. http://dx.doi.org/10.1002/(sici)1097-0177(199904)214:4<323::aid-aja5>3.0.co;2-3.

Wood, W. B., Jr. (1960). Phagocytosis, with particular reference to encapsulated bacteria. *Bacteriological Reviews*, *24*(1), 41–49.

Yang, L., Bu, L., Sun, W., Hu, L., & Zhang, S. (2014). Functional characterization of mannose-binding lectin in zebrafish: Implication for a lectin-dependent complement system in early embryos. *Developmental and Comparative Immunology*, *46*(2), 314–322. http://dx.doi.org/10.1016/j.dci.2014.05.003.

Yang, C. T., Cambier, C. J., Davis, J. M., Hall, C. J., Crosier, P. S., & Ramakrishnan, L. (2012). Neutrophils exert protection in the early tuberculous granuloma by oxidative killing of mycobacteria phagocytosed from infected macrophages. *Cell Host & Microbe*, *12*(3), 301–312. http://dx.doi.org/10.1016/j.chom.2012.07.009.

Zakrzewska, A., Cui, C., Stockhammer, O. W., Benard, E. L., Spaink, H. P., & Meijer, A. H. (2010). Macrophage-specific gene functions in Spi1-directed innate immunity. *Blood*, *116*(3), e1–e11. http://dx.doi.org/10.1182/blood-2010-01-262873.

Zheng, F., Asim, M., Lan, J., Zhao, L., Wei, S., Chen, N., … Lin, L. (2015). Molecular cloning and functional characterization of mannose receptor in zebra fish (Danio rerio) during infection with Aeromonas sobria. *International Journal of Molecular Sciences*, *16*(5), 10997–11012. http://dx.doi.org/10.3390/ijms160510997.

Zou, J., Wen, Y., Yang, X., & Wei, X. (2013). Spatial-temporal expressions of Crumbs and Nagie oko and their interdependence in zebrafish central nervous system during early development. *International Journal of Developmental Neuroscience*, *31*(8), 770–782. http://dx.doi.org/10.1016/j.ijdevneu.2013.09.005.

CHAPTER NINE

Zebrafish in Toxicology and Environmental Health

Kathryn Bambino, Jaime Chu[1]
Icahn School of Medicine at Mount Sinai, New York, United States
[1]Corresponding author: e-mail address: jaime.chu@mssm.edu

Contents

Abstract

As manufacturing processes and development of new synthetic compounds increase to keep pace with the expanding global demand, environmental health, and the effects of toxicant exposure are emerging as critical public health concerns. Additionally, chemicals that naturally occur in the environment, such as metals, have profound effects

Current Topics in Developmental Biology, Volume 124
ISSN 0070-2153
http://dx.doi.org/10.1016/bs.ctdb.2016.10.007

on human and animal health. Many of these compounds are in the news: lead, arsenic, and endocrine disruptors such as bisphenol A have all been widely publicized as causing disease or damage to humans and wildlife in recent years. Despite the widespread appreciation that environmental toxins can be harmful, there is limited understanding of how many toxins cause disease. Zebrafish are at the forefront of toxicology research; this system has been widely used as a tool to detect toxins in water samples and to investigate the mechanisms of action of environmental toxins and their related diseases. The benefits of zebrafish for studying vertebrate development are equally useful for studying teratogens. Here, we review how zebrafish are being used both to detect the presence of some toxins as well as to identify how environmental exposures affect human health and disease. We focus on areas where zebrafish have been most effectively used in ecotoxicology and in environmental health, including investigation of exposures to endocrine disruptors, industrial waste byproducts, and arsenic.

1. INTRODUCTION

Rapid growth of populations and technological advancement has resulted in innumerable pollutants and environmental toxin exposure. This has generated a vital need for toxin surveillance, identification of consequences of exposure, and understanding of the biologic, chemical, and genetic mechanisms that underlie those effects (Landrigan, 2016). The field of environmental health was established as early as the 1940s, in response to the expansion of chemical manufacturing and the occurrence of contamination of the water, soil, and air caused by widespread use of chemicals in industry and consumer products (Landrigan, 2016). Since World War II, thousands of synthetic chemical compounds have been created for industrial applications and have subsequently been introduced into consumer products. Today, approximately 70,000 chemicals are in commercial use in the United States, and 3300 of these are high production volume compounds, with annual production or importation volumes in excess of one million pounds. In addition to the risks posed by the expanding repertoire of manufactured toxins, naturally occurring chemicals, such as metals can also cause harm, as was recently brought to focus by the lead contamination of the drinking water in Flint, Michigan (Bellinger, 2016; Tong, Baghurst, McMichael, Sawyer, & Mudge, 1996).

Environmental toxins profoundly affect fish and wildlife. In particular, water pollution has damaged fish populations by affecting reproductive health, lifespan, and embryonic and larval development. This has a major effect on aquatic ecosystems and on the industries that depend on them.

Humans are exposed to environmental toxicants through fine particulate matter in the air, endocrine-disrupting chemicals (EDCs) found in food packaging, household items and personal care products, and naturally occurring compounds such as metals. Human exposure to environmental chemicals is associated with both acute toxicity and long-term consequences (Landrigan et al., 2016), which include congenital abnormalities (Swan et al., 2005), chronic diseases (Argos et al., 2010; Mazumder, 2005), cognitive disabilities (Jacobson, Muckle, Ayotte, Dewailly, & Jacobson, 2015; Muñoz-Quezada et al., 2013; Tong et al., 1996), cancer (Liu & Wu, 2010; Selikoff & Hammond, 1968; Wang, Cheng, & Zhang, 2014), and death (Argos et al., 2010). The field of environmental health is expanding to meet the demands of surveillance and prevention of consequences of environmental toxin exposure on both wildlife and human health.

There are many unanswered questions in the field of environmental health (Henn, Coull, & Wright, 2014; Landrigan, Suk, & Amler, 1999), and a surge in research effort is required to answer these. Among the most pressing are: What are the effects of low dose, cumulative exposures, and exposures to multiple toxicants? What are the developmental processes that are altered by toxicant exposure and how are these processes affected? What are the latent effects of early life exposure? Are these effects apparent in subsequent generations? Can we develop surveillance technologies to limit exposure? How can therapeutic interventions be designed and administered to reverse the effects of exposure? The barriers to addressing these questions in human populations are both practical and logistical. In terms of low-dose and cumulative exposures, the appropriate biomarkers and the ideal tissue specimens for analysis have not been identified for every toxicant or combination of toxicants. Understanding the latent and transgenerational effects of exposure is difficult in humans due out long lifespans and relatively small number of offspring. In addition, the interaction between environmental toxicants and social "exposures" including chronic stress, exposure to violence, and nutrient scarcity are only beginning to be understood. There is a critical need for in vivo animal models to study the short- and long-term effects of environmental toxins.

Zebrafish are a valuable tool for Environmental Health researchers as evidenced by a rapidly expanding body of research using zebrafish. A PubMed search using the terms "zebrafish environmental health" reveals that the use of zebrafish in this field has been steadily increasing over the past few decades (Fig. 1). In this chapter, we will highlight the unique advantages of using zebrafish embryos, larvae, and adults to address pressing issues in

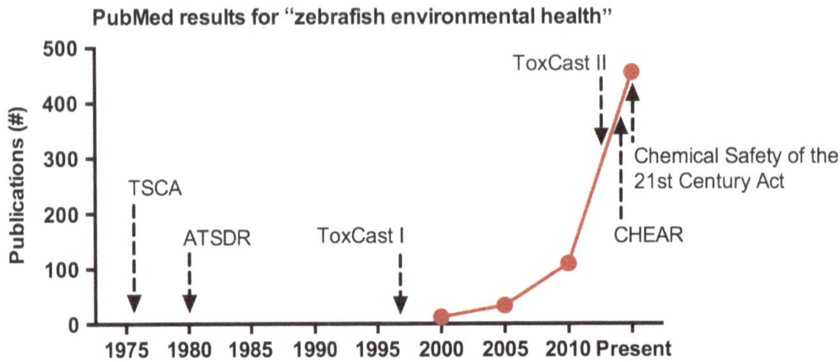

Fig. 1 Increasing use of zebrafish in environmental health studies. The use of zebrafish has steadily increased during the past three decades. *Line graph* represents the number of Pubmed articles categorized under "zebrafish environmental health." *Dashed arrows* indicate government programs and legislation focused on toxicology and environmental health from 1975 to present.

Environmental Health, including contaminant detection, environmental monitoring, toxicity/teratogenicity testing, and investigations into mechanisms of action and disease phenotypes associated with exposure to chemical compounds. In a field that rapidly changes with evolving technology and manufacturing worldwide, zebrafish can offer real-time in vivo studies to address potential hazards to human health that result from naturally occurring compounds and commercial use of new synthetics or byproducts of their production, and can improve our limited understanding of the specific effects of environmental exposures.

2. HISTORICAL PERSPECTIVE: TOXICOLOGY AND GOVERNMENT EFFORTS FOR ENVIRONMENTAL REGULATION

In 1976, the United States Congress passed the Toxic Substances Control Act, which granted the Environmental Protection Agency (EPA) the authority to require testing and reporting, and set restrictions on the manufacture and use of chemicals and mixtures (Toxic Substances Control Act, 1976). Since the passage of this law, science, and technology have made rapid progress necessitating further legislative intervention to protect the public from the health effects of chemical exposures (Birnbaum, 2010). In response, the National Toxicology Program was established in 1978 as an interagency program by the Department of Health, Education, and Welfare, which is known today as the Department of Health and Human Services, to address public, scientific, and governmental concerns that human diseases and disabilities are linked to chemical exposures (Xie, Holmgren, Andrews, & Wolfe, 2016). In 1980, as part of the Comprehensive Environmental Response, Compensation and Liability Act, more commonly referred to as the "Superfund Act," Congress formed the Agency for Toxic Substances and Disease Registry (ATSDR). ATSDR is a science-based public health agency created to study the health effects of hazardous substances in the environment and to work with communities to keep them safe from hazardous waste. To address this mission, ATSDR collects data and conducts studies in addition to using the best available scientific data to make recommendations to the EPA and other agencies to prevent and stop exposures to ensure the health of communities. Studies in zebrafish have provided critical information about the health effects of a majority of the 250+ substances on this list. Based on the strength of published data and the size of potentially exposed population, the ATSDR publishes a biannual Priority List of Hazardous Substances to select those substances that should be the subject of toxicological profiles.

With the dissolution of the National Children's Study 2014, the NIH reappropriated funds to create the Children's Health Exposure Analysis Resource (CHEAR) to focus on how children's health is shaped by the environment. Through CHEAR, researchers with NIH-funded child cohorts can apply to have biological samples analyzed for chemicals, metabolites, and biomarkers of exposure. In 2016, President Obama signed the Frank R. Lautenberg Chemical Safety of the 21st Century Act into law requiring that the EPA perform testing of chemicals currently in use, set new risk-based safety standards, require protection for vulnerable populations, and increase public transparency for chemical information. The use of zebrafish has and will continue to be a tool to provide the EPA with invaluable information regarding the short-term toxicity and long-term health effects of toxicant exposure to human health and its utility has been highlighted in a special issue of the journal *Zebrafish* that was dedicated to the use of this model in toxicology research (Gamse & Gorelick, 2016).

3. ENVIRONMENTAL AND DEVELOPMENTAL TOXICOLOGY

Zebrafish are commonly used to model human diseases using genetic modifications; applications including studies of heart, kidney, liver, hematopoietic, immune, and other systems detailed other chapters in this volume. They can similarly be used to model the health effects of environmental exposures to better understand the etiologies and mechanisms of environment-related disease in humans. The concern is growing over the persistence of chemical compounds in the environment as well as the acute and long-term health effects of exposure to environmental toxicants and contaminants and zebrafish provide an ideal model to study these effects. Chemicals can simply be added to the embryo medium and the developing and transparent zebrafish can be assessed for lethality and developmental abnormalities from fertilization through larval stages. Although juvenile and adult zebrafish are not transparent, the generation of the unpigmented Casper mutant line can be crossed to transgenic fluorescent reporters to aid observation and imaging of organ systems in older zebrafish (White et al., 2008).

The ability to observe effects of toxins in vivo allows for direct assessment of toxicity, as well as measurements of absorption, distribution, metabolism, and elimination. This can be extended for use in screening for treatments that can mitigate toxic effects in live animals as well. Zebrafish express a full range of *cytochrome P450* (*cyp*) genes required for xenobiotic metabolism and biotransformation (Goldstone et al., 2010). In the zebrafish genome assembly (GRCz10), a total of 86 *cyp* genes were identified (Saad et al., 2016) with many of the metabolic characteristics of the related human enzymes, demonstrating a strong evolutionary relationship with those found in humans. However, there remains a significant lack of information about the specific mechanisms of zebrafish xenobiotic Cyp activity.

Zebrafish have been used to study the compounds ranging from naturally occurring metals and metalloids, to synthetic components of consumer products, pesticides, and byproducts of industrial processing and waste incineration. In this chapter, we will present the zebrafish tools that have been employed for the detection of these toxicants and how zebrafish research is contributing to the understanding of the effects of these compounds on the environment and on human health.

4. ZEBRAFISH: TESTING THE WATERS FOR TOXICANTS

In 1982, George Streisinger, the founder of the zebrafish field (Streisinger, Walker, Dower, Knauber, & Singer, 1981), proposed the use of zebrafish as a vertebrate model to study the frequency of mutations in response to environmental carcinogens (Streisinger, 1983). In the following three decades, zebrafish have been used to identify teratogens, to uncover mechanisms of action of common toxicants, and to understand the tissue specificity of toxicant impact on vertebrates (Gamse & Gorelick, 2016). Zebrafish provides a unique, in vivo, medium-throughput system to expand cell culture assays to a whole vertebrate model, but are less expensive than rodents. The benefit of the large population size of zebrafish offspring is a major benefit, as studies in zebrafish allow for the rapid assessment of compound toxicity and the ability to study molecular mechanisms underlying developmental and health outcomes associated with toxicant exposure across a population of live vertebrates. In addition, large numbers of offspring enable longitudinal studies that can be done on a population scale of the developmental effects of environmental exposures at a relatively less cost than longitudinal rodent studies. Zebrafish are also easily amenable to drug discovery screens as sentinels of environmental contamination, for toxicity testing, and for investigations into the mechanisms of action of pharmaceuticals and toxicants. The zebrafish model provides the opportunity to combine the power of rapid toxicology screens with the ability to study the association of exposures with long-term outcomes in a vertebrate, making zebrafish an invaluable complementary system for research in Environmental Health.

4.1 Transgenic Zebrafish as Surveillance Tools

For nearly two decades, zebrafish have been used for biomonitoring. A major advantage of using zebrafish for this work is that the embryos and larvae are transparent and generating transgenic animals is relatively easy. This has allowed the development of transgenic lines where a fluorescent protein or other measureable readout becomes activated in the presence of contaminants or environmental stressors provide a system to assess the level of response and the tissue specificity of the response (Carvan, Dalton, Stuart, & Nebert, 2000; Gorelick, Iwanowicz, Hung, Blazer, & Halpern, 2014; Lee, Green, & Tyler, 2015). In the earliest efforts, investigators developed transgenic zebrafish lines in which expression of the luciferase or green fluorescent protein (*GFP*) gene is driven by pollutant response elements that report on the presence of aromatic hydrocarbons, electrophiles/oxidants, metals, estrogenic compounds, or retinoids (Carvan et al., 2000). Transgenic lines have been used not only to detect the presence of toxins, but can facilitate investigations into the molecular mechanisms underlying pathology associated with environmental exposures. The use

Table 1 Transgenic Zebrafish Lines for Reporting Toxicant Exposure

Transgenic Line	Reporter for	Toxicants Tested	References
Tg(cyp1a:nls-gfp)	Cytochrome p450 Cyp1a	Aromatic hydrocarbons, dioxin-like compounds	Kim et al. (2013)
Tg(cyp1a:gfp)	Cytochrome p450 Cyp1a	Aromatic hydrocarbons, dioxin-like compounds	Xu et al. (2015)
Tg(mt:egfp)	Metallothionein	Heavy metals	Liu, Yan, Wang, Wu, and Xu (2016)
Tg(huORFZ:gfp)	Human CHOP	Heavy metals, endocrine disruptors	Lee et al. (2014)
TgBAC (hspb11:GFP)	Small heat shock protein hsbp11	Pesticides	Shahid et al. (2016)
Tg(5xERE: GFP)	Estrogen receptor activity	Estradiol, xenoestrogens, environmental water samples	Gorelick et al. (2014), Gorelick and Halpern (2011), and Gorelick, Pinto, Hao, and Bondesson (2016)
Tg(cyp19a1b: GFP)	Cytochrome p450 cyp19a1b, estrogen receptor activity	BPA, environmental water samples	Cano-Nicolau et al. (2016) and Sonavane et al. (2016)

of transgenic reporter zebrafish lines to measure exposure to heavy metals, organic chemicals, endocrine disruptors, and electrophilic agents has been expertly reviewed elsewhere (Lee et al., 2015) and are outlined in Table 1. We describe how such tools are used to both detect the presence of contaminants and to understand their physiological impact.

4.2 Biosensors of Environmental Contaminants

Zebrafish have been used as sentinels to identify the effects of public water supplies. An early study examined the teratogenic effects of sediment and ground water in the Netherlands in zebrafish combined with a biochemical assay in tissue culture cells and found that the zebrafish teratogen assay was equally as sensitive in identifying the presence of toxic contaminants (Murk et al., 1996). In more recent work, several researchers have generated transgenic reporter lines in which a promoter drives expression of a fluorescent protein or other reporter that is regulated by exposure to a toxin (Table 1).

In this section, we will highlight specific transgenic lines that have been generated and used to not only identify classes of chemical contaminants in experimental settings, but with the potential to lend insights into toxicant-induced stress responses.

4.2.1 Aromatic Hydrocarbons and the Aryl Hydrocarbon Receptor (Ahr)

The aryl hydrocarbon receptor (Ahr) is a cytosolic receptor that is expressed in various tissues during development and adulthood, and signaling through this receptor has been studied in multiple developmental processes in rodents and zebrafish (Schneider, Branam, & Peterson, 2014). The Ahr is activated in response to synthetic and natural aromatic (aryl) hydrocarbons and functions as a transcription factor to bind to the dioxin-responsive element (DRE) to induce the expression of genes including those encoding the CYP enzymes, which are involved in xenobiotic metabolism. A DRE-containing fragment of the *cyp1a1* gene, which is regulated by Ahr, was used to drive expression of a nuclear-localized GFP (*Tg(cyp1a:nls-gfp)*) and this shows activation in response to 2,3,7,8-tetrachlorodibenzo-*p*-dioxin (TCDD) (Kim et al., 2013). However, more recently, the *Tg(cyp1a:gfp)* transgenic line was generated using the medaka *cyp1a* promoter, and this transgenic zebrafish provides a more sensitive biosensor for Ahr activity (Xu et al., 2015). Use of these systems has identified the kidney, liver, and gut as target tissues for TCDD and have also been shown to respond to other dioxin-like chemicals and polyaromatic hydrocarbons (Xu et al., 2015).

4.2.2 Metals

The discovery of elevated lead levels in the drinking water of Flint, Michigan in 2016 has renewed efforts to mitigate the toxic effects of human exposure to metals. In addition to lead, many other metals including copper, platinum, cadmium, and zinc have severe toxic effects on humans and animals. Zebrafish have been used extensively to study the consequences of metal exposure, and a unique transgenic animal has been developed to detect the presence of metals in water. The *Tg(mt:egfp)* transgenic zebrafish expresses enhanced green fluorescent protein (EGFP) under the transcriptional control of the metal-responsive metallothionein promoter. This line can be used as a reporter for aquatic zinc and cadmium (Liu, Yan, et al., 2016). Recent data had shown that 10 days postfertilization (dpf) zebrafish larvae did not show significant developmental abnormalities even when exposed to levels of heavy metals that exceeded current regulatory limits by 10- or 70-fold for zinc and cadmium, respectively; however, transgene activity was detected following 24 h of exposure to zinc at the current regulatory limit and cadmium at twice the current regulatory limit. Use of the *Tg(mt:egfp)* zebrafish line provides an advance in the field as it provides a more robust readout for the presence of elevated levels of heavy metals (Liu, Yan, et al., 2016). This is useful, as exposure to a number of different metals is associated with neurodevelopmental deficits. However, further work is needed to refine these tools to respond to additional environmentally relevant metals at a broader range of concentrations.

4.3 Zebrafish Transgenics Shine Light on the Mechanisms of Toxicant-Related Disease

Cellular stress is a central and conserved response to toxin exposure. Many pollutants, including metals, pesticides, and oxidative agents are known or suggested to induce endoplasmic reticulum (ER) stress (Chen, Melchior, & Guo, 2014; Kitamura, 2013). Induction of ER stress contributes to a variety of human diseases including neurodegenerative diseases, metabolic dysfunction, inflammatory diseases, and cancer, the risks for which may be compounded by underlying toxicant exposure (Wang & Kaufman, 2016). Oxidative stress is one cause of ER stress, and the metabolism of many toxic compounds, including pesticides and metals, results in the generation of reactive oxygen species. As of yet, there have been few animal models in which to investigate the consequences of toxicant exposure and metabolism, and zebrafish represent a significant advance in this field.

4.3.1 Cellular Stress Reporters

CHOP (also called DDIT3) is a transcription factor that is strongly induced and translocated to the nucleus in response to some types of ER stress (Harding, Zhang, & Ron, 1999; Palam, Baird, & Wek, 2011). The transgenic zebrafish line *huORFZ* contains a GFP transgene under the control of the upstream open reading frame of the human *CHOP* cDNA (Lee et al., 2011). There are several important features of the *huORFZ* model that make it ideal for first-line pollution monitoring: (1) exposure to different chemical stressors results in distinct patterns of GFP expression, indicating the cell types and organ systems that respond to a given toxicant; (2) the system is responsive to several pollutants at the range of concentrations enforced by current World Health Organization guidelines; and (3) GFP expression decreased following the exposure period, which suggests that expression is a direct result of the physiological response upon toxicant exposure. Lee et al. (2015) have demonstrated that this transgenic line can be used to detect the presence of environmental contaminants, including heavy metals and EDCs (Lee et al., 2014). The response of this transgenic line is not limited to a particular stressor and can be applied to a range of chemicals or toxicants that induce ER stress.

The heat shock response is a cellular strategy used to protect the cell, by the induction of a number of protein chaperones, which prevents the aggregation of unfolded and misfolded proteins that accumulate due to stress. In addition to heat stress, this survival-promoting response can be induced by aging, protein-folding diseases, and exposure to toxic chemicals (Scheff Jeremy, Stallings Jonathan, Reifman, & Rakesh, 2015). The heatshock promoter driving GFP *TgBAC(hspb11:GFP)* has been used as a surrogate marker to identify the tissue-specific effects of pesticides (Shahid et al., 2016). This recent study highlights how different cell types are impacted by exposure to the same stressor. Zebrafish embryos were exposed to a number of different pesticides from 9 to 48 hours postfertilization (hpf) and examined for induction of the *TgBAC(hspb11:GFP)* transgene and muscle integrity. This transgene largely is activated in the muscle and notochord of embryos exposed to pesticides, however, the magnitude of the response varied: azinphosmethyl had a moderate effect on induction of the *hspb11* transgene and also only modestly affected muscle integrity, whereas, galanthamine caused severe disruption of muscle integrity and strongly activated the *hspb11* promoter (Shahid et al., 2016). Interestingly, the transgene remained active in muscle tissue up to 48 h after the pesticides were removed, indicating the long-lasting effects of toxin exposure on these cells.

4.3.2 Reporters of Endocrine Activity

Activation of estrogen receptors (ERs) is important for developmental processes and sexually dimorphic behaviors. In addition to estradiol and environmental estrogens, several synthetic or exogenous compounds are known to interfere with hormone signaling and have endocrine-disrupting activity (EDCs). One of the main challenges of assessing the effects of EDC is that these compounds are typically functional at very low concentrations and exhibit nonlinear dose responses (Vandenberg et al., 2012). Because of this, the Endocrine Society recommends a "no-threshold" approach to risk assessment for EDC (Zoeller et al., 2012). Zebrafish transgenic reporters thus provide a unique system in which to detect endocrine activity in the absence of gross morphological abnormalities.

Two common zebrafish transgenic reporters that are used for the detection of estrogen receptor signaling are the *Tg(5xERE:GFP)* (Gorelick &

Halpern, 2011; Gorelick et al., 2016) and *Tg(cyp19a1b:GFP)* (Cano-Nicolau et al., 2016; Sonavane et al., 2016) zebrafish lines. These lines respond to a range of estrogenic compounds at different doses. For instance, 17α-ethynylestradiol (EE) and diethylstilbestrol (DES) induce GFP expression at the pM to nM range, whereas bisphenol A (BPA) does not induce fluorescence at exposures below the μM range (Cano-Nicolau et al., 2016; Gorelick & Halpern, 2011). In addition, reporters of endocrine activity have been used to detect environmental contamination in water samples (Gorelick et al., 2014; Sonavane et al., 2016) with similar sensitivity to the established bioluminescent yeast assay (Gorelick et al., 2014). Both of these transgenic lines have been used to detect estrogens in samples collected using the Polar Organic Chemical Integrative Sampler, which concentrates estrogens from environmental water samples. Samples extracted from the membranes are diluted in embryo water, at concentrations higher than that found directly at sampling sites, for exposure and assessment of reporter activity. Due to the nature of the sampling method, these studies determine the estrogenic effects of environmental mixtures. Although use of these transgenic reporters for detection of environmental estrogens has not yet resulted in policy change, these studies highlight the use of zebrafish not only for the detection of estrogenic activity at a single point, but reveal their use in determining variations in contamination levels over time.

While these and other reporters have been highly effective in providing both a practical tool for water quality surveillance and for studying the mechanism of toxin-mediated damage, one major limitation is that by using fluorescent proteins such as GFP, which are slow to mature and have a long half-life, these reporters cannot capture the dynamic response to toxins. A second limitation is that the detection of fluorescent reporters in high-throughput automated imaging systems may be hindered by suboptimal embryo positioning (or nonuniform transgene expression across the zebrafish), such that the brightest parts of some embryos are not imaged accurately. New approaches to surmount these challenges are currently being developed and will enhance the utility of zebrafish transgenics to uncover mechanisms underlying environmental toxicant exposures.

5. TRANSCRIPTIONAL PROFILING TO IDENTIFY CONTAMINANTS

Integration of "-omic" technologies into environmental toxicology has been occurring at a rapid pace as new advances in image processing and data analysis make the use of these applications more feasible. Connectivity mapping is a data-driven approach combining transcriptomics and machine learning technology, and has previously been used to link disease and drug-induced phenotypes on the basis of differential gene expression patterns (Lamb et al., 2006). This approach has been applied to assess exposure and toxicity to chemical groups based on mechanism(s) of action and transcriptomic changes. Software packages have been developed to allow users to compare gene expression profiles under their treatment or exposure conditions to those archived in publicly available databases (Sandmann, Kummerfeld, Gentleman, & Bourgon, 2014). Wang et al. (2016) published the first use of connectivity mapping in environmental health using zebrafish. By mining publicly available microarray datasets, the group compared the transcriptional responses to a range of chemical exposures and doses in different organs from zebrafish and fathead minnow. Mapping is more successful within species and among those samples run on the same platform. As the cost of mRNA sequencing technologies reduces, whole genome data will be available for an increasing number of model systems under different experimental conditions. Connectivity mapping also offers researchers the opportunity to generate hypotheses about the mechanisms of action for environmental pollutants or toxins for which mechanistic pathways were previously unknown.

Many toxins specifically affect the liver, as this is the primary site of xenobiotic metabolism in vertebrates. Pathologies ranging from necrosis and fatty liver, to steatohepatitis and liver cancer have been found to result from occupational and environmental exposure to chemicals and toxicants (Al-Eryani et al., 2015; Wahlang et al., 2013). Toxin-specific hepatic responses have been identified in zebrafish using gene expression analysis on a variety of platforms to determine the hepatic response to a range of toxins, including arsenic, acetaminophen, and ethanol (Xu, Lam, Shen, & Gong, 2013; Zhang, Li, & Gong, 2014). A well-defined genome, easy access to target organs, and conserved responses to toxins make zebrafish amenable to emerging genomic, proteomic, and metabolomics approaches to better understand the molecular changes caused by toxins.

6. HIGH-THROUGHPUT SCREENING FOR TOXICITY STUDIES

A major goal for toxicology studies is to be able to screen many compounds in a short amount of time and with accuracy in predicting human toxicity. In 2007, the EPA began the ToxCast program to screen chemicals in order to develop protocols that would lead to improved human toxicity prediction (Dix et al., 2007). The pilot study, ToxCast Phase I, included 310 compounds (mostly pesticides) that were screened in a large number of medium- and high-throughput screening assays (Judson et al., 2010). That same year, the National Research Council published a report titled "Toxicity testing in the 21st Century: A Vision and Strategy" (National Research Council, 2007), which prompted rapid expansion of ToxCast, and in Phase II of the ToxCast program, the chemical library was expanded to 1878 compounds for which testing concluded in 2013 (Richard et al., 2016). ToxCast Phase I and II library compounds have been tested in model organisms including *C. elegans* and zebrafish (Boyd et al., 2016; Padilla et al., 2012; Sipes, Padilla, & Knudsen, 2011). Phase III of the ToxCast program contains greater than 3800 unique chemicals and compounds under evaluation (Richard et al., 2016).

Toxicology in the 21st Century (Tox21) program is a collaboration between the NIEHS National Toxicology Program, the EPA, and the National Center for Advancing Translational Science to test more than 10,000 environmental chemicals and drugs to elucidate their toxicity in biochemical and cell-based assays (Collins, Gray, & Bucher, 2008). The FDA's ToxCast joined the collaboration in 2010 and are now jointly referred to as the "ToxCast chemical library" (Richard et al., 2016). The European community has also responded with the EU *R*egistration, *E*valuation, *A*uthorization and restriction of *CH*emical substances (REACH) legislation, requiring the collection of toxicity data for chemicals that are produced or marketed in quantities in excess of one ton per year (Selderslaghs, Blust, & Witters, 2012). Several research centers in Europe now use zebrafish as the central animal model for toxicology studies and centralized groups have issued a white paper calling for increased resources for using zebrafish for toxicology research (www.eufishbiomed.kit.edu).

Limitations to these approaches are that xenobiotic metabolism cannot be studied in vitro, determining active in vivo doses and blood concentrations from in vitro studies is not possible, understanding the effects of chronic exposure is impossible in vitro, and knowing whether or when a given genetic or signaling perturbation would result in a phenotypic change in an animal is difficult to ascertain (Tice, Austin, Kavlock, & Bucher, 2013).

Zebrafish are being used as a first-pass screen to identify chemicals with the highest likelihood of posing risk to humans and require further testing (Dix et al., 2007). Researchers at the EPA used the zebrafish developmental assay to add information to the toxicity assay database of the ToxCast Phase I library (Padilla et al., 2012). In this first large-scale screen of the effects of environmental contaminants zebrafish, embryos were exposed from 6 hpf to 5 dpf to a single dose and then a concentration range from 1 nM to 80 μM. Survival and morphological defects were assessed at 6 dpf. This was expanded in a subsequent study that analyzed the effects of several hundred chemicals from the ToxCast Phase II library on 18 different endpoints in zebrafish larvae at 5 dpf (Truong et al., 2014). More recently, 1060 compounds from the ToxCast I and II chemical libraries have been tested in a phenotype-based screen in zebrafish to predict teratogenic effects. This study showed that hypoactivity at 24 hpf in exposed zebrafish embryos is associated with an increased risk of 17 specific developmental abnormalities as assessed 5 dpf larvae (Reif et al., 2016). Interestingly, this study also identified a group of chemical compounds that caused the same degree of hypoactivity at 24 hpf, with no corresponding morphologic defect at 5 dpf, indicating that this protocol may prevent false negatives. Efforts to build databases and develop assays to predict human toxicity have capitalized on the use of zebrafish as a quick, medium throughput in vivo system to accurately predict human toxicity.

In 2009, an international group of pharmaceutical companies formed a consortium to develop a zebrafish development assay that could correctly classify a set of 10 teratogenic and 10 nonteratogenic compounds (Gustafson et al., 2012). The results of these toxicity tests were compared to mammalian data, and found to have an overall concordance of 60–70%. In a second phase of this consortium project, 38 proprietary pharmaceutical compounds were tested by two independent laboratories, and 79% of the classifications were the same between the laboratories, although the laboratories differed in their concordance with in vivo data (Ball et al., 2014). The Dechorionated Zebrafish Embryo Developmental toxicity assay was developed to identify the no-adverse-effect-level (NOAEL) and the concentration resulting in 25% lethality (LC_{25}) for a training set of 31 compounds (Brannen, Panzica-Kelly, Danberry, & Augustine-Rauch, 2010). This approach yielded 87% concordance with published in vitro teratogenicity data (Brannen et al., 2010). Improvements to this assay, including enzymatic removal of the chorion, repeating the assay with a distinct set of text compounds, and using various zebrafish strains, have been attempted to make a direct comparison between the chorion-on data published by the pharmaceutical company consortium and the chorion-off data to determine whether the presence of the chorion affected the sensitivity and specificity of the zebrafish embryo assay (Ball et al., 2014; Brannen et al., 2010; Gustafson et al., 2012; Panzica-Kelly, Zhang, & Augustine-Rauch, 2015).

New advances in the morphological assessment of toxicant-exposed zebrafish larvae allow for the determination of the effects of test compounds on developmental endpoints. Computational approaches, including the Cellomics® ArrayScan® V^{TI} high-content image analysis platform reduce the time required for analysis and reduce variability between experiments while providing an image that can be kept for permanent record or reevaluated manually (Deal et al., 2016). Bright field image analysis will identify a large number of phenotypes that may be undetectable using other methods. The use of zebrafish offers a valuable tool for high-throughput screening of compounds with demonstrated accuracy in predicting human toxicity. Further, development of these technologies and platforms will be important for identifying target organs and generating hypotheses about mechanisms of action for chemicals for which no biological data are available.

7. ASSESSING HEALTH IMPACTS OF ENVIRONMENTAL EXPOSURES USING ZEBRAFISH

The ability to assess tissues for toxin accumulation and its associated phenotypes can lead to valuable insights into disease processes and enable therapeutic compound screening. While transgenic reporter lines can monitor differentiation of distinct cell lineages and detect the induction of signaling pathways, much of the data acquisition is limited to low- and medium-throughput applications due to the time required for screening individual zebrafish embryos and larvae by fluorescence microscopy.

7.1 Automated Reporter Quantification In Vivo (ARQiv)

Automated reporter quantification in vivo (ARQiv) is a high-throughput screening platform that uses a microplate reader to detect changes in the intensity of transgenic fluorescent reporters in live zebrafish embryos and larvae over time (Walker et al., 2012). Recently, ARQiv technology has been applied to test FDA-approved drugs and their ability to increase the number of insulin-producing pancreatic β cells in a transgenic reporter zebrafish line (*Tg(ins:PhiYFP-2a-nsfB; sst2:tagRFP)lmc01*), demonstrating the feasibility of this approach for both quantification of cell number and fluorescent reporter intensity (Wang et al., 2015). An increase of as little as 10 β cells in the developing pancreas was detected, highlighting the ability of this technique to identify small changes in the development of this important organ.

Although transgenic reporters are routinely used to visualize the effects of drugs and chemical compounds on developing organ systems (Lam et al., 2011; Ma et al., 2015), this technology can also be applied to environmental exposures using the same tissue- and signaling pathway-specific transgenic reporter lines to assess toxin-induced effects on the development of cell types and organ structures. One major limitation is that it only provides quantification of reporter levels without corresponding images to allow for the analysis of morphological changes associated with changes in reporter activity (Wang et al., 2015). Data generated using ARQiv will need to be coupled with that from other imaging techniques to obtain full understanding about the phenotypic effects of a particular exposure.

7.2 Laser Ablation-Inductively Coupled Plasma-Mass Spectroscopy (LA-ICP-MS)

Laser ablation-inductively coupled plasma-mass spectroscopy (LA-ICP-MS) can be used to provide spatial information about element distribution in biological samples (Hare, Austin, & Doble, 2012). This technique can be used in calcified tissue (teeth) and soft tissue (placenta) (Arora et al., 2014; Niedzwiecki et al., 2016). We are currently optimizing the use of LA-ICP-MS to determine tissue accumulation and organ-specific distribution of elements in whole zebrafish larvae (data not shown). The technique is able to detect compounds at concentrations below parts-per-million and has spatial resolution capacity at the micrometer range allowing for detailed analysis of tissue; however, quantification of trace elements within tissue samples using LA-ICP-MS analysis is not yet reliable due to properties of the ablation process (Hare et al., 2012).

7.3 Automated Assessment of Behavior and Morphologic Phenotypes

Complex developmental effects associated with exposures can be studied in zebrafish using behavioral profiling (Rihel et al., 2010) and phenotype-driven screens (Gallardo et al., 2015). Behavioral profiling is most useful for modeling effects on brain activity and has recently been used to identify phenotypic suppressors of autism in a zebrafish genetic model of hyperactivity (Hoffman Ellen et al., 2016). Similar efforts could be used to identify environmental modifiers of genes associated with autism spectrum and other neurological disorders. Phenotype-driven chemical screening has been used to identify compounds that altered the collective migration of fluorescently marked cells (Gallardo et al., 2015).

Automation of image capture and phenotype analysis will improve the ability of researchers to screen larger libraries of compounds over wider concentration ranges, while limiting bias in the assays (Deal et al., 2016; Jeanray et al., 2015; Mikut et al., 2013). Optimized techniques for embryo immobilization will enable imaging the developing zebrafish larvae using state of the art techniques including light sheet fluorescence microscopy (Höckendorf, Thumberger, Wittbrodt, 2012; Kaufmann, Mickoleit, Weber, & Huisken, 2012). Zebrafish can facilitate analysis of developmental and structural changes over time, but require development of advanced video capabilities. Recently, an open source application for the video analysis of movement of larval zebrafish has been created for academic use (Cario, Farrell, Milanese, & Burton, 2011). The rapid advances in imaging and computational technologies to identify the morphologic consequences of toxicant exposure in zebrafish, put this model at the forefront of the field with the potential to advance the identification of teratogenic and tissue-specific effects of toxins.

8. LONG-TERM AND TRANSGENERATIONAL EFFECTS OF TOXIN EXPOSURES

Exposure to environmental toxicants during development can have both acute consequences to the embryo, leading to congenital anomalies and poor birth outcomes, as well as long-term health consequences throughout the life of an individual. In addition, exposure to low doses of environmental contaminants can have latent health effects that are not apparent for years, even after the cessation of exposure. The fetal origins hypothesis, also called Barker hypothesis, was first described by David Barker in 1986 following the observation that poor infant nutrition was associated with poor cardiovascular outcomes among men in England and Wales (Barker & Osmond, 1986). Adverse health effects in this cohort are thought to be the result of altered developmental programming or physiological changes that make an individual susceptible to disease.

A well-known examples of this phenomenon is the causal association between the development of vaginal clear cell adenocarcinoma in women who were exposed to diethylstilbestrol in utero (Hatch, Palmer, Titus-Ernstoff, et al., 1998; Herbst, Ulfelder, & Poskanzer, 1971). This hypothesis has recently been tested in zebrafish to study the latent effects of embryonic exposure to atrazine, an herbicide and suspected endocrine disruptor (Wirbisky et al., 2015, 2016), and TCDD, a persistent environmental pollutant (Baker, Peterson, & Heideman, 2013). Both male and female adult zebrafish that developed from embyros exposed to atrazine, a widely used herbicide, demonstrate altered expression of genes related to neuroendocrine function (Wirbisky et al., 2016, 2015). In another study, early life exposure to low doses of TCDD during the embryonic period caused few malformations in the fish during the exposure period; more profound were the transgenerational effects of early exposure: the offspring of adults which developed from embryos exposed to TCDD showed morphological abnormalities, including skeletal defects, and reduced reproductive success (Baker et al., 2013). Zebrafish that were exposed to TCDD during the sex determination period (3–7 weeks postfertilization) displayed skeletal anomalies in adulthood and mismatches between secondary sex characteristics and the sex of the gonads as determined by histological analysis (Baker et al., 2013). These studies exemplify the power of the zebrafish system to feasibly demonstrate early, late, and transgenerational effects of toxin exposure.

9. PATHWAYS AND MECHANISMS OF TOXICANT-INDUCED DISEASE

Zebrafish can provide a powerful tool to investigate the mechanisms of action of environmental pollutants and its related diseases, and can be used to test therapeutic candidates or intervention measures to mitigate the effects of environmental contaminants, with the goal of translation to human disease. Here, we highlight examples of mechanistic insights generated from zebrafish models of exposure, including inorganic arsenic, BPA and TCDD (Fig. 2).

9.1 Arsenic

Inorganic arsenic is a naturally occurring element that epidemiological studies have linked with multiple adverse health outcomes (Vahter, 2008). Arsenic is classified as a human carcinogen by the International Agency for Research on Cancer and long-term exposure is associated with increased risk of several cancers, including bladder, kidney, liver, and skin cancer (IARC, 2004; Wang et al., 2014). Data from the first longitudinal study of people chronically exposed to inorganic arsenic through drinking water has found that the latency for health effects can be decades (Ahsan et al.,

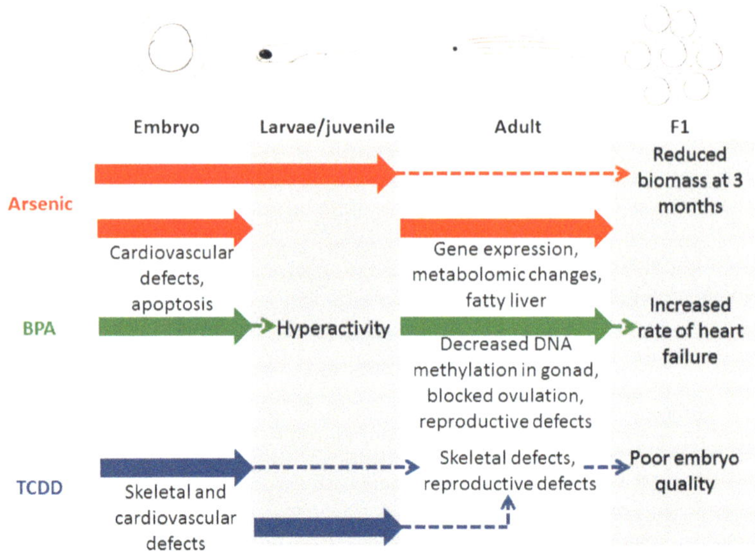

Fig. 2 Using zebrafish to model the health effects of toxicant exposure. Zebrafish exposed to a range of water-soluble environmentally relevant contaminants during different stages of the life cycle display a wide range of outcomes, from reduced reproductive success to skeletal and neurodevelopmental defects. *BPA*, bisphenol A; *TCDD*, 2,3,7,8-tetrachlorodibenzo-*p*-dioxin. Illustrations by Christopher Smith. Copyright Mount Sinai Health System 2016. Images used with permission.

2006; Argos et al., 2010). Studying the underlying mechanisms of arsenic toxicity in zebrafish can provide much needed information of how arsenic causes disease in humans.

Although arsenic is one of the most common metalloid contaminants of drinking water, and can be found at high levels in common foods such as rice and apple juice (Davis et al., 2012; Sauvé, 2014), the precise mechanism of arsenic toxicity is relatively unknown. Work in tissue culture cells and rodent models has identified oxidative stress, biotransformation and methylation, and ER stress as potential mechanisms of arsenic-induced toxicity (Gamble et al., 2005; Hughes, 2002; Vahter, 2002), although studies linking aberrant cellular processes to specific arsenic-induced disease phenotypes are lacking. Work in zebrafish is shedding light on this field.

Zebrafish express aquaglyceroporins and the trivalent arsenic specific methyltransferase (*zas3mt*), enzymes required for the uptake and metabolism of inorganic arsenic, respectively (Hamdi et al., 2009, 2012). Early studies of zebrafish models of inorganic arsenic exposure provided descriptive analysis of the effects of arsenic on the gross morphological development of zebrafish embryos and larvae, demonstrating acute toxicity and cardiovascular defects (Li et al., 2009; Seok et al., 2007). While information about the rate of arsenic metabolism by zebrafish in vivo and the production of specific metabolites are still emerging, it is clear that rodent models do not always accurately reflect the effects of human arsenic exposure. For instance, studies in rodents showed increased excretion and slower but more extensive methylation of arsenic when compared to humans so that ingested arsenic remains in the rodent blood stream for prolonged periods (Hallauer et al., 2016; States et al., 2011).

Zebrafish embryos treated with inorganic arsenic have multiple defects, and some studies implicated downregulation of Dvr1, a factor involved in mesoderm induction and the establishment of left-right asymmetry (Li et al., 2012). These processes are essential for proper cardiac morphogenesis. Interestingly, depletion of Dvr1 using morpholino knockdown led to heart defects that were similar to those seen upon exposure to 2 mM inorganic arsenic from 4 to 48 hpf (Li et al., 2012). Overexpression of human GDF1, a homolog of Dvr1, led to a reduction in the number of zebrafish that displayed morphological defects upon arsenic exposure (Li et al., 2012). Further studies from the same group showed folic acid prevented arsenic-induced toxicity in zebrafish by reducing generation of reactive oxygen species and rescuing the decrease in Dvr1 expression (Ma et al., 2015). However, the protective effect of folic acid diminished after 48 hpf (Ma et al., 2015), indicating that other mechanisms may underlie later developmental anomalies caused by arsenic exposure.

Transcriptomic and metabolomic approaches in zebrafish have also been used to gain insight into the molecular mechanisms of arsenic toxicity, particularly in the adult liver (Lam et al., 2006; Li et al., 2016; Xu et al., 2013; Yang et al., 2007). The effect of acute arsenic exposure and changes of gene expression patterns over time in the adult liver was first examined using microarray, with gene expression changes occurring as early as 8 h of exposure (Lam et al., 2006). Differentially expressed genes were grouped into categories to examine the adaptive response of the zebrafish liver to arsenic exposure, and revealed that arsenic-induced liver injury is the result of DNA and protein damage and oxidative stress resulting from the metabolism of inorganic arsenic. Gene ontology and pathway analysis of RNA-SAGE data were applied and used to identify a panel of biomarker genes to predict arsenic toxicity (Xu et al., 2013). Network analysis identified *nr2f2*, *jun*, *k-ras*, and *apoE* as four central factors that were upregulated in zebrafish liver following arsenic exposure. Each of these factors is implicated in pathways that can contribute to arsenic-induced liver disease: Jun and Kras are known oncoproteins, Nr2f2 regulates many genes involved in oxidative stress, and drug metabolism and ApoE is required for lipoprotein synthesis. Interestingly, arsenic exposure has been shown to accelerate the formation of atherosclerosis in *ApoE* deficient mice, highlighting the conservation of pathways affected by arsenic exposure in zebrafish and mammalian models (States et al., 2012; States, Srivastava, Sen, & D'Souza, 2007).

Chronic exposure of zebrafish to environmentally relevant concentrations revealed retention of arsenic in the eye, skin, and liver of 6-month-old fish and resulted in increased heart rate during larval stages and neurologic defects (Hallauer et al., 2016). Progeny of arsenic-exposed fish had reduced biomass at 3 months of age relative to the progeny of their unexposed siblings (Hallauer et al., 2016). Zebrafish studies have recapitulated the effects of arsenic on the cardiovascular system (Hallauer et al., 2016; Li et al., 2012) and have shown alterations in liver metabolism and liver function (Lam et al., 2006; Li et al., 2016; Xu et al., 2013). Our research is focused on using zebrafish to understand the mechanisms that underlie arsenic-induced liver disease in human populations (Mazumder, 2005; Santra, Das Gupta, De, Roy, & Guha Mazumder, 1999). Metabolic changes in the liver of adult zebrafish after acute arsenic exposure was investigated using gas chromatography coupled with mass spectroscopy (Li et al., 2016), identifying 34 potential metabolite markers of arsenic exposure. Additionally, histological examination of the livers of arsenic-exposed zebrafish showed cellular changes and accumulation of lipid droplets, liver function tests showed little alteration (Li et al., 2016), suggesting that metabolic changes may be a sensitive method to detect alterations in liver function induced by arsenic. Although the use of zebrafish to study arsenic toxicity is relatively recent, this model system has provided important insights into both the acute (Li et al., 2016; Xu et al., 2013) and chronic effects of arsenic exposure (Hallauer et al., 2016). Zebrafish studies of the effects of arsenic exposure will provide insight into the mechanisms of these physiological consequences and will also allow for the examination of transgenerational effects more feasibly than with rodent models.

9.2 Bisphenol A

BPA is one of the most common endocrine-disrupting environmental contaminants. It is a high production volume chemical and is present in many consumer and industrial products such as plastics. It is also present in the environment as a result of manufacturing processes and leaching from the products in which it is used. BPA is defined as an endocrine disruptor because of its ability to elicit both proestrogenic and antiestrogenic effects by binding to estrogen receptors ERα and ERβ and altering transcription in tissue- and context-specific manners (Santangeli et al., 2016). BPA binds to the zebrafish estrogen-related receptor gamma (ERRγ) in vivo (Tohmé et al., 2014). BPA has been reported to have adverse effects on reproductive health, early development, and contributes to obesity (Rochester, 2013). Both mammalian and zebrafish exposure studies have revealed related phenotypes.

Environmentally relevant doses of BPA were found to inhibit oocyte maturation by binding to the membrane estrogen receptor, Gper, and activating Egfr/Mapk3/1 signaling, which prevents resumption of meiosis (Fitzgerald, Peyton, Dong, & Thomas, 2015). Interestingly, this pathway is independent of signaling through the estrogen receptor (Fitzgerald et al., 2015). BPA is suggested to act through epigenetic mechanisms through histone modification and alteration of DNA methylation (Faulk et al., 2016; Kundakovic & Champagne, 2011; Santangeli et al., 2016), and zebrafish studies have provided key mechanistic insights. The effect of BPA on histone methylation patterns and DNA methylation has recently been shown in adult zebrafish (Laing et al., 2016; Santangeli et al., 2016). Global DNA methylation has also recently been shown to be reduced in the ovaries and testes of adult zebrafish exposed to 15 μg/L BPA for 7 days (Liu, Zhang, et al., 2016) and 1 mg/L BPA for 15 days (Laing et al., 2016). Adult female zebrafish exposed to environmentally relevant concentrations of BPA, from 5 to 20 μg/L, displayed nonmonotonic effects in that the lowest dose tested led to a complete block in ovulation, accompanied by more significant reduction in gene expression of the estrogen receptors *esr1* and *esr2a*, and induction of apoptosis markers *caspase3* and *p53* (Santangeli et al., 2016). Expression of the DNA methyltransferases *dnmt1* and *dnmt3* were upregulated in the ovaries of female zebrafish exposed to 5 μg/L BPA. Some of these gene expression changes were associated with changes in the levels of H3K4me3 and H3K27me3 levels (Santangeli et al., 2016). In contrast, adult male and female zebrafish exposed to higher doses of BPA (up to 1 mg/L) were shown to have reduced expression of *dnmt1* in the liver and ovaries when exposed to 10 μg/L, 100 μg/L, and 1 mg/L BPA, while no significant differences in expression level were observed in the testes of male zebrafish at the same exposure concentrations (Laing et al., 2016). Interestingly, the DNA methylation patterns were not strictly correlated with changes in gene expression. For instance, while no change in *dnmt1* expression was observed in the testes, analysis of 11 CpG sites in the *dnmt1* promoter revealed significant increases in some of the sites in this tissue. In the ovary, where the most consistent changes in *dnmt1* expression were observed, no significant changes in site-specific DNA methylation in the *dnmt1* promoter were found (Laing et al., 2016).

In addition to studying BPA-induced defects in reproduction, zebrafish have also been used to understand the neurotoxic effects of BPA (Cano-Nicolau et al., 2016; Kinch, Ibhazehiebo, Jeong, Habibi, & Kurrasch, 2015; Saili et al., 2012). In the zebrafish brain, BPA can activate gene expression through the canonical estrogen receptor signaling pathway (Chung, Genco, Megrelis, & Ruderman, 2011). Zebrafish exposed to BPA for a narrow (8–58 hpf) or longer (8–120 hpf) window were examined for effects on behavior. Zebrafish larvae (5 dpf) that were exposed to low dose BPA from 8 to 58 hpf demonstrated hyperactivity, and adult zebrafish that were exposed to low dose BPA from 8 to 120 hpf had behavioral and learning deficits, including larval hyperactivity and reduced ability to choose the correct arm of a T-maze to avoid an electric shock, compared to unexposed controls (Saili et al., 2012).

Using transgenic *Tg(cyp19a1b:GFP)* reporter fish, a reporter of estrogen signaling, zebrafish larvae exposed to BPA on 4 or 7 dpf resulted in activation of the estrogen-specific marker which likely occurred through activation of ERα (Cano-Nicolau et al., 2016). Reporter expression was localized to specific brain regions including the posterior telencephalon, preoptic area, and caudal hypothalamus. In the zebrafish brain, it has also recently been shown that exposure of developing zebrafish to low doses BPA caused precocious neurogenesis in the hypothalamus which resulted in hyperactivity and brain changes (Kinch et al., 2015).Together, these data show that zebrafish are capable of demonstrating not only the molecular and cellular responses to the endocrine disruptor BPA, but also provide evidence of the pathological effects of BPA exposure.

Most recently, zebrafish have been used to study the toxic effects of BPA and the products of its degradation (Makarova, Siudem, Zawada, & Kurkowiak, 2016). The degradation products of BPA were found to have lower binding affinity for both human and zebrafish estrogen receptors than BPA itself but one degradation product, 4-isopopylphenol, was predicted to have a higher binding affinity for the human ERRγ and slightly lower affinity for zebrafish ERRγA. 4-Isopopylphenol has the ability to permeate biological membranes similar to BPA, but appears to be more toxic as it caused acute lethality to zebrafish embryos while the same dose of BPA did not. These zebrafish studies emphasize that degradation products of environmental contaminants can be more toxic than their parent compounds, and that toxicity testing of intermediates may be warranted (Gamse & Gorelick, 2016).

Zebrafish have provided useful insights into the effects of BPA on the developing brain and reproductive organs. These systems are most the most likely to be affected by environmental exposures to BPA and similar compounds that rely heavily on estrogen signaling (Patisaul & Adewale, 2009; Saili et al., 2012). Studies of BPA toxicity in zebrafish have highlighted the nonmonotonic effects of this common environmental contaminant and other EDCs (Santangeli et al., 2016; Vandenberg et al., 2012).

9.3 TCDD

TCDD is one of the most widely studied environmental contaminants in zebrafish (Carney, Prasch, Heideman, & Peterson, 2006). This chemical is a polychlorinated dibenzo-*p*-dioxin, an anthropogenic, lipophilic persistent environmental contaminant, commonly found in air and soil as the result of solid waste incineration and industrial processing. Human occupational and environmental exposure may be associated with a wide range of chronic diseases, including cancer, diabetes, endometriosis, cardiovascular disease, reduced testosterone, and thyroid hormone levels (White & Birnbaum, 2009). The effects of aromatic hydrocarbon exposure on the health of wild fish populations are more difficult to assess; however, lake trout populations in regions with high levels of aromatic hydrocarbon contamination have been unable to sustain their numbers (King-Heiden et al., 2012). Zebrafish have been proposed as a model to understand not only the health effects of human exposure to aromatic hydrocarbons and dioxin-like compounds, but also to predict the effects of contamination on wild fish populations. A comprehensive review of the contributions of the zebrafish model to our understanding of the molecular mechanisms of TCDD reproductive and developmental toxicity has been published elsewhere (King-Heiden et al., 2012).

Zebrafish have been used to study TCDD-induced endocrine disruption and reproductive toxicity (Baker et al., 2013; Heiden et al., 2008), cardiovascular toxicity (Antkiewicz, Burns, Carney, Peterson, & Heideman, 2005; Goldstone & Stegeman, 2006), and skeletal abnormalities (Baker et al., 2013; Burns, Peterson, & Heideman, 2015; Henry, Spitsbergen, Hornung,

Abnet, & Peterson, 1997; Teraoka et al., 2006). Developmental malformations in zebrafish embryos and larvae exposed to TCDD are prevented by depletion of Ahr2 (Prasch et al., 2003), indicating that metabolism is required for TCDD toxicity. *Cyp1a* transgenic reporter zebrafish have been used to investigate the mechanisms of TCDD toxicity and to identify target organs for the effects of TCDD (Kim et al., 2013; Mattingly, McLachlan, & Toscano, 2001; Xu et al., 2015); however, pathways independent of Cyp1a1 also contribute to the developmental toxicity of TCDD as knockdown of zebrafish *cyp1a* does not prevent TCDD-induced phenotypes (Carney, Peterson, & Heideman, 2004). Here, we compile some of the most recent studies of zebrafish exposed to TCDD.

TCDD has been shown to affect both ovarian function and follicle maturation (Baker, Peterson, & Heideman, 2014; Heiden et al., 2008). Studies in numerous fish species have demonstrated many impairments in female reproduction that are caused by TCDD (King-Heiden et al., 2012). A microarray analysis of the zebrafish adult ovary examined the transcriptional changes that precede the physiologic dysfunction following exposure to a TCDD dose curve (Heiden et al., 2008). Exposure to TCDD resulted in downregulation of genes involved in estradiol synthesis and follicle maturation, as well as genes encoding structural proteins Krt4 and Lgals3l (Heiden et al., 2008). While this study found that gene expression changes were not dose dependent and that a majority of the differentially expressed transcripts were unknown or poorly characterized, ~40% of the differentially expressed probes contained both putative aryl hydrocarbon-response elements and estrogen response elements.

Cardiac toxicity is one of the most obvious end points of zebrafish exposure to TCDD (King-Heiden et al., 2012). Gene expression analysis over a time course was performed to understand the molecular pathways that are altered in response to TCDD exposure (Carney, Chen, et al., 2006). Within 1 h of exposure, a cluster of 42 genes involved in xenobiotic metabolism, proliferation, contractility, and regulation of heart development were induced (Carney, Chen, et al., 2006). In addition a "cell cycle gene cluster" was downregulated in zebrafish exposed to TCDD and negative regulators of cell cycle progression were upregulated, indicating that reduced cardiomyocyte number may underlie TCDD-induced cardiac toxicity (Carney, Chen, et al., 2006). Increased and ectopic expression of *Bmp4* and *Notch1b* transcripts in the region of nascent cardiac valve formation were found to be responsible for TCDD-induced failure of heart valve formation in the zebrafish (Mehta, Peterson, & Heideman, 2008). Failure to restrict these transcripts, as determined by in situ hybridization, was associated with loss of endothelial cell pattern in the region where this morphogenic process should occur. This study highlights one of the most significant advantages to using the zebrafish system, in that alterations in stereotypical developmental processes can yield insight into the cellular and molecular mechanisms underlying toxicity.

Skeletal malformation is another predominant developmental defect associated with TCDD exposure in fish and rodent species (Baker et al., 2014; Birnbaum, Harris, Stocking, Clark, & Morrissey, 1989; Henry et al., 1997; King-Heiden et al., 2012). Craniofacial malformations in TCDD-exposed zebrafish are dependent on Ahr2/Arnt1 signaling (Prasch, Tanguay, Mehta, Heideman, & Peterson, 2006; Prasch et al., 2003). A transgenic reporter *Tg(sox9b:EGFP)*, which marks perichondrial endoderm in the developing jaw, was used to demonstrate that craniofacial abnormalities in TCDD-exposed zebrafish larvae resulted from reductions in chondrocyte size and number and decreases in ossification of the jaw (Burns et al., 2015). In addition to being a marker of craniofacial and jaw development in the zebrafish, *sox9b* is required for this process. Heterozygous *sox9b* mutant zebrafish are more susceptible to TCDD-induced craniofacial malformations, and overexpression of *sox9b* in TCDD-treated zebrafish mitigated the effects of the toxicant on jaw development (Xiong, Peterson, & Heideman, 2008). Scoliosis is frequently observed in adult fish following exposure to TCDD during the embryonic or larval periods (Baker et al., 2013, 2014).

Zebrafish research into the pathways and molecular mechanisms underlying TCDD toxicity have provided information about outcomes relevant to human populations, most notably cardiac and reproductive defects. TCDD is also the most commonly studied environmental toxicant with regard to transgenerational effects, as discussed in Section 8 (Baker, King-Heiden, Peterson, & Heideman, 2014; Baker et al., 2014).

10. LIMITATIONS TO THE ZEBRAFISH MODEL SYSTEM

While zebrafish will allow researchers to answer many questions that are limited by the realities of epidemiological researchers, there are several limitations to this model. For instance, the physiological differences between zebrafish and mammals mean that disease outcomes such as asthma or placental defects are not observable in zebrafish. However, although not all disease-related phenotypes can be identified in zebrafish, many of the developmental and signaling pathways leading to these diseases are conserved between zebrafish and humans (Padilla et al., 2012).

A second consideration is that in zebrafish, some exposures may not be equivalent to the experience of human populations. In most studies, the toxicant is added directly to the water, recapitulating a dermal exposure during the early stages of zebrafish development when the embryos are not swallowing water in order to breathe. However, many toxicants are introduced into the human body via oral exposure through contaminated drinking water or food, and, as such exposure is intermittent and affects involves the gastrointestinal system. This may lead to substantial differences in the absorption, tissue distribution, metabolism, and excretion depending on the uptake and biotransformation pathways based on the route of exposure. Metabolic differences between zebrafish and mammals may also be affected by differences in the expression patterns of xenobiotic metabolism enzymes and incompletely conserved enzyme functions (Saad et al., 2016), which may also contribute to the differences in dosing required to elicit phenotypes. Urinary biomarkers of exposure and metabolism are also unavailable from zebrafish.

Studies have demonstrated that gender can play an influential role in response to toxin exposure. For example, arsenic exposure in humans leads to changes in DNA methylation in isolated cord blood cells that are different in males and females (Pilsner et al., 2012), and endocrine disruptors have been shown to have different neurobehavioral effects in boys and girls (Evans et al., 2014; Roen et al., 2015). While zebrafish have no discernible sex chromosomes and do not become sexually dimorphic until 3 weeks postfertilization (Sola & Gornung, 2001; Tong, Hsu, & Chung, 2010), toxicant exposure during this window can influence sex characteristics as seen with early TCDD exposure (Fig. 2) (Baker et al., 2013).

11. CONCLUSIONS

As manufacturing processes and development of new synthetic compounds proceed in order to keep pace with the growing world economy, environmental health, and the effects of toxicant exposure are emerging as critical areas of research. The main benefit to using zebrafish in toxicology and environmental health studies is that their unique combination of developmental features provides a system with the benefits of both in vitro and in vivo schemes. Combining the large scale of embryo production with rapid development allows for short-term assessment of toxicity in a whole animal system. In addition, the relative ease and comparatively low cost of raising large numbers of individuals allows for unprecedented investigation into latent effects and adverse outcomes in response to early life exposure to environmental contaminants. Many of the genetic, molecular, and cellular processes are conserved between zebrafish and mammals, allowing close applicability to human exposure and disease. As such, studies using zebrafish have uncovered important insights into the effects of environmental contaminants on normal development in a live vertebrate system.

ACKNOWLEDGMENTS

The authors would like to acknowledge Drs. Kirsten C. Sadler and Anjana Ramdas for critical reading of this chapter. Our research is supported by T32HD049311-01A1 (K.B.) and K08DK101340 and the Mindich Child Health and Development Institute at Mount Sinai (J.C.).

REFERENCES

Ahsan, H., Chen, Y., Parvez, F., Argos, M., Hussain, A. I., Momotaj, H., et al. (2006). Health effects of arsenic longitudinal study (HEALS): Description of a multidisciplinary epidemiologic investigation. *Journal of Exposure Science & Environmental Epidemiology*, *16*, 191–205.

Al-Eryani, L., Wahlang, B., Falkner, K. C., Guardiola, J. J., Clair, H. B., Prough, R. A., et al. (2015). Identification of environmental chemicals associated with the development of toxicant-associated fatty liver disease in rodents. *Toxicologic Pathology*, *43*, 482–497.

Antkiewicz, D. S., Burns, C. G., Carney, S. A., Peterson, R. E., & Heideman, W. (2005). Heart malformation is an early response to TCDD in embryonic zebrafish. *Toxicological Sciences*, *84*, 368–377.

Argos, M., Kalra, T., Rathouz, P. J., Chen, Y., Pierce, B., Parvez, F., et al. (2010). Arsenic exposure from drinking water, and all-cause and chronic-disease mortalities in Bangladesh (HEALS): A prospective cohort study. *Lancet*, *376*, 252–258.

Arora, M., Austin, C., Sarrafpour, B., Hernández-Ávila, M., Hu, H., Wright, R. O., et al. (2014). Determining prenatal, early childhood and cumulative long-term lead exposure using micro-spatial deciduous dentine levels. *PLoS One*, *9*, e97805.

Baker, T. R., King-Heiden, T. C., Peterson, R. E., & Heideman, W. (2014). Dioxin induction of transgenerational inheritance of disease in zebrafish. *Molecular and Cellular Endocrinology*, *398*, 36–41.

Baker, T. R., Peterson, R. E., & Heideman, W. (2013). Early dioxin exposure causes toxic effects in adult zebrafish. *Toxicological Sciences*, *135*, 241–250.

Baker, T. R., Peterson, R. E., & Heideman, W. (2014). Using zebrafish as a model system for studying the transgenerational effects of dioxin. *Toxicological Sciences*, *138*, 403–411.

Ball, J. S., Stedman, D. B., Hillegass, J. M., Zhang, C. X., Panzica-Kelly, J., Coburn, A., et al. (2014). Fishing for teratogens: A consortium effort for a harmonized zebrafish developmental toxicology assay. *Toxicological Sciences*, *139*, 210–219.

Barker, D. J. P., & Osmond, C. (1986). Infant mortality, childhood nutrition, and ischaemic heart disease in England and Wales. *The Lancet*, *327*, 1077–1081.

Bellinger, D. C. (2016). Lead contamination in Flint—An abject failure to protect public health. *New England Journal of Medicine*, *374*, 1101–1103.

Birnbaum, L. S. (2010). TSCA reform under way in Congress. *Environmental Health Perspectives, 118*, A106.
Birnbaum, L. S., Harris, M. W., Stocking, L. M., Clark, A. M., & Morrissey, R. E. (1989). Retinoic acid and 2,3,7,8-tetrachlorodibenzo-p-dioxin selectively enhance teratogenesis in C57BL/6N mice. *Toxicology and Applied Pharmacology, 98*, 487–500.
Boyd, W. A., Smith, M. V., Co, C. A., Pirone, J. R., Rice, J. R., Shockley, K. R., et al. (2016). Developmental effects of the ToxCast Phase I and II chemicals in and corresponding responses in zebrafish, rats, and rabbits. *Environmental Health Perspectives, 124*, 586–593.
Brannen, K. C., Panzica-Kelly, J. M., Danberry, T. L., & Augustine-Rauch, K. A. (2010). Development of a zebrafish embryo teratogenicity assay and quantitative prediction model. *Birth Defects Research. Part B, Developmental and Reproductive Toxicology, 89*, 66–77.
Burns, F. R., Peterson, R. E., & Heideman, W. (2015). Dioxin disrupts cranial cartilage and dermal bone development in zebrafish larvae. *Aquatic Toxicology, 164*, 52–60.
Cano-Nicolau, J., Vaillant, C., Pellegrini, E., Charlier, T. D., Kah, O., & Coumailleau, P. (2016). Estrogenic effects of several BPA analogs in the developing zebrafish brain. *Frontiers in Neuroscience, 10*, 112.
Cario, C. L., Farrell, T. C., Milanese, C., & Burton, E. A. (2011). Automated measurement of zebrafish larval movement. *The Journal of Physiology, 589*, 3703–3708.
Carney, S. A., Chen, J., Burns, C. G., Xiong, K. M., Peterson, R. E., & Heideman, W. (2006). Aryl hydrocarbon receptor activation produces heart-specific transcriptional and toxic responses in developing zebrafish. *Molecular Pharmacology, 70*, 549–561.
Carney, S. A., Peterson, R. E., & Heideman, W. (2004). 2,3,7,8-Tetrachlorodibenzo-p-dioxin activation of the aryl hydrocarbon receptor/aryl hydrocarbon receptor nuclear translocator pathway causes developmental toxicity through a CYP1A-independent mechanism in zebrafish. *Molecular Pharmacology, 66*, 512–521.
Carney, S. A., Prasch, A. L., Heideman, W., & Peterson, R. E. (2006). Understanding dioxin developmental toxicity using the zebrafish model. *Birth Defects Research. Part A, Clinical and Molecular Teratology, 76*, 7–18.
Carvan, M. J., 3rd, Dalton, T. P., Stuart, G. W., & Nebert, D. W. (2000). Transgenic zebrafish as sentinels for aquatic pollution. *Annals of the New York Academy of Sciences, 919*, 133–147.
Chen, S., Melchior, W. B., & Guo, L. (2014). Endoplasmic reticulum stress in drug- and environmental toxicant-induced liver toxicity. *Journal of Environmental Science and Health. Part C, 32*, 83–104.
Chung, E., Genco, M. C., Megrelis, L., & Ruderman, J. V. (2011). Effects of bisphenol A and triclocarban on brain-specific expression of aromatase in early zebrafish embryos. *Proceedings of the National Academy of Sciences of the United States of America, 108*, 17732–17737.
Collins, F. S., Gray, G. M., & Bucher, J. R. (2008). Toxicology. Transforming environmental health protection. *Science, 319*, 906–907.
Davis, M. A., Mackenzie, T. A., Cottingham, K. L., Gilbert-Diamond, D., Punshon, T., & Karagas, M. R. (2012). Rice consumption and urinary arsenic concentrations in U.S. children. *Environmental Health Perspectives, 120*, 1418–1424.
Deal, S., Wambaugh, J., Judson, R., Mosher, S., Radio, N., Houck, K., et al. (2016). Development of a quantitative morphological assessment of toxicant-treated zebrafish larvae using brightfield imaging and high-content analysis. *Journal of Applied Toxicology, 36*, 1214–1222.
Dix, D. J., Houck, K. A., Martin, M. T., Richard, A. M., Setzer, R. W., & Kavlock, R. J. (2007). The ToxCast program for prioritizing toxicity testing of environmental chemicals. *Toxicological Sciences, 95*, 5–12.
Evans, S. F., Kobrosly, R. W., Barrett, E. S., Thurston, S. W., Calafat, A. M., Weiss, B., et al. (2014). Prenatal bisphenol A exposure and maternally reported behavior in boys and girls. *NeuroToxicology, 45*, 91–99.

Faulk, C., Kim, J. H., Anderson, O. S., Nahar, M. S., Jones, T. R., Sartor, M. A., et al. (2016). Detection of differential DNA methylation in repetitive DNA of mice and humans perinatally exposed to bisphenol A. *Epigenetics*, *11*, 489–500.

Fitzgerald, A. C., Peyton, C., Dong, J., & Thomas, P. (2015). Bisphenol A and related alkylphenols exert nongenomic estrogenic actions through a G protein-coupled estrogen receptor 1 (Gper)/epidermal growth factor receptor (Egfr) pathway to inhibit meiotic maturation of zebrafish oocytes. *Biology of Reproduction*, *93*, 135. 131-111.

Gallardo, V. E., Varshney, G. K., Lee, M., Bupp, S., Xu, L., Shinn, P., et al. (2015). Phenotype-driven chemical screening in zebrafish for compounds that inhibit collective cell migration identifies multiple pathways potentially involved in metastatic invasion. *Disease Models & Mechanisms*, *8*, 565–576.

Gamble, M. V., Liu, X., Ahsan, H., Pilsner, J. R., Ilievski, V., Slavkovich, V., et al. (2005). Folate, homocysteine, and arsenic metabolism in arsenic-exposed individuals in Bangladesh. *Environmental Health Perspectives*, *113*, 1683–1688.

Gamse, J. T., & Gorelick, D. A. (2016). Mixtures, metabolites, and mechanisms: Understanding toxicology using zebrafish. *Zebrafish*, *13*, 377–378.

Goldstone, J. V., McArthur, A. G., Kubota, A., Zanette, J., Parente, T., Jönsson, M. E., et al. (2010). Identification and developmental expression of the full complement of Cytochrome P450 genes in Zebrafish. *BMC Genomics*, *11*, 1–21.

Goldstone, H. M. H., & Stegeman, J. J. (2006). Molecular mechanisms of 2,3,7,8-tetrachlorodibenzo-p-dioxin cardiovascular embryotoxicity. *Drug Metabolism Reviews*, *38*, 261–289.

Gorelick, D. A., & Halpern, M. E. (2011). Visualization of estrogen receptor transcriptional activation in zebrafish. *Endocrinology*, *152*, 2690–2703.

Gorelick, D. A., Iwanowicz, L. R., Hung, A. L., Blazer, V. S., & Halpern, M. E. (2014). Transgenic zebrafish reveal tissue-specific differences in estrogen signaling in response to environmental water samples. *Environmental Health Perspectives*, *122*, 356–362.

Gorelick, D. A., Pinto, C. L., Hao, R., & Bondesson, M. (2016). Use of reporter genes to analyze estrogen response: The transgenic zebrafish model. *Methods in Molecular Biology*, *1366*, 315–325.

Gustafson, A. L., Stedman, D. B., Ball, J., Hillegass, J. M., Flood, A., Zhang, C. X., et al. (2012). Inter-laboratory assessment of a harmonized zebrafish developmental toxicology assay—Progress report on phase I. *Reproductive Toxicology*, *33*, 155–164.

Hallauer, J., Geng, X., Yang, H. C., Shen, J., Tsai, K. J., & Liu, Z. (2016). The effect of chronic arsenic exposure in zebrafish. *Zebrafish*, *13*, 405–412.

Hamdi, M., Sanchez, M. A., Beene, L. C., Liu, Q., Landfear, S. M., Rosen, B. P., et al. (2009). Arsenic transport by zebrafish aquaglyceroporins. *BMC Molecular Biology*, *10*, 104.

Hamdi, M., Yoshinaga, M., Packianathan, C., Qin, J., Hallauer, J., McDermott, J. R., et al. (2012). Identification of an S-adenosylmethionine (SAM) dependent arsenic methyltransferase in Danio rerio. *Toxicology and Applied Pharmacology*, *262*, 185–193.

Harding, H. P., Zhang, Y., & Ron, D. (1999). Protein translation and folding are coupled by an endoplasmic-reticulum-resident kinase. *Nature*, *397*, 271–274.

Hare, D., Austin, C., & Doble, P. (2012). Quantification strategies for elemental imaging of biological samples using laser ablation-inductively coupled plasma-mass spectrometry. *Analyst*, *137*, 1527–1537.

Hatch, E. E., Palmer, J. R., Titus-Ernstoff, L., et al. (1998). Cancer risk in women exposed to diethylstilbestrol in utero. *JAMA*, *280*, 630–634.

Heiden, T. C., Struble, C. A., Rise, M. L., Hessner, M. J., Hutz, R. J., & Carvan, M. J., 3rd. (2008). Molecular targets of 2,3,7,8-tetrachlorodibenzo-p-dioxin (TCDD) within the zebrafish ovary: Insights into TCDD-induced endocrine disruption and reproductive toxicity. *Reproductive Toxicology*, *25*, 47–57.

Henn, B. C., Coull, B. A., & Wright, R. O. (2014). Chemical mixtures and children's health. *Current Opinion in Pediatrics*, *26*, 223–229.

Henry, T. R., Spitsbergen, J. M., Hornung, M. W., Abnet, C. C., & Peterson, R. E. (1997). Early life stage toxicity of 2,3,7,8-tetrachlorodibenzo-p-dioxin in zebrafish (Danio rerio). *Toxicology and Applied Pharmacology*, *142*, 56–68.

Herbst, A. L., Ulfelder, H., & Poskanzer, D. C. (1971). Adenocarcinoma of the Vagina. *New England Journal of Medicine*, *284*, 878–881.

Höckendorf, B., Thumberger, T., & Wittbrodt, J. (2012). Quantitative analysis of embryogenesis: A perspective for light sheet microscopy. *Developmental Cell*, *23*, 1111–1120.

Hoffman Ellen, J., Turner Katherine, J., Fernandez Joseph, M., Cifuentes, D., Ghosh, M., Ijaz, S., et al. (2016). Estrogens suppress a behavioral phenotype in zebrafish mutants of the Autism Risk Gene, CNTNAP2. *Neuron*, *89*, 725–733.

Hughes, M. F. (2002). Arsenic toxicity and potential mechanisms of action. *Toxicology Letters*, *133*, 1–16.

IARC. (2004). *Some drinking-water disinfectants and contaminants, including arsenic. Monographs on chloramine, chloral and chloral hydrate, dichloroacetic acid, trichloroacetic acid and 3-chloro-4-(dichloromethyl)-5-hydroxy-2(5H)-furanone.* Lyon: International Agency for Research on Cancer.

Jacobson, J. L., Muckle, G., Ayotte, P., Dewailly, É., & Jacobson, S. W. (2015). Relation of prenatal methylmercury exposure from environmental sources to childhood IQ. *Environmental Health Perspectives*, *123*, 827–833.

Jeanray, N., Marée, R., Pruvot, B., Stern, O., Geurts, P., Wehenkel, L., et al. (2015). Phenotype classification of zebrafish embryos by supervised learning. *PLoS One*, *10*, e0116989.

Judson, R. S., Houck, K. A., Kavlock, R. J., Knudsen, T. B., Martin, M. T., Mortensen, H. M., et al. (2010). In vitro screening of environmental chemicals for targeted testing prioritization: The ToxCast project. *Environmental Health Perspectives*, *118*, 485–492.

Kaufmann, A., Mickoleit, M., Weber, M., & Huisken, J. (2012). Multilayer mounting enables long-term imaging of zebrafish development in a light sheet microscope. *Development*, *139*, 3242–3247.

Kim, K.-H., Park, H.-J., Kim, J. H., Kim, S., Williams, D. R., Kim, M.-K., et al. (2013). Cyp1a reporter zebrafish reveals target tissues for dioxin. *Aquatic Toxicology*, *134–135*, 57–65.

Kinch, C. D., Ibhazehiebo, K., Jeong, J.-H., Habibi, H. R., & Kurrasch, D. M. (2015). Low-dose exposure to bisphenol A and replacement bisphenol S induces precocious hypothalamic neurogenesis in embryonic zebrafish. *Proceedings of the National Academy of Sciences*, *112*, 1475–1480.

King-Heiden, T. C., Mehta, V., Xiong, K. M., Lanham, K. A., Antkiewicz, D. S., Ganser, A., et al. (2012). Reproductive and developmental toxicity of dioxin in fish. *Molecular and Cellular Endocrinology*, *354*, 121–138.

Kitamura, M. (2013). The unfolded protein response triggered by environmental factors. *Seminars in Immunopathology*, *35*, 259–275.

Kundakovic, M., & Champagne, F. A. (2011). Epigenetic perspective on the developmental effects of bisphenol A. *Brain, Behavior, and Immunity*, *25*, 1084–1093.

Laing, L. V., Viana, J., Dempster, E. L., Trznadel, M., Trunkfield, L. A., Uren Webster, T. M., et al. (2016). Bisphenol A causes reproductive toxicity, decreases dnmt1 transcription, and reduces global DNA methylation in breeding zebrafish (Danio rerio). *Epigenetics*, *11*, 526–538.

Lam, S. H., Hlaing, M. M., Zhang, X., Yan, C., Duan, Z., Zhu, L., et al. (2011). Toxicogenomic and phenotypic analyses of bisphenol-A early-life exposure toxicity in zebrafish. *PLoS One*, *6*, e28273.

Lam, S. H., Winata, C. L., Tong, Y., Korzh, S., Lim, W. S., Korzh, V., et al. (2006). Transcriptome kinetics of arsenic-induced adaptive response in zebrafish liver. *Physiological Genomics*, *27*, 351–361.

Lamb, J., Crawford, E. D., Peck, D., Modell, J. W., Blat, I. C., Wrobel, M. J., et al. (2006). The Connectivity Map: Using gene-expression signatures to connect small molecules, genes, and disease. *Science*, *313*, 1929–1935.

Landrigan, P. J. (2016). Children's environmental health: A brief history. *Academic Pediatrics*, *16*, 1–9.

Landrigan, P. J., Sly, J. L., Ruchirawat, M., Silva, E. R., Huo, X., Diaz-Barriga, F., et al. (2016). Health consequences of environmental exposures: Changing global patterns of exposure and disease. *Annals of Global Health*, *82*, 10–19.

Landrigan, P. J., Suk, W. A., & Amler, R. W. (1999). Chemical wastes, children's health, and the Superfund Basic Research Program. *Environmental Health Perspectives*, *107*, 423–427.

Lee, H. C., Chen, Y. J., Liu, Y. W., Lin, K. Y., Chen, S. W., Lin, C. Y., et al. (2011). Transgenic zebrafish model to study translational control mediated by upstream open reading frame of human CHOP gene. *Nucleic Acids Research*, *39*, e139.

Lee, O., Green, J. M., & Tyler, C. R. (2015). Transgenic fish systems and their application in ecotoxicology. *Critical Reviews in Toxicology*, *45*, 124–141.

Lee, H. C., Lu, P. N., Huang, H. L., Chu, C., Li, H. P., & Tsai, H. J. (2014). Zebrafish transgenic line huORFZ is an effective living bioindicator for detecting environmental toxicants. *PLoS One*, *9*, e90160.

Li, C., Li, P., Tan, Y. M., Lam, S. H., Chan, E. C., & Gong, Z. (2016). Metabolomic characterizations of liver injury caused by acute arsenic toxicity in zebrafish. *PLoS One*, *11*, e0151225.

Li, D., Lu, C., Wang, J., Hu, W., Cao, Z., Sun, D., et al. (2009). Developmental mechanisms of arsenite toxicity in zebrafish (Danio rerio) embryos. *Aquatic Toxicology*, *91*, 229–237.

Li, X., Ma, Y., Li, D., Gao, X., Li, P., Bai, N., et al. (2012). Arsenic impairs embryo development via down-regulating Dvr1 expression in zebrafish. *Toxicology Letters*, *212*, 161–168.

Liu, Y., & Wu, F. (2010). Global burden of aflatoxin-induced hepatocellular carcinoma: A risk assessment. *Environmental Health Perspectives*, *118*, 818–824.

Liu, L., Yan, Y., Wang, J., Wu, W., & Xu, L. (2016). Generation of mt:egfp transgenic zebrafish biosensor for the detection of aquatic zinc and cadmium. *Environmental Toxicology and Chemistry*, *35*, 2066–2073.

Liu, Y., Zhang, Y., Tao, S., Guan, Y., Zhang, T., & Wang, Z. (2016). Global DNA methylation in gonads of adult zebrafish Danio rerio under bisphenol A exposure. *Ecotoxicology and Environmental Safety*, *130*, 124–132.

Ma, Y., Zhang, C., Gao, X. B., Luo, H. Y., Chen, Y., Li, H. H., et al. (2015). Folic acid protects against arsenic-mediated embryo toxicity by up-regulating the expression of Dvr1. *Scientific Reports*, *5*, 16093.

Makarova, K., Siudem, P., Zawada, K., & Kurkowiak, J. (2016). Screening of toxic effects of bisphenol A and products of its degradation: Zebrafish (Danio Rerio) embryo test and molecular docking. *Zebrafish*, *13*, 466–474.

Mattingly, C. J., McLachlan, J. A., & Toscano, W. A. (2001). Green fluorescent protein (GFP) as a marker of aryl hydrocarbon receptor (AhR) function in developing zebrafish (Danio rerio). *Environmental Health Perspectives*, *109*, 845–849.

Mazumder, D. N. (2005). Effect of chronic intake of arsenic-contaminated water on liver. *Toxicology and Applied Pharmacology*, *206*, 169–175.

Mehta, V., Peterson, R. E., & Heideman, W. (2008). 2,3,7,8-Tetrachlorodibenzo-p-dioxin exposure prevents cardiac valve formation in developing zebrafish. *Toxicological Sciences*, *104*, 303–311.

Mikut, R., Dickmeis, T., Driever, W., Geurts, P., Hamprecht, F. A., Kausler, B. X., et al. (2013). Automated processing of zebrafish imaging data: A survey. *Zebrafish*, *10*, 401–421.

Muñoz-Quezada, M. T., Lucero, B. A., Barr, D. B., Steenland, K., Levy, K., Ryan, P. B., et al. (2013). Neurodevelopmental effects in children associated with exposure to organophosphate pesticides: A systematic review. *Neurotoxicology*, *39*, 158–168.

Murk, A. J., Legler, J., Denison, M. S., Giesy, J. P., van de Guchte, C., & Brouwer, A. (1996). Chemical-activated luciferase gene expression (CALUX): A novel in vitro bioassay for Ah receptor active compounds in sediments and pore water. *Fundamental and Applied Toxicology*, *33*, 149–160.

National Research Council. (2007). *Toxicity testing in the 21st century: A vision and a strategy*. Washington, D.C.: National Academies Press.

Niedzwiecki, M. M., Austin, C., Remark, R., Merad, M., Gnjatic, S., Estrada-Gutierrez, G., et al. (2016). A multimodal imaging workflow to visualize metal mixtures in the human placenta and explore colocalization with biological response markers. *Metallomics*, *8*, 444–452.

Padilla, S., Corum, D., Padnos, B., Hunter, D. L., Beam, A., Houck, K. A., et al. (2012). Zebrafish developmental screening of the ToxCast phase I chemical library. *Reproductive Toxicology*, *33*, 174–187.

Palam, L. R., Baird, T. D., & Wek, R. C. (2011). Phosphorylation of eIF2 facilitates ribosomal bypass of an inhibitory upstream ORF to enhance CHOP translation. *Journal of Biological Chemistry*, *286*, 10939–10949.

Panzica-Kelly, J. M., Zhang, C. X., & Augustine-Rauch, K. A. (2015). Optimization and performance assessment of the Chorion-Off [Dechorinated] zebrafish developmental toxicity assay. *Toxicological Sciences*, *146*, 127–134.

Patisaul, H. B., & Adewale, H. B. (2009). Long-term effects of environmental endocrine disruptors on reproductive physiology and behavior. *Frontiers in Behavioral Neuroscience*, *3*, 10.

Pilsner, J. R., Hall, M. N., Liu, X., Ilievski, V., Slavkovich, V., Levy, D., et al. (2012). Influence of prenatal arsenic exposure and newborn sex on global methylation of cord blood DNA. *PLoS One*, 7, e37147.

Prasch, A. L., Tanguay, R. L., Mehta, V., Heideman, W., & Peterson, R. E. (2006). Identification of zebrafish ARNT1 homologs: 2,3,7,8-Tetrachlorodibenzo-p-dioxin toxicity in the developing zebrafish requires ARNT1. *Molecular Pharmacology*, *69*, 776–787.

Prasch, A. L., Teraoka, H., Carney, S. A., Dong, W., Hiraga, T., Stegeman, J. J., et al. (2003). Aryl hydrocarbon receptor 2 mediates 2,3,7,8-tetrachlorodibenzo-p-dioxin developmental toxicity in zebrafish. *Toxicological Sciences*, *76*, 138–150.

Reif, D. M., Truong, L., Mandrell, D., Marvel, S., Zhang, G., & Tanguay, R. L. (2016). High-throughput characterization of chemical-associated embryonic behavioral changes predicts teratogenic outcomes. *Archives of Toxicology*, *90*, 1459–1470.

Richard, A. M., Judson, R. S., Houck, K. A., Grulke, C. M., Volarath, P., Thillainadarajah, I., et al. (2016). ToxCast chemical landscape: Paving the road to 21st century toxicology. *Chemical Research in Toxicology*, *29*, 1225–1251.

Rihel, J., Prober, D. A., Arvanites, A., Lam, K., Zimmerman, S., Jang, S., et al. (2010). Zebrafish behavioral profiling links drugs to biological targets and rest/wake regulation. *Science*, *327*, 348–351.

Rochester, J. R. (2013). Bisphenol A and human health: A review of the literature. *Reproductive Toxicology*, *42*, 132–155.

Roen, E. L., Wang, Y., Calafat, A. M., Wang, S., Margolis, A., Herbstman, J., et al. (2015). Bisphenol A exposure and behavioral problems among inner city children at 7–9 years of age. *Environmental Research*, *142*, 739–745.

Saad, M., Cavanaugh, K., Verbueken, E., Pype, C., Casteleyn, C., Van Ginneken, C., et al. (2016). Xenobiotic metabolism in the zebrafish: A review of the spatiotemporal distribution, modulation and activity of Cytochrome P450 families 1 to 3. *The Journal of Toxicological Sciences*, *41*, 1–11.

Saili, K. S., Corvi, M. M., Weber, D. N., Patel, A. U., Das, S. R., Przybyla, J., et al. (2012). Neurodevelopmental low-dose bisphenol A exposure leads to early life-stage hyperactivity and learning deficits in adult zebrafish. *Toxicology*, *291*, 83–92.

Sandmann, T., Kummerfeld, S. K., Gentleman, R., & Bourgon, R. (2014). gCMAP: User-friendly connectivity mapping with R. *Bioinformatics*, *30*, 127–128.

Santangeli, S., Maradonna, F., Gioacchini, G., Cobellis, G., Piccinetti, C. C., Dalla Valle, L., et al. (2016). BPA-induced deregulation of epigenetic patterns: Effects on female zebrafish reproduction. *Scientific Reports*, *6*, 21982.

Santra, A., Das Gupta, J., De, B. K., Roy, B., & Guha Mazumder, D. N. (1999). Hepatic manifestations in chronic arsenic toxicity. *Indian Journal of Gastroenterology*, *18*, 152–155.

Sauvé, S. (2014). Time to revisit arsenic regulations: Comparing drinking water and rice. *BMC Public Health*, *14*, 1–5.

Scheff Jeremy, D., Stallings Jonathan, D., Reifman, J., & Rakesh, V. (2015). Mathematical modeling of the heat-shock response in HeLa cells. *Biophysical Journal*, *109*, 182–193.

Schneider, A. J., Branam, A. M., & Peterson, R. E. (2014). Intersection of AHR and Wnt signaling in development, health, and disease. *International Journal of Molecular Sciences*, *15*, 17852–17885.

Selderslaghs, I. W., Blust, R., & Witters, H. E. (2012). Feasibility study of the zebrafish assay as an alternative method to screen for developmental toxicity and embryotoxicity using a training set of 27 compounds. *Reproductive Toxicology*, *33*, 142–154.

Selikoff, I. J., & Hammond, E. C. (1968). Environmental epidemiology. 3. Community effects of nonoccupational environmental asbestos exposure. *American Journal of Public Health and the Nation's Health*, *58*, 1658–1666.

Seok, S. H., Baek, M. W., Lee, H. Y., Kim, D. J., Na, Y. R., Noh, K. J., et al. (2007). Quantitative GFP fluorescence as an indicator of arsenite developmental toxicity in mosaic heat shock protein 70 transgenic zebrafish. *Toxicology and Applied Pharmacology*, *225*, 154–161.

Shahid, M., Takamiya, M., Stegmaier, J., Middel, V., Gradl, M., Klüver, N., et al. (2016). Zebrafish biosensor for toxicant induced muscle hyperactivity. *Scientific Reports*, *6*, 23768.

Sipes, N. S., Padilla, S., & Knudsen, T. B. (2011). Zebrafish—As an integrative model for twenty-first century toxicity testing. *Birth Defects Research. Part C, Embryo Today: Reviews*, *93*, 256–267.

Sola, L., & Gornung, E. (2001). Classical and molecular cytogenetics of the zebrafish, Danio rerio (Cyprinidae, Cypriniformes): An overview. *Genetica*, *111*, 397–412.

Sonavane, M., Creusot, N., Maillot-Maréchal, E., Péry, A., Brion, F., & Aït-Aïssa, S. (2016). Zebrafish-based reporter gene assays reveal different estrogenic activities in river waters compared to a conventional human-derived assay. *Science of the Total Environment*, *550*, 934–939.

States, J. C., Barchowsky, A., Cartwright, I. L., Reichard, J. F., Futscher, B. W., & Lantz, R. C. (2011). Arsenic toxicology: Translating between experimental models and human pathology. *Environmental Health Perspectives*, *119*, 1356–1363.

States, J. C., Singh, A. V., Knudsen, T. B., Rouchka, E. C., Ngalame, N. O., Arteel, G. E., et al. (2012). Prenatal arsenic exposure alters gene expression in the adult liver to a proinflammatory state contributing to accelerated atherosclerosis. *PLoS One*, 7, e38713.

States, J. C., Srivastava, S., Sen, U., & D'Souza, S. E. (2007). Early onset of atherosclerosis in ApoE-knockout mice is induced by in utero arsenic exposure. *The FASEB Journal*, *21*, A810.

Streisinger, G. (1983). Extrapolations from species to species and from various cell types in assessing risks from chemical mutagens. *Mutation Research*, *114*, 93–105.

Streisinger, G., Walker, C., Dower, N., Knauber, D., & Singer, F. (1981). Production of clones of homozygous diploid zebra fish (Brachydanio rerio). *Nature*, *291*, 293–296.

Swan, S. H., Main, K. M., Liu, F., Stewart, S. L., Kruse, R. L., Calafat, A. M., et al. (2005). Decrease in anogenital distance among male infants with prenatal phthalate exposure. *Environmental Health Perspectives*, *113*, 1056–1061.

Teraoka, H., Dong, W., Okuhara, Y., Iwasa, H., Shindo, A., Hill, A. J., et al. (2006). Impairment of lower jaw growth in developing zebrafish exposed to 2,3,7,8-tetrachlorodibenzo-p-dioxin and reduced hedgehog expression. *Aquatic Toxicology*, *78*, 103–113.

Tice, R. R., Austin, C. P., Kavlock, R. J., & Bucher, J. R. (2013). Improving the human hazard characterization of chemicals: A Tox21 update. *Environmental Health Perspectives*, *121*, 756–765.

Tohmé, M., Prud'homme, S. M., Boulahtouf, A., Samarut, E., Brunet, F., Bernard, L., et al. (2014). Estrogen-related receptor γ is an in vivo receptor of bisphenol A. *The FASEB Journal*, *28*, 3124–3133.

Tong, S., Baghurst, P., McMichael, A., Sawyer, M., & Mudge, J. (1996). Lifetime exposure to environmental lead and children's intelligence at 11-13 years: The Port Pirie cohort study. *BMJ: British Medical Journal*, *312*, 1569–1575.

Tong, S. K., Hsu, H. J., & Chung, B. C. (2010). Zebrafish monosex population reveals female dominance in sex determination and earliest events of gonad differentiation. *Developmental Biology*, *344*, 849–856.

Toxic Substances Control Act, 15 U.S.C. §2601 et seq., 1976.

Truong, L., Reif, D. M., St Mary, L., Geier, M. C., Truong, H. D., & Tanguay, R. L. (2014). Multidimensional in vivo hazard assessment using zebrafish. *Toxicological Sciences*, *137*, 212–233.

Vahter, M. (2002). Mechanisms of arsenic biotransformation. *Toxicology*, *181–182*, 211–217.

Vahter, M. (2008). Health effects of early life exposure to arsenic. *Basic & Clinical Pharmacology & Toxicology*, *102*, 204–211.

Vandenberg, L. N., Colborn, T., Hayes, T. B., Heindel, J. J., Jacobs, D. R., Lee, D.-H., et al. (2012). Hormones and endocrine-disrupting chemicals: Low-dose effects and nonmonotonic dose responses. *Endocrine Reviews*, *33*, 378–455.

Wahlang, B., Beier, J. I., Clair, H. B., Bellis-Jones, H. J., Falkner, K. C., McClain, C. J., et al. (2013). Toxicant-associated steatohepatitis. *Toxicologic Pathology*, *41*, 343–360.

Walker, S. L., Ariga, J., Mathias, J. R., Coothankandaswamy, V., Xie, X., Distel, M., et al. (2012). Automated reporter quantification in vivo: High-throughput screening method for reporter-based assays in zebrafish. *PLoS One*, *7*, e29916.

Wang, R. L., Biales, A. D., Garcia-Reyero, N., Perkins, E. J., Villeneuve, D. L., Ankley, G. T., et al. (2016). Fish connectivity mapping: Linking chemical stressors by their mechanisms of action-driven transcriptomic profiles. *BMC Genomics*, *17*, 84.

Wang, W., Cheng, S., & Zhang, D. (2014). Association of inorganic arsenic exposure with liver cancer mortality: A meta-analysis. *Environmental Research*, *135*, 120–125.

Wang, M., & Kaufman, R. J. (2016). Protein misfolding in the endoplasmic reticulum as a conduit to human disease. *Nature*, *529*, 326–335.

Wang, G., Rajpurohit, S. K., Delaspre, F., Walker, S. L., White, D. T., Ceasrine, A., et al. (2015). First quantitative high-throughput screen in zebrafish identifies novel pathways for increasing pancreatic beta-cell mass. *eLife*, *4*, e08261.

White, S. S., & Birnbaum, L. S. (2009). An overview of the effects of dioxins and dioxin-like compounds on vertebrates, as documented in human and ecological epidemiology. *Journal of Environmental Science and Health. Part C, Environmental Carcinogenesis & Ecotoxicology Reviews*, *27*, 197–211.

White, R. M., Sessa, A., Burke, C., Bowman, T., LeBlanc, J., Ceol, C., et al. (2008). Transparent adult zebrafish as a tool for in vivo transplantation analysis. *Cell Stem Cell*, *2*, 183–189.

Wirbisky, S. E., Sepúlveda, M. S., Weber, G. J., Jannasch, A. S., Horzmann, K. A., & Freeman, J. L. (2016). Embryonic atrazine exposure elicits alterations in genes associated with neuroendocrine function in adult male zebrafish. *Toxicological Sciences*, *153*, 149–164.

Wirbisky, S. E., Weber, G. J., Sepúlveda, M. S., Xiao, C., Cannon, J. R., & Freeman, J. L. (2015). Developmental origins of neurotransmitter and transcriptome alterations in adult female zebrafish exposed to atrazine during embryogenesis. *Toxicology*, *333*, 156–167.

Xie, Y., Holmgren, S., Andrews, D. M., & Wolfe, M. S. (2016). Evaluating the impact of the U.S. National toxicology program: A case study on hexavalent chromium. *Environmental Health Perspectives*. http://dx.doi.org/10.1289/EHP21, Epub ahead of print.

Xiong, K. M., Peterson, R. E., & Heideman, W. (2008). Aryl hydrocarbon receptor-mediated down-regulation of Sox9b causes jaw malformation in zebrafish embryos. *Molecular Pharmacology*, *74*, 1544–1553.

Xu, H., Lam, S. H., Shen, Y., & Gong, Z. (2013). Genome-wide identification of molecular pathways and biomarkers in response to arsenic exposure in zebrafish liver. *PLoS One*, *8*, e68737.

Xu, H., Li, C., Li, Y., Ng, G. H., Liu, C., Zhang, X., et al. (2015). Generation of Tg(cyp1a: gfp) transgenic zebrafish for development of a convenient and sensitive in vivo assay for aryl hydrocarbon receptor activity. *Marine Biotechnology (New York, N.Y.)*, *17*, 831–840.

Yang, L., Kemadjou, J. R., Zinsmeister, C., Bauer, M., Legradi, J., Muller, F., et al. (2007). Transcriptional profiling reveals barcode-like toxicogenomic responses in the zebrafish embryo. *Genome Biology*, *8*, R227.

Zhang, X., Li, C., & Gong, Z. (2014). Development of a convenient in vivo hepatotoxin assay using a transgenic zebrafish line with liver-specific DsRed expression. *PLoS One*, *9*, e91874.

Zoeller, R. T., Brown, T. R., Doan, L. L., Gore, A. C., Skakkebaek, N. E., Soto, A. M., et al. (2012). Endocrine-disrupting chemicals and public health protection: A statement of principles from the endocrine society. *Endocrinology*, *153*, 4097–4110.

FSC
www.fsc.org
MIX
Papier | Fördert gute Waldnutzung
FSC® C083411

Kolibri 360 GmbH
Ferdinand-Jühlke-Straße 7
99095 Erfurt, Deutschland
produktsicherheit@kolibri360.de